A Comprehensive Course in Number Theory

Developed from the author's popular text, *A Concise Introduction to the Theory of Numbers*, this book provides a comprehensive initiation to all the major branches of number theory. Beginning with the rudiments of the subject, the author proceeds to more advanced topics, including elements of cryptography and primality testing; an account of number fields in the classical vein including properties of their units, ideals and ideal classes; aspects of analytic number theory including studies of the Riemann zeta-function, the prime-number theorem and primes in arithmetical progressions; a description of the Hardy–Littlewood and sieve methods from, respectively, additive and multiplicative number theory; and an exposition of the arithmetic of elliptic curves.

The book includes many worked examples, exercises and, as with the earlier volume, there is a guide to further reading at the end of each chapter. Its wide coverage and versatility make this book suitable for courses extending from the elementary to the graduate level.

ALAN BAKER, FRS, is Emeritus Professor of Pure Mathematics in the University of Cambridge and Fellow of Trinity College, Cambridge. His many distinctions include the Fields Medal (1970) and the Adams Prize (1972).

A COMPREHENSIVE COURSE IN NUMBER THEORY

ALAN BAKER
University of Cambridge

CAMBRIDGE
UNIVERSITY PRESS

University Printing House, Cambridge CB2 8BS, United Kingdom

One Liberty Plaza, 20th Floor, New York, NY 10006, USA

477 Williamstown Road, Port Melbourne, VIC 3207, Australia

314-321, 3rd Floor, Plot 3, Splendor Forum, Jasola District Centre, New Delhi - 110025, India

79 Anson Road, #06-04/06, Singapore 079906

Cambridge University Press is part of the University of Cambridge.

It furthers the University's mission by disseminating knowledge in the pursuit of education, learning and research at the highest international levels of excellence.

www.cambridge.org
Information on this title: www.cambridge.org/9781107603790

First published 2012
Reprinted 2013

A catalogue record for this publication is available from the British Library

Library of Congress Cataloging in Publication data
Baker, Alan, 1939–
A comprehensive course in number theory / Alan Baker.
p.   cm.
Includes bibliographical references and index.
ISBN 978-1-107-01901-0 (hardback)
1. Number theory – Textbooks.   I. Title.
QA241.B237   2012
512.7–dc23
2012013414

ISBN 978-1-107-01901-0 Hardback
ISBN 978-1-107-60379-0 Paperback

Contents

Preface

This is a sequel to my earlier book, *A Concise Introduction to the Theory of Numbers*. The latter was based on a short preparatory course of the kind traditionally taught in Cambridge at around the time of publication about 25 years ago. Clearly it was in need of updating, and it was originally intended that a second edition be produced. However, on looking through, it became apparent that the work would blend well with more advanced material arising from my lecture courses in Cambridge at a higher level, and it was decided accordingly that it would be more appropriate to produce a substantially new book. The now much expanded text covers elements of cryptography and primality testing. It also provides an account of number fields in the classical vein including properties of their units, ideals and ideal classes. In addition it covers various aspects of analytic number theory including studies of the Riemann zeta-function, the prime-number theorem, primes in arithmetical progressions and a brief exposition of the Hardy–Littlewood and sieve methods. Many worked examples are given and, as with the earlier volume, there are guides to further reading at the ends of the chapters.

The following remarks, taken from the *Concise Introduction*, apply even more appropriately here:

The theory of numbers has a long and distinguished history, and indeed the concepts and problems relating to the field have been instrumental in the foundation of a large part of mathematics. It is very much to be hoped that our exposition will serve to stimulate the reader to delve into the rich literature associated with the subject and thereby to discover some of the deep and beautiful theories that have been created as a result of numerous researches over the centuries. By way of introduction, there is a short account of the *Disquisitiones Arithmeticae* of Gauss, and, to begin with, the reader can scarcely do better than to consult this famous work.

To complete the text there is a chapter on elliptic curves; here my main source has been lecture notes by Dr Tom Fisher of a course that he has given regularly in Cambridge in recent times. I am indebted to him for generously providing me with a copy of the notes and for further expert advice. I am grateful also to Mrs Michèle Bailey for her invaluable secretarial assistance with my lectures over many years and to Dr David Tranah of Cambridge University Press for his constant encouragement in the production of this book.

Cambridge 2012   A.B.

Introduction

Gauss and Number Theory[†]

Without doubt the theory of numbers was Gauss' favourite subject. Indeed, in a much quoted dictum, he asserted that Mathematics is the Queen of the Sciences and the Theory of Numbers is the Queen of Mathematics. Moreover, in the introduction to Eisenstein's *Mathematische Abhandlungen*, Gauss wrote:

The Higher Arithmetic presents us with an inexhaustible storehouse of interesting truths – of truths, too, which are not isolated but stand in the closest relation to one another, and between which, with each successive advance of the science, we continually discover new and sometimes wholly unexpected points of contact. A great part of the theories of Arithmetic derive an additional charm from the peculiarity that we easily arrive by induction at important propositions which have the stamp of simplicity upon them but the demonstration of which lies so deep as not to be discovered until after many fruitless efforts; and even then it is obtained by some tedious and artificial process while the simpler methods of proof long remain hidden from us.

All this is well illustrated by what is perhaps Gauss' most profound publication, namely his *Disquisitiones Arithmeticae*. It has been described, quite justifiably I believe, as the Magna Carta of Number Theory, and the depth and originality of thought manifest in this work are particularly remarkable considering that it was written when Gauss was only about 18 years of age. Of course, as Gauss said himself, not all of the subject matter was new at the time of writing, and Gauss acknowledged the considerable debt that he owed to earlier scholars, in particular Fermat, Euler, Lagrange and Legendre. But the *Disquisitiones Arithmeticae* was the first systematic treatise on the Higher Arithmetic and it provided the foundations and stimulus for a great volume

---

[†] This article was originally prepared for a meeting of the British Society for the History of Mathematics held in Cambridge in 1977 to celebrate the bicentenary of Gauss' birth.

of subsequent research which is in fact continuing to this day. The importance of the work was recognized as soon as it was published in 1801 and the first edition quickly became unobtainable; indeed many scholars of the time had to resort to taking handwritten copies. But it was generally regarded as a rather impenetrable work and it was probably not widely understood; perhaps the formal Latin style contributed in this respect. Now, however, after numerous reformulations, most of the material is very well known, and the earlier sections at least are included in every basic course on number theory.

The text begins with the definition of a congruence, namely two numbers are said to be congruent modulo n if their difference is divisible by n. This is plainly an equivalence relation in the now familiar terminology. Gauss proceeds to the discussion of linear congruences and shows that they can in fact be treated somewhat analogously to linear equations. He then turns his attention to power residues and introduces, amongst other things, the concepts of primitive roots and indices; and he notes, in particular, the resemblance between the latter and the ordinary logarithms. There follows an exposition of the theory of quadratic congruences, and it is here that we meet, more especially, the famous law of quadratic reciprocity; this asserts that if p, q are primes, not both congruent to 3 (mod 4), then p is a residue or non-residue of q according as q is a residue or non-residue of p, while in the remaining case the opposite occurs. As is well known, Gauss spent a great deal of time on this result and gave several demonstrations; and it has subsequently stimulated much excellent research. In particular, following works of Jacobi, Eisenstein and Kummer, Hilbert raised as the ninth of his famous list of problems presented at the Paris Congress of 1900 the question of obtaining higher reciprocity laws, and this led to the celebrated studies of Furtwängler, Artin and others in the context of class field theory.

By far the largest section of the *Disquisitiones Arithmeticae* is concerned with the theory of binary quadratic forms. Here Gauss describes how quadratic forms with a given discriminant can be divided into classes so that two forms belong to the same class if and only if there exists an integral unimodular substitution relating them, and how the classes can be divided into genera, so that two forms are in the same genus if and only if they are rationally equivalent. He proceeds to apply these concepts so as, for instance, to throw light on the difficult question of the representation of integers by binary forms. It is a remarkable and beautiful theory with many important ramifications. Indeed, after re-interpretation in terms of quadratic fields, it became apparent that it could be applied much more widely, and in fact it can be regarded as having provided the foundations for the whole of algebraic number theory. The term 'Gaussian field', meaning the field generated over the rationals by i, is a reminder of Gauss' pioneering work in this area.

The remainder of the *Disquisitiones Arithmeticae* contains results of a more miscellaneous character, relating, for instance, to the construction of 17-sided polygons, which was clearly of particular appeal to Gauss, and to what is now termed the cyclotomic field, that is, the field generated by a primitive root of unity. And especially noteworthy here is the discussion of certain sums involving roots of unity, now referred to as Gaussian sums, which play a fundamental role in the analytic theory of numbers.

I conclude this introduction with some words of Mordell. In an essay published in 1917 he wrote 'The theory of numbers is unrivalled for the number and variety of its results and for the beauty and wealth of its demonstrations. The Higher Arithmetic seems to include most of the romance of mathematics. As Gauss wrote to Sophie Germain, the enchanting beauties of this sublime study are revealed in their full charm only to those who have the courage to pursue it.' And Mordell added 'We are reminded of the folk-tales, current amongst all peoples, of the Prince Charming who can assume his proper form as a handsome prince only because of the devotedness of the faithful heroine.'

1
Divisibility

1.1 Foundations

The set $1, 2, 3, \ldots$ of all natural numbers will be denoted by $\mathbb{N}$. There is no need to enter here into philosophical questions concerning the existence of $\mathbb{N}$. It will suffice to assume that it is a given set for which the Peano axioms are satisfied. They imply that addition and multiplication can be defined on $\mathbb{N}$ such that the commutative, associative and distributive laws are valid. Further, an ordering on $\mathbb{N}$ can be introduced so that either $m < n$ or $n < m$ for any distinct elements m, n in $\mathbb{N}$. Furthermore, it is evident from the axioms that the principle of mathematical induction holds and that every non-empty subset of $\mathbb{N}$ has a least member. We shall frequently appeal to these properties.

As customary, we shall denote by $\mathbb{Z}$ the set of integers $0, \pm1, \pm2, \ldots$, and by $\mathbb{Q}$ the set of rationals, that is, the numbers p/q with p in $\mathbb{Z}$ and q in $\mathbb{N}$. The construction, commencing with $\mathbb{N}$, of $\mathbb{Z}$, $\mathbb{Q}$ and then, through Cauchy sequences and ordered pairs, the real and complex numbers $\mathbb{R}$ and $\mathbb{C}$ forms the basis of mathematical analysis and it is assumed known.

1.2 Division algorithm

Suppose that a, b are elements of $\mathbb{N}$. One says that b divides a (written $b|a$) if there exists an element c of $\mathbb{N}$ such that $a = bc$. In this case b is referred to as a divisor of a, and a is called a multiple of b. The relation $b|a$ is reflexive and transitive but not symmetric; in fact if $b|a$ and $a|b$ then $a = b$. Clearly also if $b|a$ then $b \leq a$ and so a natural number has only finitely many divisors. The concept of divisibility is readily extended to $\mathbb{Z}$; if a, b are elements of $\mathbb{Z}$, with $b \neq 0$, then b is said to divide a if there exists c in $\mathbb{Z}$ such that $a = bc$.

We shall frequently appeal to the division algorithm. This asserts that for any a, b in $\mathbb{Z}$, with $b > 0$, there exist q, r in $\mathbb{Z}$ such that $a = bq + r$ and $0 \leq r < b$. The proof is simple; indeed if bq is the largest multiple of b that does not exceed a then the integer $r = a - bq$ is certainly non-negative and, since $b(q + 1) > a$, we have $r < b$. The result plainly remains valid for any integer $b \neq 0$ provided that the bound $r < b$ is replaced by $r < |b|$.

1.3 Greatest common divisor

By the greatest common divisor of natural numbers a, b we mean an element d of $\mathbb{N}$ such that $d|a, d|b$ and every common divisor of a and b also divides d. We proceed to prove that a number d with these properties exists; plainly it will be unique, for any other such number d' would divide a, b and so also d, and since similarly $d|d'$ we have $d = d'$.

Accordingly consider the set of all natural numbers of the form $ax + by$ with x, y in $\mathbb{Z}$. The set is not empty since, for instance, it contains a and b; hence there is a least member d, say. Now $d = ax + by$ for some integers x, y, whence every common divisor of a and b certainly divides d. Further, by the division algorithm, we have $a = dq + r$ for some q, r in $\mathbb{Z}$ with $0 \leq r < d$; this gives $r = ax' + by'$, where $x' = 1 - qx$ and $y' = -qy$. Thus, from the minimal property of d, it follows that $r = 0$, whence $d|a$. Similarly we have $d|b$, as required.

It is customary to signify the greatest common divisor of a, b by (a, b). Clearly, for any n in $\mathbb{N}$, the equation $ax + by = n$ is soluble in integers x, y if and only if (a, b) divides n. In the case $(a, b) = 1$ we say that a and b are relatively prime or coprime (or that a is prime to b). Then the equation $ax + by = n$ is always soluble.

Obviously one can extend these concepts to more than two numbers. In fact one can show that any elements $a_1, \ldots, a_m$ of $\mathbb{N}$ have a greatest common divisor $d = (a_1, \ldots, a_m)$ such that $d = a_1 x_1 + \cdots + a_m x_m$ for some integers $x_1, \ldots, x_m$. Further, if $d = 1$, we say that $a_1, \ldots, a_m$ are relatively prime and then the equation $a_1 x_1 + \cdots + a_m x_m = n$ is always soluble.

1.4 Euclid's algorithm

A method for finding the greatest common divisor d of a, b was described by Euclid. It proceeds as follows.

By the division algorithm there exist integers q_1, r_1 such that $a = bq_1 + r_1$ and $0 \leq r_1 < b$. If $r_1 \neq 0$ then there exist integers q_2, r_2 such that $b = r_1 q_2 + r_2$ and $0 \leq r_2 < r_1$. If $r_2 \neq 0$ then there exist integers q_3, r_3 such that $r_1 = r_2 q_3 + r_3$ and $0 \le r_3 < r_2$. Continuing thus, one obtains a decreasing sequence $r_1, r_2, \ldots$ satisfying $r_{j-2} = r_{j-1} q_j + r_j$. The sequence terminates when $r_{k+1} = 0$ for some k, that is, when $r_{k-1} = r_k q_{k+1}$. It is then readily verified that $d = r_k$. Indeed it is evident from the equations

$$
\begin{aligned}
a &= bq_1 + r_1, & 0 &< r_1 < b; \\
b &= r_1 q_2 + r_2, & 0 &< r_2 < r_1; \\
r_1 &= r_2 q_3 + r_3, & 0 &< r_3 < r_2; \\
&\cdots \\
r_{k-2} &= r_{k-1} q_k + r_k, & 0 &< r_k < r_{k-1}; \\
r_{k-1} &= r_k q_{k+1}
\end{aligned}
$$

that every common divisor of a and b divides $r_1, r_2, \ldots, r_k$; and, moreover, viewing the equations in the reverse order, it is clear that r_k divides each r_j and so also b and a.

Euclid's algorithm furnishes another proof of the existence of integers x, y satisfying $d = ax + by$, and furthermore it enables these x, y to be explicitly calculated. For we have $d = r_k$ and $r_j = r_{j-2} - r_{j-1} q_j$, whence the required values can be obtained by successive substitution. Let us take, for example, $a = 187$ and $b = 35$. Then, following Euclid, we have

$$187 = 35 \cdot 5 + 12, \quad 35 = 12 \cdot 2 + 11, \quad 12 = 11 \cdot 1 + 1.$$

Thus we see that $(187, 35) = 1$ and moreover

$$1 = 12 - 11 \cdot 1 = 12 - (35 - 12 \cdot 2) = 3(187 - 35 \cdot 5) - 35.$$

Hence a solution of the equation $187x + 35y = 1$ in integers x, y is given by $x = 3$, $y = -16$.

For another example let us take $a = 1000$ and $b = 45$; then we get

$$1000 = 45 \cdot 22 + 10, \quad 45 = 10 \cdot 4 + 5, \quad 10 = 5 \cdot 2$$

and so $d = 5$. The solutions to $ax + by = d$ can then be calculated from

$$5 = 45 - 10 \cdot 4 = 45 - (1000 - 45 \cdot 22)4 = 45 \cdot 89 - 1000 \cdot 4$$

which gives $x = -4$, $y = 89$. Note that the process is very efficient: if $a > b$ then a solution x, y can be found in $O((\log a)^3)$ bit operations.

There is a close connection between Euclid's algorithm and the theory of continued fractions; this will be discussed in Chapter 6.

1.5 Fundamental theorem

A natural number, other than 1, is called a prime if it is divisible only by itself and 1. The smallest primes are therefore given by $2, 3, 5, 7, 11, \ldots$.

Let n be any natural number other than 1. The least divisor of n that exceeds 1 is plainly a prime, say p_1. If $n \neq p_1$ then, similarly, there is a prime p_2 dividing n/p_1. If $n \neq p_1 p_2$ then there is a prime p_3 dividing $n/p_1 p_2$; and so on. After a finite number of steps we obtain $n = p_1 \cdots p_m$; and by grouping together we get the standard factorization (or canonical decomposition) $n = p_1^{j_1} \cdots p_k^{j_k}$, where $p_1, \ldots, p_k$ denote distinct primes and $j_1, \ldots, j_k$ are elements of $\mathbb{N}$.

The fundamental theorem of arithmetic asserts that the above factorization is unique except for the order of the factors. To prove the result, note first that if a prime p divides a product mn of natural numbers then either p divides m or p divides n. Indeed if p does not divide m then $(p, m) = 1$, whence there exist integers x, y such that $px + my = 1$; thus we have $pnx + mny = n$ and hence p divides n. More generally we conclude that if p divides $n_1 n_2 \cdots n_k$ then p divides n_l for some l. Now suppose that, apart from the factorization $n = p_1^{j_1} \cdots p_k^{j_k}$ derived above, there is another decomposition and that p' is one of the primes occurring therein. From the preceding conclusion we obtain $p' = p_l$ for some l. Hence we deduce that, if the standard factorization for n/p' is unique, then so also is that for n. The fundamental theorem follows by induction.

It is simple to express the greatest common divisor (a, b) of elements a, b of $\mathbb{N}$ in terms of the primes occurring in their decompositions. In fact we can write $a = p_1^{\alpha_1} \cdots p_k^{\alpha_k}$ and $b = p_1^{\beta_1} \cdots p_k^{\beta_k}$, where $p_1, \ldots, p_k$ are distinct primes and the α s and β s are non-negative integers; then $(a, b) = p_1^{\gamma_1} \cdots p_k^{\gamma_k}$, where $\gamma_l = \min(\alpha_l, \beta_l)$. With the same notation, the lowest common multiple of a, b is defined by $\{a, b\} = p_1^{\delta_1} \cdots p_k^{\delta_k}$, where $\delta_l = \max(\alpha_l, \beta_l)$. The identity $(a, b)\{a, b\} = ab$ is readily verified.

1.6 Properties of the primes

There exist infinitely many primes, for if $p_1, \ldots, p_n$ is any finite set of primes then $p_1 \cdots p_n + 1$ is divisible by a prime different from $p_1, \ldots, p_n$; the argument is due to Euclid. It follows that, if p_n is the nth prime in ascending order of magnitude, then p_m divides $p_1 \cdots p_n + 1$ for some $m \geq n + 1$; from this we deduce by induction that $p_n > 2^{2^n}$. In fact a much stronger result is known; indeed $p_n \sim n \log n$ as $n \to \infty$.[†] The result is equivalent to the assertion that

---

[†] The notation $f \sim g$ means that $f/g \to 1$; and one says that f is asymptotic to g.

the number $\pi(x)$ of primes $p \leq x$ satisfies $\pi(x) \sim x/\log x$ as $x \to \infty$. This is called the prime-number theorem and it was proved by Hadamard and de la Vallée Poussin independently in 1896. Their proofs were based on properties of the Riemann zeta-function about which we shall speak in Chapter 2. In 1737 Euler proved that the series $\sum 1/p_n$ diverges and he noted that this gives another demonstration of the existence of infinitely many primes. In fact it can be shown by elementary arguments that, for some number c,

$$\sum_{p \leq x} 1/p = \log \log x + c + O(1/\log x).$$

Fermat conjectured that the numbers $2^{2^n} + 1$ $(n = 1, 2, \ldots)$ are all primes; this is true for $n = 1, 2, 3$ and 4 but false for $n = 5$, as was proved by Euler. In fact 641 divides $2^{32} + 1$. Numbers of the above form that are primes are called Fermat primes. They are closely connected with the existence of a construction of a regular plane polygon with ruler and compasses only. In fact the regular plane polygon with p sides, where p is a prime, is capable of construction if and only if p is a Fermat prime. It is not known at present whether the number of Fermat primes is finite or infinite.

Numbers of the form $2^n - 1$ that are primes are called Mersenne primes. In this case n is a prime, for plainly $2^m - 1$ divides $2^n - 1$ if m divides n. Mersenne primes are of particular interest in providing examples of large prime numbers; for instance it is known that $2^{44\,497} - 1$ is the 27th Mersenne prime, a number with 13 395 digits.

It is easily seen that no polynomial $f(n)$ with integer coefficients can be prime for all n in $\mathbb{N}$, or even for all sufficiently large n, unless f is constant. Indeed by Taylor's theorem, $f(mf(n) + n)$ is divisible by $f(n)$ for all m in $\mathbb{N}$. On the other hand, the remarkable polynomial $n^2 - n + 41$ is prime for $n = 1, 2, \ldots, 40$. Furthermore one can write down a polynomial $f(n_1, \ldots, n_k)$ with the property that, as the n_j run through the elements of $\mathbb{N}$, the set of positive values assumed by f is precisely the sequence of primes. The latter result arises from studies in logic relating to Hilbert's tenth problem (see Chapter 8).

The primes are well distributed in the sense that, for every $n > 1$, there is always a prime between n and $2n$. This result, which is commonly referred to as Bertrand's postulate, can be regarded as the forerunner of extensive researches on the difference $p_{n+1} - p_n$ of consecutive primes. In fact estimates of the form $p_{n+1} - p_n = O(p_n^\kappa)$ are known with values of κ just a little greater than $\frac{1}{2}$; but, on the other hand, the difference is certainly not bounded, since the consecutive integers $n! + m$ with $m = 2, 3, \ldots, n$ are all composite. A famous theorem of Dirichlet asserts that any arithmetical progression $a, a + q$, $a + 2q, \ldots$, where $(a, q) = 1$, contains infinitely many primes. Some special cases, for instance the existence of infinitely many primes of the form $4n + 3$, can be deduced simply by modifying Euclid's argument given at the beginning, but the general result lies quite deep. Indeed Dirichlet's proof involved, amongst other things, the concepts of characters and L-functions, and of class numbers of quadratic forms, and it has been of far-reaching significance in the history of mathematics.

Two notorious unsolved problems in prime-number theory are the Goldbach conjecture, mentioned in a letter to Euler of 1742, to the effect that every even integer (> 2) is the sum of two primes, and the twin-prime conjecture, to the effect that there exist infinitely many pairs of primes, such as 3, 5 and 17, 19, that differ by 2. By ingenious work on sieve methods, Chen showed in 1974 that these conjectures are valid if one of the primes is replaced by a number with at most two prime factors (assuming, in the Goldbach case, that the even integer is sufficiently large). The oldest known sieve, incidentally, is due to Eratosthenes. He observed that if one deletes from the set of integers $2, 3, \ldots, n$, first all multiples of 2, then all multiples of 3, and so on up to the largest integer not exceeding $\sqrt{n}$, then only primes remain. Studies on Goldbach's conjecture gave rise to the Hardy–Littlewood circle method of analysis and, in particular, to the celebrated theorem of Vinogradov to the effect that every sufficiently large odd integer is the sum of three primes.

1.7 Further reading[‡]

For a good account of the Peano axioms see the book by E. Landau, *Foundations of Analysis* (Chelsea Publishing, 1951).

The division algorithm, Euclid's algorithm and the fundamental theorem of arithmetic are discussed in every elementary text on number theory. The tracts are too numerous to list here but for many years the book by G. H. Hardy and E. M. Wright, *An Introduction to the Theory of Numbers* (Oxford University Press, 2008) has been regarded as a standard work in the field. The books of similar title by T. Nagell (Wiley, 1951) and H. M. Stark (MIT Press, 1978) are also to be recommended, as well as the volume by E. Landau, *Elementary Number Theory* (Chelsea Publishing, 1958).

For properties of the primes, see the book by Hardy and Wright mentioned above and, for more advanced reading, see, for instance, H. Davenport, *Multiplicative Number Theory* (Springer, 2000) and H. Halberstam and H. E. Richert, *Sieve Methods* (Academic Press, 1974). The latter contains, in particular, a proof of Chen's theorem. The result referred to on a polynomial in several

---

[‡] For full publication details please refer to the Bibliography on page 240.

variables representing primes arose from work of Davis, Robinson, Putnam and Matiyasevich on Hilbert's tenth problem; see, for instance, the article by J. P. Jones *et al.* in *American Math. Monthly* **83** (1976), 449–464, where it is shown that 12 variables suffice. The best result to date, due to Matiyasevich, is 10 variables; a proof is given in the article by J. P. Jones in *J. Symbolic Logic* **47** (1982), 549–571.

1.8 Exercises

(i) Find integers x, y such that $22x + 37y = 1$.
(ii) Find integers x, y such that $95x + 432y = 1$.
(iii) Find integers x, y, z such that $6x + 15y + 10z = 1$.
(iv) Find integers x, y, z such that $35x + 55y + 77z = 1$.
(v) Prove that $1 + \frac{1}{2} + \cdots + 1/n$ is not an integer for $n > 1$.
(vi) Prove that

$$(\{a, b\}, \{b, c\}, \{c, a\}) = \{(a, b), (b, c), (c, a)\}.$$

(vii) Prove that if $g_1, g_2, \ldots$ are integers > 1 then every natural number can be expressed uniquely in the form $a_0 + a_1 g_1 + a_2 g_1 g_2 + \cdots + a_k g_1 \cdots g_k$, where the a_j are integers satisfying $0 \le a_j < g_{j+1}$.
(viii) Show that there exist infinitely many primes of the form $4n + 3$.
(ix) Show that, if $2^n + 1$ is a prime, then it is in fact a Fermat prime.
(x) Show that, if $m > n$, then $2^{2^n} + 1$ divides $2^{2^m} - 1$ and so $(2^{2^m} + 1, 2^{2^n} + 1) = 1$.
(xi) Deduce that $p_{n+1} \le 2^{2^n} + 1$, whence $\pi(x) \ge \log\log x$ for $x \ge 2$.

2

Arithmetical functions

2.1 The function [x]

For any real x, one signifies by $[x]$ the largest integer $\leq x$, that is, the unique integer such that $x - 1 < [x] \leq x$. The function is called 'the integral part of x'. It is readily verified that $[x + y] \geq [x] + [y]$ and that, for any positive integer n, $[x + n] = [x] + n$ and $[x/n] = [[x]/n]$. The difference $x - [x]$ is called 'the fractional part of x'; it is written $\{x\}$ and satisfies $0 \leq \{x\} < 1$.

Let now p be a prime. The largest integer l such that p^l divides $n!$ can be neatly expressed in terms of the above function. In fact, on noting that $[n/p]$ of the numbers $1, 2, \ldots, n$ are divisible by p, that $[n/p^2]$ are divisible by p^2, and so on, we obtain

$$l = \sum_{\substack{m=1 \\ p^j | m}}^{n} \sum_{j=1}^{\infty} 1 = \sum_{j=1}^{\infty} \sum_{\substack{m=1 \\ p^j | m}}^{n} 1 = \sum_{j=1}^{\infty} [n/p^j].$$

It follows easily that $l \leq [n/(p-1)]$; for the latter sum is at most $n(1/p + 1/p^2 + \cdots)$. The result also shows at once that the binomial coefficient

$$\binom{m}{n} = \frac{m!}{n!(m-n)!}$$

is an integer; for we have

$$[m/p^j] \geq [n/p^j] + [(m-n)/p^j].$$

Indeed, more generally, if $n_1, \ldots, n_k$ are positive integers such that $n_1 + \cdots + n_k = m$ then the expression $m!/(n_1! \cdots n_k!)$ is an integer.

8

2.2 Multiplicative functions

A real function f defined on the positive integers is said to be multiplicative if $f(m)f(n) = f(mn)$ for all m, n with $(m, n) = 1$. We shall meet many examples. Plainly if f is multiplicative and does not vanish identically then $f(1) = 1$. Further, if $n = p_1{}^{j_1} \cdots p_k{}^{j_k}$ in standard form then

$$f(n) = f(p_1{}^{j_1}) \cdots f(p_k{}^{j_k}).$$

Thus to evaluate f it suffices to calculate its values on the prime powers; we shall appeal to this property frequently.

We shall also use the fact that if f is multiplicative and if

$$g(n) = \sum_{d \mid n} f(d),$$

where the sum is over all divisors d of n, then g is a multiplicative function. Indeed, if $(m, n) = 1$, we have

$$g(mn) = \sum_{d \mid m} \sum_{d' \mid n} f(dd') = \sum_{d \mid m} f(d) \sum_{d' \mid n} f(d')$$

$$= g(m)g(n).$$

2.3 Euler's (totient) function $\phi(n)$

By $\phi(n)$ we mean the number of numbers $1, 2, \ldots, n$ that are relatively prime to n. Thus, in particular, $\phi(1) = \phi(2) = 1$ and $\phi(3) = \phi(4) = 2$.

We shall show, in the next chapter, from properties of congruences, that ϕ is multiplicative. Now, as is easily verified, $\phi(p^j) = p^j - p^{j-1}$ for all prime powers p^j. It follows at once that

$$\phi(n) = n \prod_{p \mid n} (1 - 1/p).$$

We proceed to establish this formula directly without assuming that ϕ is multiplicative. In fact the formula furnishes another proof of this property.

Let $p_1, \ldots, p_k$ be the distinct prime factors of n. Then it suffices to show that $\phi(n)$ is given by

$$n - \sum_{r} (n/p_r) + \sum_{r>s} n/(p_r p_s) - \sum_{r>s>t} n/(p_r p_s p_t) + \cdots .$$

But n/p_r is the number of numbers $1, 2, \ldots, n$ that are divisible by p_r; $n/(p_r p_s)$ is the number that are divisible by $p_r p_s$; and so on. Hence the above expression is

$$\sum_{m=1}^{n} \left(1 - \sum_{\substack{r \\ p_r | m}} 1 + \sum_{\substack{r > s \\ p_r p_s | m}} 1 - \cdots \right) = \sum_{m=1}^{n} \left(1 - \binom{l}{1} + \binom{l}{2} - \cdots \right),$$

where $l = l(m)$ is the number of primes $p_1, \ldots, p_k$ that divide m. Now the summand on the right is $(1-1)^l = 0$ if $l > 0$, and it is 1 if $l = 0$. The required result follows. The demonstration is a particular example of an argument due to Sylvester. Note that the result can be obtained alternatively as an immediate application of the inclusion–exclusion principle. For the respective sums in the required expression for $\phi(n)$ give the number of elements in the set $1, 2, \ldots, n$ that possess precisely $1, 2, 3, \ldots$ of the properties of divisibility by p_j for $1 \le j \le k$ and the principle (or rather the complement of it) gives the analogous expression for the number of elements in an arbitrary set of n objects that possess none of k possible properties.

It is a simple consequence of the multiplicative property of ϕ that

$$\sum_{d|n} \phi(d) = n.$$

In fact the expression on the left is multiplicative and, when $n = p^j$, it becomes

$$\phi(1) + \phi(p) + \cdots + \phi(p^j) = 1 + (p-1) + \cdots + (p^j - p^{j-1}) = p^j.$$

2.4 The Möbius function $\mu(n)$

This is defined, for any positive integer n, as 0 if n contains a squared factor, and as $(-1)^k$ if $n = p_1 \cdots p_k$ as a product of k distinct primes. Further, by convention, $\mu(1) = 1$.

It is clear that μ is multiplicative. Thus the function

$$\nu(n) = \sum_{d|n} \mu(d)$$

is also multiplicative. Now for all prime powers p^j with $j > 0$ we have $\nu(p^j) = \mu(1) + \mu(p) = 0$. Hence we obtain the basic property, namely $\nu(n) = 0$ for $n > 1$ and $\nu(1) = 1$. We proceed to use this property to establish the Möbius inversion formulae.

Let f be any arithmetical function, that is, a function defined on the positive integers, and let

$$g(n) = \sum_{d \mid n} f(d).$$

Then we have

$$f(n) = \sum_{d \mid n} \mu(d) g(n/d).$$

In fact the right-hand side is

$$\sum_{d \mid n} \sum_{d' \mid n/d} \mu(d) f(d') = \sum_{d' \mid n} f(d') v(n/d'),$$

and the result follows since $v(n/d') = 0$ unless $d' = n$. The converse also holds, for we can write the second equation in the form

$$f(n) = \sum_{d' \mid n} \mu(n/d') g(d')$$

and then

$$\sum_{d \mid n} f(d) = \sum_{d \mid n} f(n/d) = \sum_{d \mid n} \sum_{d' \mid n/d} \mu(n/dd') g(d')$$

$$= \sum_{d' \mid n} g(d') v(n/d').$$

Again we have $v(n/d') = 0$ unless $d' = n$, whence the expression on the right is $g(n)$.

The Euler and Möbius functions are related by the equation

$$\phi(n) = n \sum_{d \mid n} \mu(d)/d.$$

This can be seen directly from the formula for ϕ established in Section 2.3, and it also follows at once by Möbius inversion from the property of ϕ recorded at the end of Section 2.3. Indeed the relation is clear from the multiplicative properties of ϕ and μ.

There is an analogue of Möbius inversion for functions defined over the reals, namely if

$$g(x) = \sum_{n \leq x} f(x/n)$$

then

$$f(x) = \sum_{n \leq x} \mu(n) g(x/n).$$

In fact the last sum is

$$\sum_{n \leq x} \sum_{m \leq x/n} \mu(n) f(x/mn) = \sum_{l \leq x} f(x/l) v(l)$$

and the result follows since $v(l) = 0$ for $l > 1$. We shall give several applications of Möbius inversion in the examples at the end of the chapter.

2.5 The functions $\tau(n)$ and $\sigma(n)$

For any positive integer n, we denote by $\tau(n)$ the number of divisors of n (in some books, in particular in that of Hardy and Wright, the function is written $d(n)$). By $\sigma(n)$ we denote the sum of the divisors of n. Thus

$$\tau(n) = \sum_{d|n} 1, \qquad \sigma(n) = \sum_{d|n} d.$$

It is plain that both $\tau(n)$ and $\sigma(n)$ are multiplicative. Further, for any prime power p^j we have $\tau(p^j) = j + 1$ and

$$\sigma(p^j) = 1 + p + \cdots + p^j = (p^{j+1} - 1)/(p - 1).$$

Thus if p^j is the highest power of p that divides n then

$$\tau(n) = \prod_{p|n}(j + 1), \qquad \sigma(n) = \prod_{p|n}(p^{j+1} - 1)/(p - 1).$$

It is easy to give rough estimates for the sizes of $\tau(n)$ and $\sigma(n)$. Indeed we have $\tau(n) < cn^\delta$ for any $\delta > 0$, where c is a number depending only on δ; for the function $f(n) = \tau(n)/n^\delta$ is multiplicative and satisfies $f(p^j) = (j + 1)/p^{j\delta} < 1$ for all but a finite number of values of p and j, the exceptions being bounded in terms of δ. Further, we have

$$\sigma(n) = n \sum_{d|n} 1/d \leq n \sum_{d \leq n} 1/d < n(1 + \log n).$$

The last estimate implies that $\phi(n) > \frac{1}{4}n/\log n$ for $n > 1$. In fact the function $f(n) = \sigma(n)\phi(n)/n^2$ is multiplicative and, for any prime power p^j, we have

$$f(p^j) = 1 - p^{-j-1} \geq 1 - 1/p^2;$$

hence, since

$$\prod_{p|n}(1 - 1/p^2) \geq \prod_{m=2}^{\infty}(1 - 1/m^2) = \frac{1}{2},$$

it follows that $\sigma(n)\phi(n) \geq \frac{1}{2}n^2$, and this together with $\sigma(n) < 2n\log n$ for $n > 2$ gives the estimate for ϕ.

2.6 Average orders

It is often of interest to determine the magnitude 'on average' of arithmetical functions f, that is, to find estimates for sums of the form $\sum f(n)$ with $n \leq x$, where x is a large real number. We shall obtain such estimates when f is τ, σ and ϕ.

First we observe that

$$\sum_{n\leq x} \tau(n) = \sum_{n\leq x}\sum_{d|n} 1 = \sum_{d\leq x}\sum_{m\leq x/d} 1 = \sum_{d\leq x}[x/d].$$

Now we have

$$\sum_{d\leq x} 1/d = \log x + O(1),$$

and hence

$$\sum_{n\leq x} \tau(n) = x\log x + O(x).$$

This implies that $(1/x)\sum \tau(n) \sim \log x$ as $x \to \infty$. The argument can be refined to give

$$\sum_{n\leq x} \tau(n) = x\log x + (2\gamma - 1)x + O(\sqrt{x}),$$

where γ is Euler's constant. Note that although one can say that the 'average order' of $\tau(n)$ is $\log n$ (since $\sum \log n \sim x\log x$), it is not true that 'almost all' numbers have about $\log n$ divisors; here almost all numbers are said to have a certain property if the proportion $\leq x$ not possessing the property is $o(x)$. In fact 'almost all' numbers have about $(\log n)^{\log 2}$ divisors, that is, for any $\varepsilon > 0$ and for almost all n, the function $\tau(n)/(\log n)^{\log 2}$ lies between $(\log n)^\varepsilon$ and $(\log n)^{-\varepsilon}$.

To determine the average order of $\sigma(n)$ we observe that

$$\sum_{n\leq x} \sigma(n) = \sum_{n\leq x}\sum_{d|n}(n/d) = \sum_{d\leq x}\sum_{m\leq x/d} m.$$

The last sum is

$$\frac{1}{2}[x/d]([x/d]+1) = \frac{1}{2}(x/d)^2 + O(x/d).$$

Now

$$\sum_{d\leq x} 1/d^2 = \sum_{d=1}^{\infty} 1/d^2 + O(1/x),$$

and thus we obtain

$$\sum_{n\leq x} \sigma(n) = \frac{1}{12}\pi^2 x^2 + O(x\log x).$$

This implies that the 'average order' of $\sigma(n)$ is $\frac{1}{6}\pi^2 n$ (since $\sum n \sim \frac{1}{2}x^2$).
Finally we derive an average estimate for ϕ. We have

$$\sum_{n\leq x} \phi(n) = \sum_{n\leq x}\sum_{d\mid n} \mu(d)(n/d) = \sum_{d\leq x} \mu(d) \sum_{m\leq x/d} m.$$

The last sum is

$$\frac{1}{2}(x/d)^2 + O(x/d).$$

Now

$$\sum_{d\leq x} \mu(d)/d^2 = \sum_{d=1}^{\infty} \mu(d)/d^2 + O(1/x),$$

and the infinite series here has sum $6/\pi^2$, as will be clear from Section 2.8.
Hence we obtain

$$\sum_{n\leq x} \phi(n) = (3/\pi^2)x^2 + O(x\log x).$$

This implies that the 'average order' of $\phi(n)$ is $6n/\pi^2$. Moreover the result
shows that the probability that two integers are relatively prime is $6/\pi^2$. For
there are $\frac{1}{2}n(n+1)$ pairs of integers p, q with $1 \leq p \leq q \leq n$, and precisely
$\phi(1)+\cdots+\phi(n)$ of the corresponding fractions p/q are in their lowest terms.

2.7 Perfect numbers

A natural number n is said to be perfect if $\sigma(n)=2n$, that is, if n is equal to
the sum of its divisors other than itself. Thus, for instance, 6 and 28 are perfect
numbers.

Whether there exist any odd perfect numbers is a notorious unresolved problem. By contrast, however, the even perfect numbers can be specified precisely.
Indeed an even number is perfect if and only if it has the form $2^{p-1}(2^p - 1)$,

where both p and $2^p - 1$ are primes. It suffices to prove the necessity, for it is readily verified that numbers of this form are certainly perfect. Suppose therefore that $\sigma(n) = 2n$ and that $n = 2^k m$, where k and m are positive integers with m odd. We have $(2^{k+1} - 1)\sigma(m) = 2^{k+1}m$ and hence $\sigma(m) = 2^{k+1}l$ and $m = (2^{k+1} - 1)l$ for some positive integer l. If now l were greater than 1 then m would have distinct divisors l, m and 1, whence we would have $\sigma(m) \geq l + m + 1$. But $l + m = 2^{k+1}l = \sigma(m)$, and this gives a contradiction. Thus $l = 1$ and $\sigma(m) = m + 1$, which implies that m is a prime. In fact m is a Mersenne prime and hence $k + 1$ is a prime p, say (cf. Section 1.6). This shows that n has the required form.

2.8 The Riemann zeta-function

In a classic memoir of 1860 Riemann showed that questions concerning the distribution of the primes are intimately related to properties of the zeta-function

$$\zeta(s) = \sum_{n=1}^{\infty} 1/n^s,$$

where s denotes a complex variable. It is clear that the series converges absolutely for $\sigma > 1$, where $s = \sigma + it$ with σ, t real, and indeed that it converges uniformly for $\sigma > 1 + \delta$ for any $\delta > 0$. Riemann showed that $\zeta(s)$ can be continued analytically throughout the complex plane and that it is regular there except for a simple pole at $s = 1$ with residue 1. He showed moreover that it satisfies the functional equation $\Xi(s) = \Xi(1 - s)$, where

$$\Xi(s) = \pi^{-\frac{1}{2}s}\Gamma(\tfrac{1}{2}s)\zeta(s).$$

The fundamental connection between the zeta-function and the primes is given by the Euler product

$$\zeta(s) = \prod_p (1 - 1/p^s)^{-1},$$

valid for $\sigma > 1$. The relation is readily verified; in fact it is clear that, for any positive integer N,

$$\prod_{p \leq N} (1 - 1/p^s)^{-1} = \prod_{p \leq N} (1 + p^{-s} + p^{-2s} + \cdots) = \sum_m m^{-s},$$

where m runs through all the positive integers that are divisible only by primes $\leq N$, and

$$\left| \sum_m m^{-s} - \sum_{n \leq N} n^{-s} \right| \leq \sum_{n > N} n^{-\sigma} \to 0 \quad \text{as} \quad N \to \infty.$$

The Euler product shows that $\zeta(s)$ has no zeros for $\sigma > 1$. In view of the functional equation it follows that $\zeta(s)$ has no zeros for $\sigma < 0$ except at the points $s = -2, -4, -6, \ldots$; these are termed the 'trivial zeros'. All other zeros of $\zeta(s)$ must lie in the 'critical strip' given by $0 \leq \sigma \leq 1$, and Riemann conjectured that they in fact lie on the line $\sigma = \frac{1}{2}$. This is the famous Riemann hypothesis and it remains unproved to this day. There is much evidence in favour of the hypothesis; in particular Hardy proved in 1915 that infinitely many zeros of $\zeta(s)$ lie on the critical line, and extensive computations have verified that at least the first trillion, that is, 10^{12}, zeros above the real axis do so. It has been shown that, if the hypothesis is true, then, for instance, there is a refinement of the prime-number theorem to the effect that

$$\pi(x) = \int_2^x \frac{dt}{\log t} + O(\sqrt{x} \log x),$$

and that the difference between consecutive primes satisfies $p_{n+1} - p_n = O(p_n^{\frac{1}{2}+\varepsilon})$. In fact it has been shown that there is a narrow zero-free region for $\zeta(s)$ to the left of the line $\sigma = 1$, and this implies that results as above are indeed valid but with weaker error terms. It is also known that the Riemann hypothesis is equivalent to the assertion that, for any $\varepsilon > 0$,

$$\sum_{n \leq x} \mu(n) = O(x^{\frac{1}{2}+\varepsilon}).$$

The basic relation between the Möbius function and the Riemann zeta-function is given by

$$1/\zeta(s) = \sum_{n=1}^{\infty} \mu(n)/n^s.$$

This is clearly valid for $\sigma > 1$ since the product of the series on the right with $\sum 1/n^s$ is $\sum v(n)/n^s$. In fact if the Riemann hypothesis holds then the equation remains true for $\sigma > \frac{1}{2}$. There is a similar equation for the Euler function, valid for $\sigma > 2$, namely

$$\zeta(s-1)/\zeta(s) = \sum_{n=1}^{\infty} \phi(n)/n^s.$$

This is readily verified from the result at the end of Section 2.3. Likewise there are equations for $\tau(n)$ and $\sigma(n)$, valid respectively for $\sigma > 1$ and $\sigma > 2$, namely

$$(\zeta(s))^2 = \sum_{n=1}^{\infty} \tau(n)/n^s, \qquad \zeta(s)\zeta(s-1) = \sum_{n=1}^{\infty} \sigma(n)/n^s.$$

2.9 Further reading

The elementary arithmetical functions are discussed in every introductory text on number theory; again Hardy and Wright's *An Introduction to the Theory of Numbers* (Oxford University Press, 2008) is a good reference. Other books to be recommended are those of T. M. Apostol (Springer, 1976) and K. Chandrasekharan (Springer, 1968), both with the title *Introduction to Analytic Number Theory*; see also Chandrasekharan's *Arithmetical Functions* (Springer, 1970).

As regards the last section, the classic text on the subject is that of E. C. Titchmarsh, *The Theory of the Riemann Zeta-Function* (Oxford University Press, 1986). There are substantial books covering more recent ground by A. Ivić (Wiley, 1985) and by A. A. Karatsuba and S. M. Voronin (de Gruyter, 1992), both with the title *The Riemann Zeta-Function*. The volumes of similar title by H. M. Edwards (Academic Press, 1974) and S. J. Patterson (Cambridge University Press, 1988) provide accessible introductions to the topic.

2.10 Exercises

(i) Evaluate $\sum_{d|n} \mu(d)\sigma(d)$ in terms of the distinct prime factors of n.

(ii) Let $\Lambda(n) = \log p$ if n is a power of a prime p and let $\Lambda(n) = 0$ otherwise (Λ is called von Mangoldt's function). Evaluate $\sum_{d|n} \Lambda(d)$. Express $\sum \Lambda(n)/n^s$ in terms of $\zeta(s)$.

(iii) Let a run through all the integers with $1 \le a \le n$ and $(a, n) = 1$. Show that $f(n) = (1/n) \sum a$ satisfies $\sum_{d|n} f(d) = \frac{1}{2}(n+1)$. Hence prove that $f(n) = \frac{1}{2}\phi(n)$ for $n > 1$.

(iv) Let a run through the integers as in Exercise (iii). Prove that

$$(1/n^3) \sum a^3 = \frac{1}{4}\phi(n)(1 + (-1)^k p_1 \cdots p_k/n^2),$$

where $p_1, \ldots, p_k$ are the distinct prime factors of $n(>1)$.

(v) Show that the product of all the integers a in Exercise (iii) is given by $n^{\phi(n)} \prod_{d|n} (d!/d^d)^{\mu(n/d)}$.

(vi) Show that $\sum_{n \leq x} \mu(n)[x/n] = 1$. Hence prove that $|\sum_{n \leq x} \mu(n)/n| \leq 1$.

(vii) Let m, n be positive integers and let d run through all divisors of (m, n). Prove that $\sum d\mu(n/d) = \mu(n/(m, n))\phi(n)/\phi(n/(m, n))$. (The sum here is called Ramanujan's sum.)

(viii) Prove that if n has k distinct prime factors then $\sum_{d|n} |\mu(d)| = 2^k$.

(ix) Prove that

$$\sum_{d|n}(\mu(d))^2/\phi(d) = n/\phi(n), \quad \sum_{d|2n}\mu(d)\phi(d) = 0.$$

(x) Find all positive integers n such that

(a) $\phi(n)|n$,   (b) $\phi(n) = \frac{1}{2}n$,   (c) $\phi(n) = \phi(2n)$,   (d) $\phi(n) = 12$.

(xi) Prove that $\sum_{n=1}^{\infty} \phi(n)x^n/(1-x^n) = x/(1-x)^2$. (Series of this kind are called Lambert series.)

(xii) Prove that $\sum_{n \leq x} \phi(n)/n = (6/\pi^2)x + O(\log x)$.

3
Congruences

3.1 Definitions

Suppose that a, b are integers and that n is a natural number. By $a \equiv b \pmod{n}$ one means n divides $b - a$; and one says that a is congruent to b modulo n. If $0 \le b < n$ then one refers to b as the residue of $a \pmod{n}$. It is readily verified that the congruence relation is an equivalence relation; the equivalence classes are called residue classes or congruence classes. By a complete set of residues $\pmod{n}$ one means a set of n integers, one from each residue class $\pmod{n}$.

It is clear that if $a \equiv a' \pmod{n}$ and $b \equiv b' \pmod{n}$ then $a + b \equiv a' + b'$ and $a - b \equiv a' - b' \pmod{n}$. Further, we have $ab \equiv a'b' \pmod{n}$, since n divides $(a - a')b + a'(b - b')$. Furthermore, if $f(x)$ is any polynomial with integer coefficients, then $f(a) \equiv f(a') \pmod{n}$.

Note also that if $ka \equiv ka' \pmod{n}$ for some natural number k with $(k, n) = 1$ then $a \equiv a' \pmod{n}$; thus if $a_1, \ldots, a_n$ is a complete set of residues $\pmod{n}$ then so is $ka_1, \ldots, ka_n$. More generally, if k is any natural number such that $ka \equiv ka' \pmod{n}$ then $a \equiv a' \pmod{n/(k, n)}$, since obviously $k/(k, n)$ and $n/(k, n)$ are relatively prime.

3.2 Chinese remainder theorem

Let a, n be natural numbers and let b be any integer. We prove first that the linear congruence $ax \equiv b \pmod{n}$ is soluble for some integer x if and only if (a, n) divides b. The condition is certainly necessary, for (a, n) divides both a and n. To prove the sufficiency, suppose that $d = (a, n)$ divides b. Put $a' = a/d$, $b' = b/d$ and $n' = n/d$. Then it suffices to solve $a'x \equiv b' \pmod{n'}$. But this has precisely one solution $\pmod{n'}$, since $(a', n') = 1$ and so $a'x$ runs through a complete set of residues $\pmod{n'}$ as x runs through such a set. It is clear that

19

if x' is any solution of $a'x' \equiv b'$ (mod n') then the complete set of solutions (mod n) of $ax \equiv b$ (mod n) is given by $x = x' + mn'$, where $m = 1, 2, \ldots, d$. Hence, when d divides b, the congruence $ax \equiv b$ (mod n) has precisely d solutions (mod n).

It follows from the last result that if p is a prime and if a is not divisible by p then the congruence $ax \equiv b$ (mod p) is always soluble; in fact there is a unique solution (mod p). This implies that the residues $0, 1, \ldots, p - 1$ form a field under addition and multiplication (mod p); for indeed every non-zero element has a unique inverse in the multiplicative group. We shall denote the field of residues mod p by $\mathbb{F}_p$.[†] Plainly the field has characteristic p. Since any other finite field with characteristic p is a vector space over $\mathbb{F}_p$, it must have $q = p^e$ elements for some e; an essentially unique field with q elements actually exists but we shall not be concerned with the theory relating to it here.

We turn now to simultaneous linear congruences and prove the Chinese remainder theorem; the result was apparently known to the Chinese at least 1500 years ago. Let $n_1, \ldots, n_k$ be natural numbers and suppose that they are coprime in pairs, that is, $(n_i, n_j) = 1$ for $i \neq j$. The theorem asserts that, for any integers $c_1, \ldots, c_k$, the congruences $x \equiv c_j$ (mod n_j), with $1 \leq j \leq k$, are soluble simultaneously for some integer x; in fact there is a unique solution modulo $n = n_1 \cdots n_k$. For the proof, let $m_j = n/n_j (1 \leq j \leq k)$. Then $(m_j, n_j) = 1$ and thus there is an integer x_j such that $m_j x_j \equiv c_j$ (mod n_j). Now it is readily seen that $x = m_1 x_1 + \cdots + m_k x_k$ satisfies $x \equiv c_j$ (mod n_j), as required. The uniqueness is clear, for if x, y are two solutions then $x \equiv y$ (mod n_j) for $1 \leq j \leq k$, whence, since the n_j are coprime in pairs, we have $x \equiv y$ (mod n). Plainly the Chinese remainder theorem together with the first result of this section implies that if $n_1, \ldots, n_k$ are coprime in pairs then the congruences $a_j x \equiv b_j$ (mod n_j), with $1 \leq j \leq k$, are soluble simultaneously if and only if (a_j, n_j) divides b_j for all j.

As an example, consider the congruences $x \equiv 2$ (mod 5), $x \equiv 3$ (mod 7), $x \equiv 4$ (mod 11). In this case a solution is given by $x = 77x_1 + 55x_2 + 35x_3$, where x_1, x_2, x_3 satisfy $2x_1 \equiv 2$ (mod 5), $6x_2 \equiv 3$ (mod 7), $2x_3 \equiv 4$ (mod 11). Thus we can take $x_1 = 1, x_2 = 4, x_3 = 2$, and these give $x = 367$. The complete solution is $x \equiv -18$ (mod 385). As another example, consider the congruences $x \equiv 1$ (mod 3), $x \equiv 2$ (mod 10), $x \equiv 3$ (mod 11). A solution is given by $x = 110x_1 + 33x_2 + 30x_3$, where x_1, x_2, x_3 satisfy $2x_1 \equiv 1$ (mod 3), $3x_2 \equiv 2$ (mod 10), $8x_3 \equiv 3$ (mod 11). Again solving by inspection, we get $x_1 = 2, x_2 = 4, x_3 = 10$, which gives $x = 652$. The complete solution is $x \equiv -8$ (mod 330).

[†] This is currently the most common of several standard notations; they include $\mathbb{Z}/p\mathbb{Z}$, $\mathbb{Z}/p$ and GF(p) (the Galois field with p elements). The notation $\mathbb{Z}_p$, which was used in the *Concise Introduction*, also commonly occurs but it is open to objection since it clashes with notation customarily adopted in the context of p-adic numbers.

Note that, when (a, n) divides b, an explicit solution to the congruence $ax \equiv b \pmod{n}$ can always be obtained from Euclid's algorithm although, as in the examples above, a simple observation often suffices.

3.3 The theorems of Fermat and Euler

First we introduce the concept of a reduced set of residues (mod n). By this we mean a set of $\phi(n)$ numbers, one from each of the $\phi(n)$ residue classes that consist of numbers relatively prime to n. In particular, the numbers a with $1 \le a \le n$ and $(a, n) = 1$ form a reduced set of residues (mod n).

We proceed now to establish the multiplicative property of ϕ, referred to in Section 2.3, using the above concept. Accordingly let n, n' be natural numbers with $(n, n') = 1$. Further, let a and a' run through reduced sets of residues (mod n) and (mod n') respectively. Then it suffices to prove that $an' + a'n$ runs through a reduced set of residues (mod nn'); for this implies that $\phi(n)\phi(n') = \phi(nn')$, as required. Now clearly, since $(a, n) = 1$ and $(a', n') = 1$, the number $an' + a'n$ is relatively prime to n and to n' and so to nn'. Furthermore any two distinct numbers of the form are incongruent (mod nn'). Thus we have only to prove that if $(b, nn') = 1$ then $b \equiv an' + a'n \pmod{nn'}$ for some a, a' as above. But since $(n, n') = 1$ there exist integers m, m' satisfying $mn' + m'n = 1$. Plainly $(bm, n) = 1$ and so $a \equiv bm \pmod{n}$ for some a; similary $a' \equiv bm' \pmod{n'}$ for some a', and now it is easily seen that a, a' have the required property.

Fermat's theorem states that if a is any natural number and if p is any prime then $a^p \equiv a \pmod{p}$. In particular, if $(a, p) = 1$ then $a^{p-1} \equiv 1 \pmod{p}$. The theorem was announced by Fermat in 1640 but without proof. Euler gave the first demonstration about a century later and, in 1760, he established a more general result to the effect that, if a, n are natural numbers with $(a, n) = 1$ then $a^{\phi(n)} \equiv 1 \pmod{n}$. For the proof of Euler's theorem, we observe simply that as x runs through a reduced set of residues (mod n) so also ax runs through such a set. Hence $\prod(ax) \equiv \prod(x) \pmod{n}$, where the products are taken over all x in the reduced set, and the theorem follows on cancelling $\prod(x)$ from both sides.

3.4 Wilson's theorem

This asserts that $(p - 1)! \equiv -1 \pmod{p}$ for any prime p. Though the result is attributed to Wilson, the statement was apparently first published by Waring in his *Meditationes Algebraicae* of 1770 and a proof was furnished a little later by Lagrange.

For the demonstration, it suffices to assume that p is odd. Now to every integer a with $0 < a < p$ there is a unique integer a' with $0 < a' < p$ such that $aa' \equiv 1 \pmod{p}$. Further, if $a = a'$ then $a^2 \equiv 1 \pmod{p}$, whence $a = 1$ or $a = p - 1$. Thus the set $2, 3, \ldots, p - 2$ can be divided into $\frac{1}{2}(p - 3)$ pairs a, a' with $aa' \equiv 1 \pmod{p}$. Hence we have $2 \cdot 3 \cdots (p - 2) \equiv 1 \pmod{p}$, and so $(p - 1)! \equiv p - 1 \equiv -1 \pmod{p}$, as required.

Wilson's theorem admits a converse and so yields a criterion for primes. Indeed an integer $n > 1$ is a prime if and only if $(n - 1)! \equiv -1 \pmod{n}$. To verify the sufficiency note that any divisor of n, other than itself, must divide $(n - 1)!$.

As an immediate deduction from Wilson's theorem we see that if p is a prime with $p \equiv 1 \pmod{4}$ then the congruence $x^2 \equiv -1 \pmod{p}$ has solutions $x = \pm (r!)$, where $r = \frac{1}{2}(p - 1)$. This follows on replacing $a + r$ in $(p - 1)!$ by the congruent integer $a - r - 1$ for each a with $1 \leq a \leq r$. Note that the congruence has no solutions when $p \equiv 3 \pmod{4}$, for otherwise we would have $x^{p-1} = x^{2r} \equiv (-1)^r = -1 \pmod{p}$, contrary to Fermat's theorem.

3.5 Lagrange's theorem

Let $f(x)$ be a polynomial with integer coefficients and with degree n. Suppose that p is a prime and that the leading coefficient of f, that is, the coefficient of x^n, is not divisible by p. Lagrange's theorem states that the congruence $f(x) \equiv 0 \pmod{p}$ has at most n solutions $\pmod{p}$.

The theorem certainly holds for $n = 1$ by the first result in Section 2.2. We assume that it is valid for polynomials with degree $n - 1$ and proceed inductively to prove the theorem for polynomials with degree n. Now, for any integer a we have $f(x) - f(a) = (x - a)g(x)$, where g is a polynomial with degree $n - 1$, with integer coefficients and with the same leading coefficient as f. Thus if $f(x) \equiv 0 \pmod{p}$ has a solution $x = a$ then all solutions of the congruence satisfy $(x - a)g(x) \equiv 0 \pmod{p}$. But, by the inductive hypothesis, the congruence $g(x) \equiv 0 \pmod{p}$ has at most $n - 1$ solutions $\pmod{p}$. The theorem follows. It is customary to write $f(x) \equiv g(x) \pmod{p}$ to signify that the coefficients of like powers of x in the polynomials f, g are congruent $\pmod{p}$; and it is clear that if the congruence $f(x) \equiv 0 \pmod{p}$ has its full complement $a_1, \ldots, a_n$ of solutions $\pmod{p}$ then

$$f(x) \equiv c(x - a_1) \cdots (x - a_n) \pmod{p},$$

where c is the leading coefficient of f. In particular, by Fermat's theorem, we have

$$x^{p-1} - 1 \equiv (x - 1) \cdots (x - p + 1) \pmod{p},$$

and, on comparing constant coefficients, we obtain another proof of Wilson's theorem.

Plainly, instead of speaking of congruences, we can express the above succinctly in terms of polynomials defined over $\mathbb{F}_p$. Thus Lagrange's theorem asserts that the number of zeros in $\mathbb{F}_p$ of a polynomial defined over this field cannot exceed its degree. The proof proceeds in this instance by supposing that $f(x)$ is a polynomial over $\mathbb{F}_p$ with degree n and with at least one zero a in $\mathbb{F}_p$; then $f(x) = f(x) - f(a) = (x - a)g(x)$, where $g(x)$ is a polynomial over $\mathbb{F}_p$ with degree $n - 1$ and as before, by induction on n, the result follows. As a corollary we deduce that the polynomial $x^d - 1$ has precisely d zeros in $\mathbb{F}_p$ for each divisor d of $p - 1$. For we have $x^{p-1} - 1 = (x^d - 1)g(x)$, where $g(x)$ has degree $p - 1 - d$. But, by Fermat's theorem, $x^{p-1} - 1$ has $p - 1$ zeros in $\mathbb{F}_p$ and, by Lagrange's theorem, $g(x)$ has at most $p - 1 - d$ zeros in $\mathbb{F}_p$. Thus $x^d - 1$ has at least $(p - 1) - (p - 1 - d) = d$ zeros in $\mathbb{F}_p$, whence the assertion. In particular, on taking $d = 4$, we deduce that $x^2 + 1$ has precisely two zeros in $\mathbb{F}_p$ when $p \equiv 1 \pmod 4$, a result related to both Section 3.4 and Section 4.2.

Lagrange's theorem does not remain true for composite moduli. In fact it is readily verified from the Chinese remainder theorem that if $m_1, \ldots, m_k$ are natural numbers coprime in pairs, if $f(x)$ is a polynomial with integer coefficients, and if the congruence $f(x) \equiv 0 \pmod{m_j}$ has s_j solutions $\pmod{m_j}$, then the congruence $f(x) \equiv 0 \pmod m$, where $m = m_1 \cdots m_k$, has $s = s_1 \cdots s_k$ solutions $\pmod m$. Lagrange's theorem is still false for prime power moduli; for example $x^2 \equiv 1 \pmod 8$ has four solutions. But if the prime p does not divide the discriminant of f then the theorem holds for all powers p^j; indeed the number of solutions of $f(x) \equiv 0 \pmod{p^j}$ is, in this case, the same as the number of solutions of $f(x) \equiv 0 \pmod p$. This can be seen at once when, for instance, $f(x) = x^2 - a$; for if p is any odd prime that does not divide a, then from a solution y of $f(y) \equiv 0 \pmod{p^j}$ we obtain a solution $x = y + p^j z$ of $f(x) \equiv 0 \pmod{p^{j+1}}$ by solving the congruence $2yz + f(y)/p^j \equiv 0 \pmod p$ for z, as is possible since $(2y, p) = 1$.

3.6 Primitive roots

Let a, n be natural numbers with $(a, n) = 1$. The least natural number d such that $a^d \equiv 1 \pmod n$ is called the order of $a \pmod n$, and a is said to belong to

$d \pmod{n}$. By Euler's theorem, the order d exists and it divides $\phi(n)$. In fact d divides every integer k such that $a^k \equiv 1 \pmod{n}$, for, by the division algorithm, $k = dq + r$ with $0 \le r < d$, whence $a^r \equiv 1 \pmod{n}$ and so $r = 0$.

By a primitive root $\pmod{n}$ we mean a number that belongs to $\phi(n) \pmod{n}$. Thus, for a prime p, a primitive root $\pmod{p}$ is an integer g, not divisible by p, such that $p - 1$ is the smallest exponent with $g^{p-1} \equiv 1 \pmod{p}$. In other words, a primitive root $\pmod{p}$ can be defined as a generator g of the multiplicative group of the field $\mathbb{F}_p$. It is relatively easy to obtain examples of primitive roots $\pmod{p}$. Thus, if we take $p = 17$, then, by testing sequentially, we find that the smallest primitive root is $g = 3$; in fact the respective powers of 3 $\pmod{17}$ are 3, 9, 10, 13, 5, 15, 11, 16, 14, 8, 7, 4, 12, 2, 6, 1.

We proceed to prove that for every odd prime p there exists a primitive root $\pmod{p}$ and indeed that there are precisely $\phi(p-1)$ primitive roots $\pmod{p}$. Now each of the numbers $1, 2, \ldots, p-1$ belongs $\pmod{p}$ to some divisor d of $p - 1$; let $\psi(d)$ be the number that belongs to $d \pmod{p}$ so that

$$\sum_{d|(p-1)} \psi(d) = p - 1.$$

It will suffice to prove that if $\psi(d) \ne 0$ then $\psi(d) = \phi(d)$. For, by Section 2.3, we have

$$\sum_{d|(p-1)} \phi(d) = p - 1,$$

whence $\psi(d) \ne 0$ for all d and so $\psi(p-1) = \phi(p-1)$ as required.

To verify the assertion concerning ψ, suppose that $\psi(d) \ne 0$ and let a be a number that belongs to $d \pmod{p}$. Then $a, a^2, \ldots, a^d$ are mutually incongruent solutions of $x^d \equiv 1 \pmod{p}$ and thus, by Lagrange's theorem, they represent all the solutions (in fact we showed in Section 2.5 that the congruence has precisely d solutions $\pmod{p}$). It is now easily seen that the numbers a^m with $1 \le m \le d$ and $(m, d) = 1$ represent all the numbers that belong to $d \pmod{p}$; indeed each has order d, for if $a^{md'} \equiv 1$ then $d|d'$, and if b is any number that belongs to $d \pmod{p}$ then $b \equiv a^m$ for some m with $1 \le m \le d$, and we have $(m, d) = 1$ since $b^{d/(m,d)} \equiv (a^d)^{m/(m,d)} \equiv 1 \pmod{p}$. This gives $\psi(d) = \phi(d)$, as asserted.

As noted before, arguments of this kind can be expressed alternatively by referring to the field $\mathbb{F}_p$. In this context, by a primitive root $\pmod{p}$ we mean a generator g of the multiplicative group of $\mathbb{F}_p$ and by the order of a non-zero element a of $\mathbb{F}_p$ we mean the least positive integer d such that $a^d = 1$. Let $\psi(d)$ be the number of elements in $\mathbb{F}_p$ with order d. Supposing that $\psi(d) \ne 0$

and a is any element of $\mathbb{F}_p$ with order d, we show that the $\phi(d)$ elements a^m with $1 \le m \le d$ and $(m, d) = 1$ are precisely those with order d; this gives $\psi(d) = \phi(d)$ as required. Now certainly the a^m with $1 \le m \le d$ are distinct zeros of the polynomial $x^d - 1$ and thus, by Lagrange's theorem, they are all the zeros. Hence any element with order d is given by a^m for some m and, since $(a^m)^{d/(m,d)} = (a^d)^{m/(m,d)} = 1$, we must have $(m, d) = 1$. Further, each of the a^m with $(m, d) = 1$ has order d since $a^{md} = 1$ and md is the smallest multiple of m divisible by d. The result follows.

Let g be a primitive root (mod p). We prove now that there exists an integer x such that $g' = g + px$ is a primitive root (mod p^j) for all prime powers p^j. We have $g^{p-1} = 1 + py$ for some integer y and so, by the binomial theorem, $g'^{p-1} = 1 + pz$, where

$$z \equiv y + (p-1)g^{p-2}x \ (\text{mod } p).$$

The coefficient of x is not divisible by p and so we can choose x such that $(z, p) = 1$. Then g' has the required property. For suppose that g' belongs to d (mod p^j). Then d divides $\phi(p^j) = p^{j-1}(p-1)$. But g' is a primitive root (mod p) and thus $p - 1$ divides d. Hence $d = p^k(p-1)$ for some $k < j$. Further, since p is odd, we have

$$(1 + pz)^{p^k} = 1 + p^{k+1}z_k,$$

where $(z_k, p) = 1$. Now since $g'^d \equiv 1 \ (\text{mod } p^j)$ it follows that $j = k + 1$ and this gives $d = \phi(p^j)$, as required.

Finally we deduce that, for any natural number n, there exists a primitive root (mod n) if and only if n has the form 2, 4, p^j or $2p^j$, where p is an odd prime. Clearly 1 and 3 are primitive roots (mod 2) and (mod 4). Further, if g is a primitive root (mod p^j) then the odd element of the pair $g, g + p^j$ is a primitive root (mod $2p^j$), since $\phi(2p^j) = \phi(p^j)$. Hence it remains only to prove the necessity of the assertion. Now if $n = n_1 n_2$, where $(n_1, n_2) = 1$ and $n_1 > 2, n_2 > 2$, then there is no primitive root (mod n). For $\phi(n_1)$ and $\phi(n_2)$ are even and thus for any natural number a we have

$$a^{\frac{1}{2}\phi(n)} = (a^{\phi(n_1)})^{\frac{1}{2}\phi(n_2)} \equiv 1 \ (\text{mod } n_1);$$

similarly $a^{\frac{1}{2}\phi(n)} \equiv 1 \ (\text{mod } n_2)$, whence $a^{\frac{1}{2}\phi(n)} \equiv 1 \ (\text{mod } n)$. Further, there are no primitive roots (mod 2^j) for $j > 2$, since, by induction, we have $a^{2^{j-2}} \equiv 1 \ (\text{mod } 2^j)$ for all odd numbers a. This proves the theorem.

3.7 Indices

Let g be a primitive root (mod n). The numbers g^l with $l = 0, 1, \ldots,$ $\phi(n) - 1$ form a reduced set of residues (mod n). Hence, for every integer a with $(a, n) = 1$ there is a unique l such that $g^l \equiv a$ (mod n). The exponent l is called the index of a with respect to g and it is denoted by ind a. Plainly we have

$$\text{ind } a + \text{ind } b \equiv \text{ind } (ab) \ (\text{mod } \phi(n)),$$

and ind $1 = 0$, ind $g = 1$. Further, for every natural number m, we have ind $(a^m) \equiv m$ ind a (mod $\phi(n)$). These properties of the index are clearly analogous to the properties of logarithms. We also have ind $(-1) = \frac{1}{2}\phi(n)$ for $n > 2$ since $g^{2 \text{ ind}(-1)} \equiv 1$ (mod n) and 2 ind $(-1) < 2\phi(n)$.

As an example of the use of indices, consider the congruence $x^n \equiv a$ (mod p), where p is a prime. We have n ind $x \equiv$ ind a (mod $(p - 1)$) and thus if $(n, p - 1) = 1$ then there is just one solution. Consider, in particular, $x^5 \equiv 2$ (mod 7). It is readily verified that 3 is a primitive root (mod 7) and we have $3^2 \equiv 2$ (mod 7). Thus 5 ind $x \equiv 2$ (mod 6), which gives ind $x = 4$ and $x \equiv 3^4 \equiv 4$ (mod 7).

Note that although there is no primitive root (mod 2^j) for $j > 2$, the number 5 belongs to 2^{j-2} (mod 2^j) and every odd integer a is congruent (mod 2^j) to just one integer of the form $(-1)^l 5^m$, where $l = 0$, 1 and $m = 0, 1, \ldots, 2^{j-2} - 1$. The pair l, m has similar properties to the index defined above.

3.8 Further reading

A good account of the elementary theory of congruences is given by T. Nagell, *Introduction to Number Theory* (Wiley, 1951); this contains, in particular, a table of primitive roots. There is another and in fact more extensive table in I. M. Vinogradov's *An Introduction to the Theory of Numbers* (Pergamon Press, 1961). Again Hardy and Wright's book of the same title (Oxford University Press, 2008) covers the subject well.

3.9 Exercises

(i) Find an integer x such that $2x \equiv 1$ (mod 3), $3x \equiv 1$ (mod 5), $5x \equiv 1$ (mod 7).

(ii) Find an integer x such that $3x \equiv 1$ (mod 5), $5x \equiv 1$ (mod 17), $7x \equiv 1$ (mod 23).

(iii) Find integers a, b, c, d, e such that the congruences $x \equiv a \pmod 2$, $x \equiv b \pmod 3$, $x \equiv c \pmod 4$, $x \equiv d \pmod 6$, $x \equiv e \pmod{12}$ overlap, that is, such that at least one is soluble for every x.

(iv) Show that $a^{kp-k+1} \equiv a \pmod p$ for all primes p, integers a and positive integers k. Deduce that 798 divides $a^{19} - a$ for all integers a.

(v) Suppose that $a_1, \ldots, a_p$ and $b_1, \ldots, b_p$ are each complete sets of residues $\pmod p$ for a prime p. Is it possible that $a_1 b_1, \ldots, a_p b_p$ is also a complete set of residues $\pmod p$?

(vi) Show that, for an odd prime p, the congruence $x^2 \equiv (-1)^{\frac{1}{2}(p+1)} \pmod p$ has the solution $x = (\frac{1}{2}(p-1))!$.

(vii) Show that, for composite n, the congruence $(n-1)! \equiv 0 \pmod n$ holds with one exception. Show further that $(n-1)! + 1$ is not a power of n.

(viii) Prove that, for any positive integers a, n with $(a, n) = 1$, $\sum \{ax/n\} = \frac{1}{2}\phi(n)$, where the summation is over all x in a reduced set of residues $\pmod n$.

(ix) The integers a and $n > 1$ satisfy $a^{n-1} \equiv 1 \pmod n$ but $a^m \not\equiv 1 \pmod n$ for each divisor m of $n-1$, other than itself. Prove that n is a prime.

(x) Show that the congruence $x^{p-1} - 1 \equiv 0 \pmod{p^j}$ has just $p-1$ solutions $\pmod{p^j}$ for every prime power p^j.

(xii) Prove that, for every natural number n, either there is no primitive root $\pmod n$ or there are $\phi(\phi(n))$ primitive roots $\pmod n$.

(xiii) Prove that, for any prime p, the sum of all the distinct primitive roots $\pmod p$ is congruent to $\mu(p-1) \pmod p$.

(xiv) Prove that, for a prime $p > 3$, the product of all the distinct primitive roots $\pmod p$ is congruent to $1 \pmod p$.

(xv) Prove that if p is a prime and k is a positive integer then $\sum_{n=1}^{p} n^k$ is congruent $\pmod p$ to -1 if $p-1$ divides k and to 0 otherwise.

(xvi) Determine all the solutions of the congruence $y^2 \equiv 5x^3 \pmod 7$ in integers x, y.

(xvii) Prove that, for any prime $p > 3$, the numerator of $1 + \frac{1}{2} + \cdots + 1/(p-1)$ is divisible by p^2 (Wolstenholme's theorem).

4

Quadratic residues

4.1 Legendre's symbol

In the last chapter we discussed the linear congruence $ax \equiv b \pmod{n}$. Here we shall study the quadratic congruence $x^2 \equiv a \pmod{n}$; in fact this amounts to the study of the general quadratic congruence $ax^2 + bx + c \equiv 0 \pmod{n}$, since on writing $d = b^2 - 4ac$ and $y = 2ax + b$, the latter gives $y^2 \equiv d \pmod{4an}$.

Let a be any integer, let n be a natural number and suppose that $(a, n) = 1$. Then a is called a quadratic residue $\pmod{n}$ if the congruence $x^2 \equiv a \pmod{n}$ is soluble; otherwise it is called a quadratic non-residue $\pmod{n}$. The Legendre symbol $\left(\frac{a}{p}\right)$, where p is a prime and $(a, p) = 1$, is defined as 1 if a is a quadratic residue $\pmod{p}$ and as -1 if a is a quadratic non-residue $\pmod{p}$. The symbol is customarily extended to the case when p divides a by defining it as 0 in this instance. Clearly, if $a \equiv a' \pmod{p}$, we have

$$\left(\frac{a}{p}\right) = \left(\frac{a'}{p}\right).$$

4.2 Euler's criterion

This states that if p is an odd prime then

$$\left(\frac{a}{p}\right) \equiv a^{\frac{1}{2}(p-1)} \pmod{p}.$$

For the proof we write, for brevity, $r = \frac{1}{2}(p - 1)$ and we note first that if a is a quadratic residue $\pmod{p}$ then for some x in $\mathbb{N}$ we have $x^2 \equiv a \pmod{p}$, whence, by Fermat's theorem, $a^r \equiv x^{p-1} \equiv 1 \pmod{p}$. Thus it suffices to show that if a is a quadratic non-residue $\pmod{p}$ then $a^r \equiv -1 \pmod{p}$. Now in any reduced set of residues $\pmod{p}$ there are r quadratic residues $\pmod{p}$ and r quadratic non-residues $\pmod{p}$; for the numbers $1^2, 2^2, \ldots, r^2$ are mutually incongruent (mod p) and since, for any integer k, $(p - k)^2 \equiv k^2$ (mod p), the numbers represent all the quadratic residues (mod p). Each of the numbers satisfies $x^r \equiv 1$ (mod p), and, by Lagrange's theorem, the congruence has at most r solutions (mod p). Hence if a is a quadratic non-residue (mod p) then a is not a solution of the congruence. But, by Fermat's theorem, $a^{p-1} \equiv 1$ (mod p), whence $a^r \equiv \pm 1$ (mod p). The required result follows.

It will be seen that the proof given above can be expressed briefly in terms of the field $\mathbb{F}_p$. In fact it is enough to observe that, from Fermat's theorem, every element of $\mathbb{F}_p$ other than 0 is a zero of one of the polynomials $x^{\frac{1}{2}(p-1)} \pm 1$ and, from Lagrange's theorem, $x^{\frac{1}{2}(p-1)} - 1$ has precisely the zeros $1^2, 2^2, \ldots,$ $(\frac{1}{2}(p - 1))^2$, which is a complete set of quadratic residues. Note also that one can argue alternatively in terms of a primitive root (mod p), say g; indeed it is clear that the quadratic residues (mod p) are given by $1, g^2, \ldots, g^{2(r-1)}$.

As an immediate corollary to Euler's criterion we have the multiplicative property of the Legendre symbol, namely

$$\left(\frac{a}{p}\right)\left(\frac{b}{p}\right) = \left(\frac{ab}{p}\right)$$

for all integers a, b not divisible by p; here equality holds since both sides are ± 1. Similarly we have

$$\left(\frac{-1}{p}\right) = (-1)^{\frac{1}{2}(p-1)};$$

in other words, -1 is a quadratic residue of all primes $\equiv 1$ (mod 4) and a quadratic non-residue of all primes $\equiv 3$ (mod 4). It will be recalled from Section 3.4 that when $p \equiv 1$ (mod 4) the solutions of $x^2 \equiv -1$ (mod p) are given by $x = \pm (r!)$.

4.3 Gauss' lemma

For any integer a and any natural number n we define the numerically least residue of a (mod n) as that integer a' for which $a \equiv a'$ (mod n) and $-\frac{1}{2}n < a' \leq \frac{1}{2}n$.

Let now p be an odd prime and suppose that $(a, p) = 1$. Further, let a_j be the numerically least residue of aj (mod p) for $j = 1, 2, \ldots$. Then Gauss' lemma states that

$$\left(\frac{a}{p}\right) = (-1)^l,$$

where l is the number of $j \leq \frac{1}{2}(p - 1)$ for which $a_j < 0$.

For the proof we observe that the numbers $|a_j|$ with $1 \leq j \leq r$, where $r = \frac{1}{2}(p-1)$, are simply the numbers $1, 2, \ldots, r$ in some order. For certainly we have $1 \leq |a_j| \leq r$, and the $|a_j|$ are distinct since $a_j = -a_k$, with $k \leq r$, would give $a(j+k) \equiv 0 \pmod{p}$ with $0 < j+k < p$, which is impossible, and $a_j = a_k$ gives $a_j \equiv ak \pmod{p}$, whence $j = k$. Hence we have $a_1 \cdots a_r = (-1)^l r!$. But $a_j \equiv aj \pmod{p}$ and so $a_1 \cdots a_r \equiv a^r r! \pmod{p}$. Thus $a^r \equiv (-1)^l \pmod{p}$, and the result now follows from Euler's criterion.

As a corollary we obtain

$$\left(\frac{2}{p}\right) = (-1)^{\frac{1}{8}(p^2-1)},$$

that is, 2 is a quadratic residue of all primes $\equiv \pm 1 \pmod{8}$ and a quadratic non-residue of all primes $\equiv \pm 3 \pmod{8}$. To verify this result, note that, when $a = 2$, we have $a_j = 2j$ for $1 \leq j \leq [\frac{1}{4}p]$ and $a_j = 2j - p$ for $[\frac{1}{4}p] < j \leq \frac{1}{2}(p-1)$. Hence in this case $l = \frac{1}{2}(p-1) - [\frac{1}{4}p]$, and it is readily checked that $l \equiv \frac{1}{8}(p^2-1) \pmod 2$.

4.4 Law of quadratic reciprocity

We come now to the famous theorem stated by Euler in 1783 and first proved by Gauss in 1796. Apparently Euler, Legendre and Gauss each discovered the theorem independently and Gauss worked on it intensively for a year before establishing the result; he subsequently gave no fewer than eight demonstrations.

The law of quadratic reciprocity asserts that if p, q are distinct odd primes then

$$\left(\frac{p}{q}\right)\left(\frac{q}{p}\right) = (-1)^{\frac{1}{4}(p-1)(q-1)}.$$

Thus if p, q are not both congruent to 3 (mod 4) then

$$\left(\frac{p}{q}\right) = \left(\frac{q}{p}\right),$$

and in the exceptional case

$$\left(\frac{p}{q}\right) = -\left(\frac{q}{p}\right).$$

For the proof we observe that, by Gauss' lemma, $\left(\frac{p}{q}\right) = (-1)^l$, where l is the number of lattice points (x, y) (that is, pairs of integers) satisfying $0 < x < \frac{1}{2}q$ and $-\frac{1}{2}q < px - qy < 0$. Now these inequalities give $y < (px/q) + \frac{1}{2} < \frac{1}{2}(p+1)$. Hence, since y is an integer, we see that l is the number of lattice

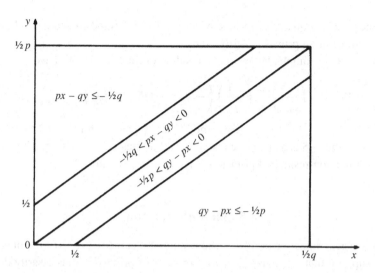

Fig. 4.1   The rectangle R in the proof of the law of quadratic reciprocity.

points in the rectangle R defined by $0 < x < \frac{1}{2}q$, $0 < y < \frac{1}{2}p$, satisfying $-\frac{1}{2}q < px - qy < 0$ (see Fig. 4.1). Similarly

$$\left(\frac{q}{p}\right) = (-1)^m,$$

where m is the number of lattice points in R satisfying $-\frac{1}{2}p < qy - px < 0$. Now it suffices to prove that $\frac{1}{4}(p-1)(q-1) - (l+m)$ is even. But $\frac{1}{4}(p-1)(q-1)$ is just the number of lattice points in R, and thus the latter expression is the number of lattice points in R satisfying either $px - qy \le -\frac{1}{2}q$ or $qy - px \le -\frac{1}{2}p$. The regions in R defined by these inequalities are disjoint and they contain the same number of lattice points since, as is readily verified, the substitution

$$x = \tfrac{1}{2}(q+1) - x', \; y = \tfrac{1}{2}(p+1) - y'$$

furnishes a one–one correspondence between them. The theorem follows.

The law of quadratic reciprocity is useful in the calculation of Legendre symbols. For example, we have

$$\left(\frac{15}{71}\right) = \left(\frac{3}{71}\right)\left(\frac{5}{71}\right) = -\left(\frac{71}{3}\right)\left(\frac{71}{5}\right) = -\left(\frac{2}{3}\right)\left(\frac{1}{5}\right) = 1.$$

Further, for instance, we obtain

$$\left(\frac{-3}{p}\right) = \left(\frac{-1}{p}\right)\left(\frac{3}{p}\right) = (-1)^{\frac{1}{2}(p-1)}\left(\frac{3}{p}\right) = \left(\frac{p}{3}\right),$$

whence -3 is a quadratic residue of all primes $\equiv 1 \pmod 6$ and a quadratic non-residue of all primes $\equiv -1 \pmod 6$.

As another example, let us use the result to evaluate $\left(\frac{-5}{p}\right)$. We have

$$\left(\frac{-5}{p}\right) = \left(\frac{-1}{p}\right)\left(\frac{5}{p}\right) = (-1)^{\frac{1}{2}(p-1)}\left(\frac{p}{5}\right),$$

whence, since $\left(\frac{p}{5}\right) = 1$ for $p \equiv \pm 1 \pmod 5$ and $\left(\frac{p}{5}\right) = -1$ for $p \equiv \pm 2 \pmod 5$, it follows that -5 is a quadratic residue of all primes $\equiv 1, 3, 7, 9 \pmod{20}$ and a quadratic non-residue of primes $\equiv -1, -3, -7, -9 \pmod{20}$.

4.5 Jacobi's symbol

This is a generalization of the Legendre symbol. Let n be a positive odd integer and suppose that $n = p_1 p_2 \cdots p_k$ as a product of primes, not necessarily distinct. Then, for any integer a with $(a, n) = 1$, the Jacobi symbol is defined by

$$\left(\frac{a}{n}\right) = \left(\frac{a}{p_1}\right) \cdots \left(\frac{a}{p_k}\right),$$

where the factors on the right are Legendre symbols. When $n = 1$ the Jacobi symbol is defined as 1 and when $(a, n) > 1$ it is defined as 0. Clearly, if $a \equiv a' \pmod n$ then

$$\left(\frac{a}{n}\right) = \left(\frac{a'}{n}\right).$$

It should be noted at once that

$$\left(\frac{a}{n}\right) = 1$$

does not imply that a is a quadratic residue $\pmod n$. Indeed a is a quadratic residue $\pmod n$ if and only if a is a quadratic residue $\pmod p$ for each prime divisor p of n (see Section 3.5). But

$$\left(\frac{a}{n}\right) = -1$$

does imply that a is a quadratic non-residue $\pmod n$. Thus, for example, since

$$\left(\frac{6}{35}\right) = \left(\frac{6}{5}\right)\left(\frac{6}{7}\right) = \left(\frac{1}{5}\right)\left(\frac{-1}{7}\right) = -1,$$

we conclude that 6 is a quadratic non-residue $\pmod{35}$.

The Jacobi symbol is multiplicative, like the Legendre symbol; that is,

$$\left(\frac{ab}{n}\right) = \left(\frac{a}{n}\right)\left(\frac{b}{n}\right)$$

for all integers a, b relatively prime to n. Further, if m, n are odd and $(a, mn) = 1$ then

$$\left(\frac{a}{mn}\right) = \left(\frac{a}{m}\right)\left(\frac{a}{n}\right).$$

Furthermore we have

$$\left(\frac{-1}{n}\right) = (-1)^{\frac{1}{2}(n-1)}, \qquad \left(\frac{2}{n}\right) = (-1)^{\frac{1}{8}(n^2-1)},$$

and the analogue of the law of quadratic reciprocity holds, namely if m, n are odd and $(m, n) = 1$ then

$$\left(\frac{m}{n}\right)\left(\frac{n}{m}\right) = (-1)^{\frac{1}{4}(m-1)(n-1)}.$$

These results are readily verified from the corresponding theorems for the Legendre symbol on noting that, if $n = n_1 n_2$, then

$$\tfrac{1}{2}(n-1) \equiv \tfrac{1}{2}(n_1 - 1) + \tfrac{1}{2}(n_2 - 1) \ (\text{mod } 2),$$

since $\tfrac{1}{2}(n_1 - 1)(n_2 - 1) \equiv 0 \ (\text{mod } 2)$, and that a similar congruence holds for $\tfrac{1}{8}(n^2 - 1)$.

Jacobi symbols can be used to facilitate the calculation of Legendre symbols. We have, for example,

$$\left(\frac{335}{2999}\right) = -\left(\frac{2999}{335}\right) = -\left(\frac{-16}{335}\right) = -\left(\frac{-1}{335}\right) = 1,$$

whence, since 2999 is a prime, it follows that 335 is a quadratic residue (mod 2999).

For another illustration of the use of the Jacobi symbol consider the equations

$$\left(\frac{21}{275}\right) = \left(\frac{275}{21}\right) = \left(\frac{2}{21}\right) = -1.$$

Now if $\left(\frac{a}{n}\right) = -1$ then $\left(\frac{a}{p}\right) = -1$ for some prime factor p of n and, since $x^2 \equiv a$ (mod n) implies $x^2 \equiv a$ (mod p), it follows that a is a quadratic non-residue of n; hence 21 is a quadratic non-residue of 275. But the converse is not true. For instance, though $\left(\frac{3}{275}\right) = -\left(\frac{2}{3}\right) = 1$, we cannot conclude that 3 is a quadratic residue of 275; indeed $\left(\frac{3}{5}\right) = -1$ and so 3 is a quadratic non-residue of 275.

4.6 Further reading

The theories here date back to the *Disquisitiones Arithmeticae* of Gauss, and they are covered by numerous texts. An excellent account of the history relating to the law of quadratic reciprocity is given by Bachmann, *Niedere Zahlentheorie*, Vol. 1 (Teubner, 1902). In particular he gives references to some 40 different proofs. For an account of modern developments associated with the law of quadratic reciprocity see Artin and Tate, *Class Field Theory* (W. A. Benjamin, 1967) and Cassels and Fröhlich, eds, *Algebraic Number Theory* (Academic Press, 1967).

The study of higher congruences, that is, congruences of the form $f(x_1, \ldots, x_n) \equiv 0 \pmod{p^j}$, where f is a polynomial with integer coefficients, leads to the concept of p-adic numbers and to deep theories in the realm of algebraic geometry; see, for example, Borevich and Shafarevich, *Number Theory* (Academic Press, 1966), and Weil, 'Numbers of solutions of equations in finite fields', *Bull. American Math. Soc.* **55** (1949), 497–508.

4.7 Exercises

(i) Determine the primes p for which 5 is a quadratic residue (mod p).

(ii) Show that if p is a prime $\equiv 3 \pmod 4$ and if $p' = 2p + 1$ is a prime then $2^p \equiv 1 \pmod{p'}$. Deduce that $2^{251} - 1$ is not a Mersenne prime.

(iii) Show that if p is an odd prime then the product P of all the quadratic residues (mod p) satisfies $P \equiv (-1)^{\frac{1}{2}(p+1)} \pmod p$. Show further that, if $p > 3$, then their sum S satisfies $S \equiv 0 \pmod p$. Deduce analogous results for the product and sum of all the quadratic non-residues (mod p).

(iv) Prove that if p is a prime $\equiv 1 \pmod 4$ then $\sum r = \frac{1}{4}p(p-1)$, where the summation is over all quadratic residues r with $1 \le r \le p - 1$.

(v) Use Euler's criterion to show that the primitive roots (mod p) for a prime $p = 2^n + 1$ are precisely the quadratic non-residues (mod p). Deduce that
(a) if $n > 1$ then 3 is a primitive root (mod p),
(b) if $n = 2^k$ with $k > 1$ then 5 is a primitive root (mod p).

(vi) Show that the prime factors of $n^2 + 4$, where n is a positive odd integer, are congruent to 1 or 5 (mod 8). Deduce that there are infinitely many primes congruent to 5 (mod 8). By considering $n^2 + 2$ and $n^2 - 2$, show further that there are infinitely many primes congruent to 3 (mod 8) and to 7 (mod 8).

(vii) Find the least integer $n > 1$ such that $a^n \equiv a \pmod{12\,121}$ for all integers a.

(viii) Let p be an odd prime and let a be an integer not divisible by p. Prove that, if a is a quadratic residue (mod p), then it is a quadratic residue (mod p^k) for all positive integers k.

(ix) Show that, for $p > 3$, the latter holds also for cubic residues; by a cubic residue (mod n), one means an integer a with $(a, n) = 1$ such that $x^3 \equiv a$ (mod n) is soluble.

(x) Evaluate the Jacobi symbol $\left(\frac{123}{917}\right)$.

(xi) Evaluate the Jacobi symbols $\left(\frac{103}{2773}\right)$ and $\left(\frac{117}{3553}\right)$. Are 103 and 117 quadratic residues mod 2773 and mod 3553 respectively?

(xii) Let $f(x) = ax^2 + bx + c$, where a, b, c are integers, and let p be an odd prime that does not divide a. Prove that the number of solutions of the congruence $f(x) \equiv 0$ (mod p) is $1 + \left(\frac{d}{p}\right)$, where $d = b^2 - 4ac$ and $\left(\frac{d}{p}\right) = 0$ if p divides d.

(xiii) Find the number of solutions (mod 997) of
(a) $x^2 + x + 1 \equiv 0$, (b) $x^2 + x - 2 \equiv 0$, (c) $x^2 + 25x - 93 \equiv 0$.

(xiv) With the notation of Exercise (xii), show that, if p does not divide d, then

$$\sum_{x=1}^{p} \left(\frac{f(x)}{p}\right) = -\left(\frac{a}{p}\right).$$

Evaluate the sum when p divides d.

(xv) Prove that if p' is a prime $\equiv 1$ (mod 4) and if $p = 2p' + 1$ is a prime then 2 is a primitive root (mod p). For which primes p' with $p = 2p' + 1$ prime is 5 a primitive root (mod p)?

(xvi) Show that if p is a prime and a, b, c are integers not divisible by p then there are integers x, y such that $ax^2 + by^2 \equiv c$ (mod p).

(xvii) Let $f = f(x_1, \ldots, x_n)$ be a polynomial with integer coefficients that vanishes at the origin and let p be a prime. Prove that if the congruence $f \equiv 0$ (mod p) has only the trivial solution then the polynomial

$$1 - f^{p-1} - (1 - x_1^{p-1}) \cdots (1 - x_n^{p-1})$$

is divisible by p for all integers $x_1, \ldots, x_n$. Deduce that if f has total degree less than n then the congruence $f \equiv 0$ (mod p) has a non-trivial solution (Chevalley's theorem).

(xviii) Prove that if $f = f(x_1, \ldots, x_n)$ is a quadratic form with integer coefficients, if $n \geq 3$ and if p is a prime then the congruence $f \equiv 0$ (mod p) has a non-trivial solution.

5

Quadratic forms

5.1 Equivalence

We shall consider binary quadratic forms

$$f(x, y) = ax^2 + bxy + cy^2,$$

where a, b, c are integers. By the discriminant of f we mean the number $d = b^2 - 4ac$. Plainly $d \equiv 0 \pmod 4$ if b is even and $d \equiv 1 \pmod 4$ if b is odd. The forms $x^2 - \frac{1}{4}dy^2$ for $d \equiv 0 \pmod 4$ and $x^2 + xy + \frac{1}{4}(1-d)y^2$ for $d \equiv 1 \pmod 4$ are called the principal forms with discriminant d. We have

$$4af(x, y) = (2ax + by)^2 - dy^2,$$

whence if $d < 0$ the values taken by f are all of the same sign (or zero); f is called positive or negative definite accordingly. If $d > 0$ then f takes values of both signs and it is called indefinite.

We say that two quadratic forms are equivalent if one can be transformed into the other by an integral unimodular substitution, that is, a substitution of the form

$$x = px' + qy', \qquad y = rx' + sy',$$

where p, q, r, s are integers with $ps - qr = 1$. It is readily verified that this relation is reflexive, symmetric and transitive. Further, it is clear that the set of values assumed by equivalent forms as x, y run through the integers are the same, and indeed they assume the same set of values as the pair x, y runs through all relatively prime integers; for $(x, y) = 1$ if and only if $(x', y') = 1$. Furthermore equivalent forms have the same discriminant. For the substitution takes f into

$$f'(x', y') = a'x'^2 + b'x'y' + c'y'^2,$$

36

where

$$a' = f(p, r), \quad b' = 2apq + b(ps + qr) + 2crs, \quad c' = f(q, s),$$

and it is readily checked that $b'^2 - 4a'c' = d(ps - qr)^2$. Alternatively, in matrix notation, we can write f as $X^T F X$ and the substitution as $X = U X'$, where

$$X = \begin{pmatrix} x \\ y \end{pmatrix}, \quad X' = \begin{pmatrix} x' \\ y' \end{pmatrix}, \quad F = \begin{pmatrix} a & \frac{1}{2}b \\ \frac{1}{2}b & c \end{pmatrix}, \quad U = \begin{pmatrix} p & q \\ r & s \end{pmatrix};$$

then f is transformed into $X'^T F' X'$, where $F' = U^T F U$, and, since the determinant of U is 1, it follows that the determinants of F and F' are equal.

5.2 Reduction

There is an elegant theory of reduction relating to positive definite quadratic forms which we shall now describe. Accordingly we shall assume henceforth that $d < 0$ and that $a > 0$; then we have also $c > 0$.

We begin by observing that by a finite sequence of unimodular substitutions of the form $x = y'$, $y = -x'$ and $x = x' \pm y'$, $y = y'$, f can be transformed into another binary form for which $|b| \leq a \leq c$. For the first of these substitutions interchanges a and c, whence it allows one to replace $a > c$ by $a < c$; and the second has the effect of changing b to $b \pm 2a$, leaving a unchanged, whence, by finitely many applications it allows one to replace $|b| > a$ by $|b| \leq a$. The process must terminate since whenever the first substitution is applied it results in a smaller value of a. In fact we can transform f into a binary form for which either

$$-a < b \leq a < c \quad \text{or} \quad 0 \leq b \leq a = c.$$

For if $b = -a$ then the second of the above substitutions allows one to take $b = a$, leaving c unchanged, and if $a = c$ then the first substitution allows one to take $0 \leq b$. A binary form for which one or other of the above conditions on a, b, c holds is said to be reduced.

There are only finitely many reduced forms with a given discriminant d; for if f is reduced then $-d = 4ac - b^2 \geq 3ac$, whence a, c and $|b|$ cannot exceed $\frac{1}{3}|d|$. The number of reduced forms with discriminant d is called the class number and is denoted by $h(d)$. To calculate the class number when $d = -4$, for example, we note that the inequality $3ac \leq 4$ gives $a = c = 1$, whence $b = 0$ and $h(-4) = 1$. The number $h(d)$ is actually the number of inequivalent classes of binary quadratic forms with discriminant d since, as we shall now prove, any two reduced forms are not equivalent.

Let $f(x, y)$ be a reduced form. Then if x, y are non-zero integers and $|x| \geq |y|$ we have

$$f(x, y) \geq |x|(a|x| - |by|) + c|y|^2$$
$$\geq |x|^2(a - |b|) + c|y|^2 \geq a - |b| + c.$$

Similarly if $|y| \geq |x|$ we have $f(x, y) \geq a - |b| + c$. Hence the smallest values assumed by f for relatively prime integers x, y are a, c and $a - |b| + c$ in that order; these values are taken at $(1, 0)$, $(0, 1)$ and either $(1, 1)$ or $(1, -1)$. Now the sequences of values assumed by equivalent forms for relatively prime x, y are the same, except for a rearrangement, and thus if f' is a form, as in Section 5.1, equivalent to f, and if also f' is reduced, then $a = a'$, $c = c'$ and $b = \pm b'$. It remains therefore to prove that if $b = -b'$ then in fact $b = 0$. We can assume here that $-a < b < a < c$, for, since f' is reduced, we have $-a < -b$, and if $a = c$ then we have $b \geq 0$, $-b \geq 0$, whence $b = 0$. It follows that $f(x, y) \geq a - |b| + c > c > a$ for all non-zero integers x, y. But, with the notation of Section 5.1 for the substitution taking f to f', we have $a = f(p, r)$. Thus $p = \pm 1$, $r = 0$, and from $ps - qr = 1$ we obtain $s = \pm 1$. Further, we have $c = f(q, s)$, whence $q = 0$. Hence the only substitutions taking f to f' are $x = x'$, $y = y'$ and $x = -x'$, $y = -y'$. These give $b = 0$, as required.

5.3 Representations by binary forms

A number n is said to be properly represented by a binary form f if $n = f(x, y)$ for some integers x, y with $(x, y) = 1$. There is a useful criterion in connection with such representations, namely n is properly represented by some binary form with discriminant d if and only if the congruence $x^2 \equiv d \pmod{4n}$ is soluble.

For the proof, suppose first that the congruence is soluble and let $x = b$ be a solution. Define c by $b^2 - 4nc = d$ and put $a = n$. Then the form f, as in Section 5.1, has discriminant d and it properly represents n; in fact $f(1, 0) = n$. Conversely suppose that f has discriminant d and that $n = f(p, r)$ for some integers p, r with $(p, r) = 1$. Then there exist integers q, s with $ps - qr = 1$ and f is equivalent to a form f' as in Section 5.1 with $a' = n$. But f and f' have the same discriminants and so $b'^2 - 4nc' = d$. Hence the congruence $x^2 \equiv d \pmod{4n}$ has a solution $x = b'$.

The ideas here can be developed to furnish, in the case $(n, d) = 1$, the number of proper representations of n by all reduced forms with a given discriminant d. Indeed the quantity in question is given by ws, where s is the number of solutions of the congruence $x^2 \equiv d$ (mod $4n$) with $0 \le x < 2n$ and w is the number of automorphs of a reduced form; by an automorph of f we mean an integral unimodular substitution that takes f into itself. The number w is related to the solutions of the Pell equation (see Section 7.3); it is given by 2 for $d < -4$, by 4 for $d = -4$ and by 6 for $d = -3$. In fact the only automorphs, for $d < -4$, are $x = x'$, $y = y'$ and $x = -x'$, $y = -y'$.

5.4 Sums of two squares

Let n be a natural number. We proceed to prove that n can be expressed in the form $x^2 + y^2$ for some integers x, y if and only if every prime divisor p of n with $p \equiv 3$ (mod 4) occurs to an even power in the standard factorization of n. The result dates back to Fermat and Euler.

The necessity is easily verified, for suppose that $n = x^2 + y^2$ and that n is divisible by a prime $p \equiv 3$ (mod 4). Then $x^2 \equiv -y^2$ (mod p) and, since -1 is a quadratic non-residue (mod p), we see that p divides x and y. Thus we have $(x/p)^2 + (y/p)^2 = n/p^2$, and it follows by induction that p divides n to an even power.

To prove the converse it will suffice to assume that n is square-free and to show that if each odd prime divisor p of n satisfies $p \equiv 1$ (mod 4) then n can be represented by $x^2 + y^2$; for clearly if $n = x^2 + y^2$ then $nm^2 = (xm)^2 + (ym)^2$. Now the quadratic form $x^2 + y^2$ is a reduced form with discriminant -4, and it was proved in Section 5.2 that $h(-4) = 1$. Hence it is the only such reduced form. It follows from Section 5.3 that n is properly represented by $x^2 + y^2$ if and only if the congruence $x^2 \equiv -4$ (mod $4n$) is soluble. But, by hypothesis, -1 is a quadratic residue (mod p) for each prime divisor p of n. Hence -1 is a quadratic residue (mod n) and the result follows.

It will be noted that the argument involves the Chinese remainder theorem; but this can be avoided by appeal to the identity

$$(x^2 + y^2)(x'^2 + y'^2) = (xx' + yy')^2 + (xy' - yx')^2,$$

which enables one to consider only prime values of n. In fact there is a well-known proof of the theorem based on this identity alone, similar to Section 5.5 below.

The demonstration here can be refined to furnish the number of representations of n as $x^2 + y^2$. The number is given by $4 \sum \left(\frac{-1}{m}\right)$, where the summation is over all odd divisors m of n. Thus, for instance, each prime $p \equiv 1$ (mod 4) can be expressed in precisely eight ways as the sum of two squares.

5.5 Sums of four squares

We prove now the famous theorem stated by Bachet in 1621 and first demonstrated by Lagrange in 1770 to the effect that every natural number can be expressed as the sum of four integer squares. Our proof will be based on the identity

$$(x^2 + y^2 + z^2 + w^2)(x'^2 + y'^2 + z'^2 + w'^2)$$
$$= (xx' + yy' + zz' + ww')^2 + (xy' - yx' + wz' - zw')^2$$
$$+ (xz' - zx' + yw' - wy')^2 + (xw' - wx' + zy' - yz')^2,$$

which is related to the theory of quaternions.

In view of the identity and the trivial representation $2 = 1^2 + 1^2 + 0^2 + 0^2$, it will suffice to prove the theorem for odd primes p. Now the numbers x^2 with $0 \le x \le \frac{1}{2}(p-1)$ are mutually incongruent (mod p), and the same holds for the numbers $-1 - y^2$ with $0 \le y \le \frac{1}{2}(p-1)$. Thus we have $x^2 \equiv -1 - y^2$ (mod p) for some x, y satisfying $x^2 + y^2 + 1 < 1 + 2(\frac{1}{2}p)^2 < p^2$. Hence we obtain $mp = x^2 + y^2 + 1$ for some integer m with $0 < m < p$.

Let l be the least positive integer such that $lp = x^2 + y^2 + z^2 + w^2$ for some integers x, y, z, w. Then $l \le m < p$. Further, l is odd, for if l were even then an even number of x, y, z, w would be odd and we could assume that $x + y$, $x - y$, $z + w$, $z - w$ are even; but

$$\tfrac{1}{2}lp = (\tfrac{1}{2}(x+y))^2 + (\tfrac{1}{2}(x-y))^2 + (\tfrac{1}{2}(z+w))^2 + (\tfrac{1}{2}(z-w))^2$$

and this is inconsistent with the minimal choice of l. To prove the theorem we have to show that $l = 1$; accordingly we suppose that $l > 1$ and obtain a contradiction. Let x', y', z', w' be the numerically least residues of x, y, z, w (mod l) and put

$$n = x'^2 + y'^2 + z'^2 + w'^2.$$

Then $n \equiv 0 \pmod{l}$ and we have $n > 0$, for otherwise l would divide p. Further, since l is odd, we have $n < 4(\frac{1}{2}l)^2 = l^2$. Thus $n = kl$ for some integer k with $0 < k < l$. Now by the identity we see that $(kl)(lp)$ is expressible as a sum of four integer squares, and moreover it is clear that each of these squares is divisible by l^2. Thus kp is expressible as a sum of four integer squares. But this contradicts the definition of l and the theorem follows. The argument here is an illustration of Fermat's method of infinite descent.

There is a result dating back to Legendre and Gauss to the effect that a natural number is the sum of three squares if and only if it is not of the form

$4^j(8k+7)$ with j, k non-negative integers. Here the necessity is obvious since a square is congruent to 0, 1 or 4 (mod 8) but the sufficiency depends on the theory of ternary quadratic forms.

Waring conjectured in 1770 that every natural number can be represented as the sum of 4 squares, 9 cubes, 19 biquadrates 'and so on'. One interprets the latter to mean that, for every integer $k \geq 2$ there exists an integer $s = s(k)$ such that every natural number n can be expressed in the form $x_1^k + \cdots + x_s^k$ with $x_1, \ldots, x_s$ non-negative integers; and it is customary to denote the least such s by $g(k)$. Thus we have $g(2) = 4$. Waring's conjecture was proved by Hilbert in 1909. Another, quite different proof was given by Hardy and Littlewood in 1920 and it was here that they described for the first time their famous 'circle method'. The work depends on the identity

$$\sum_{n=0}^{\infty} r(n)z^n = (f(z))^s,$$

where $r(n)$ denotes the number of representations of n in the required form and $f(z) = 1 + z^{1^k} + z^{2^k} + \cdots$. Thus we have

$$r(n) = \frac{1}{2\pi i} \int_C \frac{(f(z))^s}{z^{n+1}} \, dz$$

for a suitable contour C. The argument now involves a delicate division of the contour into 'major and minor' arcs, and the analysis leads to an asymptotic expression for $r(n)$ and to precise estimates for $g(k)$.

5.6 Further reading

A careful account of the theory of binary quadratic forms is given in Landau, *Elementary Number Theory* (Chelsea Publishing, 1958); see also Davenport, *The Higher Arithmetic* (Cambridge University Press, 2008). As there, we have used the classical definition of equivalence in terms of substitutions with determinant 1; however, there is an analogous theory involving substitutions with determinant ± 1 and this is described in Niven, Zuckerman and Montgomery, *An Introduction to the Theory of Numbers* (Wiley, 1991).

For a comprehensive account of the general theory of quadratic forms see Cassels, *Rational Quadratic Forms* (Academic Press, 1978). For an account of the analysis appertaining to Waring's problem see R. C. Vaughan, *The Hardy–Littlewood Method* (Cambridge University Press, 1997).

5.7 Exercises

(i) Prove that $h(d) = 1$ when $d = -3, -4, -7, -8, -11, -19, -43, -67$ and -163.

(ii) Determine all the odd primes that can be expressed in the form $x^2 + xy + 5y^2$.

(iii) Determine all the positive integers that can be expressed in the form $x^2 + 2y^2$.

(iv) Determine all the positive integers that can be expressed in the form $x^2 - y^2$.

(v) Show that there are precisely two reduced forms with discriminant -20. Hence prove that the primes that can be represented by $x^2 + 5y^2$ are 5 and those congruent to 1 or 9 (mod 20).

(vi) Calculate $h(-31)$.

(vii) Find the least positive integer that can be represented by $4x^2 + 17xy + 20y^2$.

(viii) Prove that n and $2n$, where n is any positive integer, have the same number of representations as the sum of two squares.

(ix) Find the least integer s such that $n = 2^k[(3/2)^k] - 1$ can be expressed in the form $n = x_1^k + \cdots + x_s^k$ with $x_1, \ldots, x_s$ positive integers.

6

Diophantine approximation

6.1 Dirichlet's theorem

Diophantine approximation is concerned with the solubility of inequalities in integers. The simplest result in this field was obtained by Dirichlet in 1842. He showed that, for any real θ and any integer $Q > 1$, there exist integers p, q with $0 < q < Q$ such that $|q\theta - p| \leq 1/Q$.

The result can be derived at once from the so-called 'box' or 'pigeon-hole' principle. This asserts that if there are n holes containing $n + 1$ pigeons then there must be at least two pigeons in some hole. Consider in fact the $Q + 1$ numbers $0, 1, \{\theta\}, \{2\theta\}, \ldots, \{(Q - 1)\theta\}$, where $\{x\}$ denotes the fractional part of x as in Chapter 2. These numbers all lie in the interval $[0, 1]$, and if one divides the latter, as clearly one can, into Q disjoint sub-intervals, each of length $1/Q$, then it follows that two of the $Q + 1$ numbers must lie in one of the Q sub-intervals. The difference between the two numbers has the form $q\theta - p$, where p, q are integers with $0 < q < Q$, and we have $|q\theta - p| \leq 1/Q$, as required.

Dirichlet's theorem holds more generally for any real $Q > 1$; the result for non-integral Q follows from the theorem just established with Q replaced by $[Q] + 1$. Further, it is clear that the integers p, q referred to in the theorem can be chosen to be relatively prime. When θ is irrational we have the important corollary that there exist infinitely many rationals $p/q\,(q > 0)$ such that $|\theta - p/q| < 1/q^2$. Indeed, for $Q > 1$, there is a rational p/q with $|\theta - p/q| \leq 1/(Qq) < 1/q^2$; moreover, if θ is irrational then, for any Q' exceeding $1/|q\theta - p|$, the rational corresponding to Q' will be different from p/q. Note that the corollary does not remain valid for rational θ; for if $\theta = a/b$ with a, b integers and $b > 0$ then, when $\theta \neq p/q$, we have $|\theta - p/q| \geq 1/(qb)$ and so there are only finitely many rationals p/q such that $|\theta - p/q| < 1/q^2$.

6.2 Continued fractions

The continued-fraction algorithm sets up a one–one correspondence between all irrational θ and all infinite sets of integers $a_0, a_1, a_2, \ldots$ with $a_1, a_2, \ldots$ positive. It also sets up a one–one correspondence between all rational θ and all finite sets of integers $a_0, a_1, \ldots, a_n$ with $a_1, a_2, \ldots, a_{n-1}$ positive and with $a_n \geq 2$.

To describe the algorithm, let θ be any real number. We put $a_0 = [\theta]$. If $a_0 \neq \theta$ we write $\theta = a_0 + 1/\theta_1$, so that $\theta_1 > 1$, and we put $a_1 = [\theta_1]$. If $a_1 \neq \theta_1$ we write $\theta_1 = a_1 + 1/\theta_2$, so that $\theta_2 > 1$, and we put $a_2 = [\theta_2]$. The process continues indefinitely unless $a_n = \theta_n$ for some n. It is clear that if the latter occurs then θ is rational; in fact we have

$$\theta = a_0 + \cfrac{1}{a_1 + \cfrac{1}{a_2 + \cfrac{1}{\ddots \cfrac{}{a_n}}}}.$$

Conversely, as will be clear in a moment, if θ is rational then the process terminates. The expression above is called the continued fraction for θ; it is customary to write the equation briefly as

$$\theta = a_0 + \frac{1}{a_1+} \frac{1}{a_2+} \cdots \frac{1}{a_n}$$

or, more briefly, as

$$\theta = [a_0, a_1, a_2, \ldots, a_n].$$

If $a_n \neq \theta_n$ for all n, so that the process does not terminate, then θ is irrational. We proceed to show that one can then write

$$\theta = a_0 + \frac{1}{a_1+} \frac{1}{a_2+} \cdots ,$$

or briefly

$$\theta = [a_0, a_1, a_2, \ldots].$$

The integers $a_0, a_1, a_2, \ldots$ are known as the partial quotients of θ; the numbers $\theta_1, \theta_2, \ldots$ are referred to as the complete quotients of θ. We shall prove that the rationals

$$p_n/q_n = [a_0, a_1, \ldots, a_n],$$

where p_n, q_n denote relatively prime integers, tend to θ as $n \to \infty$; they are in fact known as the convergents to θ.

First we show that the p_n, q_n are generated recursively by the equations

$$p_n = a_n p_{n-1} + p_{n-2}, \qquad q_n = a_n q_{n-1} + q_{n-2},$$

where $p_0 = a_0$, $q_0 = 1$ and $p_1 = a_0 a_1 + 1$, $q_1 = a_1$. The recurrences plainly hold for $n = 2$; we assume that they hold for $n = m - 1 \geq 2$ and we proceed to verify them for $n = m$. We define relatively prime integers p'_j, q'_j ($j = 0, 1, \ldots$) by

$$p'_j/q'_j = [a_1, a_2, \ldots, a_{j+1}],$$

and we apply the recurrences to p'_{m-1}, q'_{m-1}; they give

$$p'_{m-1} = a_m p'_{m-2} + p'_{m-3}, \qquad q'_{m-1} = a_m q'_{m-2} + q'_{m-3}.$$

But we have $p_j/q_j = a_0 + q'_{j-1}/p'_{j-1}$, whence

$$p_j = a_0 p'_{j-1} + q'_{j-1}, \qquad q_j = p'_{j-1}.$$

Thus, on taking $j = m$, we obtain

$$p_m = a_m(a_0 p'_{m-2} + q'_{m-2}) + a_0 p'_{m-3} + q'_{m-3},$$
$$q_m = a_m p'_{m-2} + p'_{m-3},$$

and, on taking $j = m - 1$ and $j = m - 2$, it follows that

$$p_m = a_m p_{m-1} + p_{m-2}, \qquad q_m = a_m q_{m-1} + q_{m-2},$$

as required.

Now by the definition of $\theta_1, \theta_2, \ldots$ we have

$$\theta = [a_0, a_1, \ldots, a_n, \theta_{n+1}],$$

where $0 < 1/\theta_{n+1} \leq 1/a_{n+1}$; hence θ lies between p_n/q_n and p_{n+1}/q_{n+1}. It is readily seen by induction that the above recurrences give

$$p_n q_{n+1} - p_{n+1} q_n = (-1)^{n+1},$$

and thus we have

$$|p_n/q_n - p_{n+1}/q_{n+1}| = 1/(q_n q_{n+1}).$$

It follows that the convergents p_n/q_n to θ satisfy

$$|\theta - p_n/q_n| \leq 1/(q_n q_{n+1}),$$

and so certainly $p_n/q_n \to \theta$ as $n \to \infty$.

In view of the latter inequality and the remarks at the end of Section 6.1, it is now clear that when θ is rational the continued-fraction process terminates.

Indeed, for rational θ, the process is closely related to Euclid's algorithm as described in Chapter 1. In fact if we take $\theta = a/b$ then, with the notation of Section 1.4, the partial quotients $a_0, a_1, a_2, \ldots$ of θ are just $q_1, q_2, q_3, \ldots, q_{k+1}$ and the complete quotients $\theta_1, \theta_2, \ldots$ are given by $b/r_1, r_1/r_2, \ldots, r_{k-1}/r_k$. In other words, on defining $a_j = q_{j+1} (0 \le j \le k)$, we have

$$\theta = [a_0, a_1, \ldots, a_k];$$

thus, for example, $\frac{187}{35} = [5, 2, 1, 11]$.

6.3 Rational approximations

It follows from the results of Section 6.2 that, for any real θ, each convergent p/q satisfies $|\theta - p/q| < 1/q^2$. We observe now that, of any two consecutive convergents, say p_n/q_n and p_{n+1}/q_{n+1}, one at least satisfies $|\theta - p/q| < 1/(2q^2)$. Indeed, since $\theta - p_n/q_n$ and $\theta - p_{n+1}/q_{n+1}$ have opposite signs, we have

$$|\theta - p_n/q_n| + |\theta - p_{n+1}/q_{n+1}| = |p_n/q_n - p_{n+1}/q_{n+1}|$$
$$= 1/(q_n q_{n+1});$$

but, for any real α, β with $\alpha \ne \beta$, we have $\alpha\beta < \frac{1}{2}(\alpha^2 + \beta^2)$, whence

$$1/(q_n q_{n+1}) < 1/(2q_n^2) + 1/(2q_{n+1}^2),$$

and this gives the required result. We observe further that, of any three consecutive convergents, say p_n/q_n, p_{n+1}/q_{n+1} and p_{n+2}/q_{n+2}, one at least satisfies $|\theta - p/q| < 1/(\sqrt{5} q^2)$. In fact, if the result were false, then the equations above would give

$$1/(\sqrt{5} q_n^2) + 1/(\sqrt{5} q_{n+1}^2) \le 1/(q_n q_{n+1}),$$

that is, $\lambda + 1/\lambda \le \sqrt{5}$, where $\lambda = q_{n+1}/q_n$. Since λ is rational it follows that strict inequality holds and so

$$(\lambda - \tfrac{1}{2}(1 + \sqrt{5}))(\lambda + \tfrac{1}{2}(1 - \sqrt{5})) < 0,$$

whence $\lambda < \frac{1}{2}(1 + \sqrt{5})$. Similarly, on writing $\mu = q_{n+2}/q_{n+1}$, we would have $\mu < \frac{1}{2}(1 + \sqrt{5})$. But, by Section 6.2, we have $q_{n+2} = a_{n+2} q_{n+1} + q_n$, and thus $\mu \ge 1 + 1/\lambda$; this gives a contradiction, for if $\lambda < \frac{1}{2}(1 + \sqrt{5})$ then $1/\lambda > \frac{1}{2}(-1 + \sqrt{5})$.

The latter result confirms a theorem of Hurwitz to the effect that, for any irrational θ, there exist infinitely many rational p/q such that $|\theta - p/q| < 1/(\sqrt{5}q^2)$. The constant $1/\sqrt{5}$ is best possible, as can be verified (see Section 6.5) by taking

$$\theta = \tfrac{1}{2}(1 + \sqrt{5}) = [1, 1, 1, \ldots].$$

However if one excludes all irrationals equivalent to θ, that is, those whose continued fractions have all but finitely many partial quotients equal to 1, then Hurwitz's theorem holds with $1/\sqrt{8}$ in place of $1/\sqrt{5}$, and this is again best possible. There is an infinite sequence of such results, with constants tending to $1/3$, and they constitute the so-called Markoff chain.

We note next that the convergents give successively closer approximations to θ. In fact we have the stronger result that $|q_n\theta - p_n|$ decreases as n increases. To verify this, we observe that the recurrences in Section 6.2 hold for any indeterminates $a_0, a_1, \ldots$, whence, for $n \geq 1$, we have

$$\theta = \frac{p_n\theta_{n+1} + p_{n-1}}{q_n\theta_{n+1} + q_{n-1}};$$

thus we obtain

$$|q_n\theta - p_n| = 1/(q_n\theta_{n+1} + q_{n-1}),$$

and the assertion follows since, for $n > 1$, the denominator on the right exceeds

$$q_n + q_{n-1} = (a_n + 1)q_{n-1} + q_{n-2} > q_{n-1}\theta_n + q_{n-2},$$

and, for $n = 1$, it exceeds θ_1. The argument here shows, incidentally, that the convergents to θ satisfy

$$\frac{1}{(a_{n+1} + 2)q_n^2} < \left|\theta - \frac{p_n}{q_n}\right| < \frac{1}{a_{n+1}q_n^2}.$$

The convergents are indeed the best approximations to θ in the sense that, if p, q are integers with $0 < q < q_{n+1}$, then $|q\theta - p| \geq |q_n\theta - p_n|$. For if we define integers u, v by

$$p = up_n + vp_{n+1}, \qquad q = uq_n + vq_{n+1},$$

then it is easily seen that $u \neq 0$ and that, if $v \neq 0$, then u, v have opposite signs; hence, since $q_n\theta - p_n$ and $q_{n+1}\theta - p_{n+1}$ have opposite signs, we obtain

$$|q\theta - p| = |u(q_n\theta - p_n) + v(q_{n+1}\theta - p_{n+1})|$$
$$\geq |q_n\theta - p_n|,$$

as required. As a corollary, we deduce that if a rational p/q satisfies $|\theta - p/q| < 1/(2q^2)$ then it is a convergent to θ. In fact we have $p/q = p_n/q_n$, where $q_n \leq q < q_{n+1}$; for clearly

$$|p/q - p_n/q_n| \leq |\theta - p/q| + |\theta - p_n/q_n|$$
$$\leq (1/q + 1/q_n)|q\theta - p|,$$

and, since $q \geq q_n$ and $|q\theta - p| < 1/(2q)$, the number on the right is less than $1/(qq_n)$; hence the number on the left vanishes, as required.

To conclude this section we remark that, for almost all real θ in the sense of Lebesgue measure, the inequality $|\theta - p/q| < 1/(q^2 \log q)$ has infinitely many rational solutions p/q; in fact the same applies to the inequality $|\theta - p/q| < f(q)/q$, where f is any monotonically decreasing function such that $\sum f(q)$ diverges. However, almost no θ have the property if $\sum f(q)$ converges, for instance if $f(q) = 1/(q(\log q)^{1+\delta})$ with $\delta > 0$.

6.4 Quadratic irrationals

By a quadratic irrational we mean a zero of a polynomial $ax^2 + bx + c$, where a, b, c are integers and the discriminant $d = b^2 - 4ac$ is positive and not a perfect square. One of the most remarkable results in the theory of numbers, known since the time of Lagrange, is that a continued fraction represents a quadratic irrational if and only if it is ultimately periodic, that is, if and only if the partial quotients $a_0, a_1, \ldots$ satisfy $a_{m+n} = a_n$ for some positive integer m and for all sufficiently large n. Thus a continued fraction θ is a quadratic irrational if and only if it has the form

$$\theta = [a_0, a_1, \ldots, a_{k-1}, \overline{a_k, \ldots, a_{k+m-1}}],$$

where the bar indicates that the block of partial quotients is repeated indefinitely. As examples, we have $\sqrt{2} = [1, \overline{2}]$, $\frac{1}{3}(3 + \sqrt{3}) = [1, 1, \overline{1, 2}]$ and $\frac{1}{2}(3 + \sqrt{2}) = [2, 4, \overline{1, 4}]$. In the latter instance, the quantity is a root of the equation $4x^2 - 12x + 7 = 0$ and, to generate the continued fraction, one expresses it as $2 + \frac{1}{2}(\sqrt{2} - 1)$ and notes that

$$1/(\tfrac{1}{2}(\sqrt{2} - 1)) = 4 + 2(\sqrt{2} - 1), \quad 1/(2(\sqrt{2} - 1)) = 1 + \tfrac{1}{2}(\sqrt{2} - 1).$$

Similarly, for example, one obtains $\sqrt{20} = [4, \overline{2, 8}]$ and $\sqrt{22} = [4, \overline{1, 2, 4, 2, 1, 8}]$.

It is easy to see that if the continued fraction for θ has the above form then θ is a quadratic irrational. For the number

$$\phi = [\overline{a_k, \ldots, a_{k+m-1}}]$$

is a complete quotient of θ and so, by Section 6.3, we have, for $k \geq 2$,

$$\theta = \frac{p_{k-1}\phi + p_{k-2}}{q_{k-1}\phi + q_{k-2}},$$

where $p_n/q_n (n = 0, 1, \ldots)$ are the convergents to θ; further, we have, for $m \geq 2$,

$$\phi = \frac{p'_{m-1}\phi + p'_{m-2}}{q'_{m-1}\phi + q'_{m-2}},$$

where $p'_n/q'_n (n = 0, 1, \ldots)$ are the convergents to ϕ. It is clear from the latter equation that ϕ is quadratic and hence, by the preceding equation, so also is θ; and this plainly remains valid for $k = 0$ and 1, and for $m = 1$. Since the continued fraction for θ does not terminate, it follows that θ is a quadratic irrational, as required.

To prove the converse, suppose that θ is a quadratic irrational so that θ satisfies an equation $ax^2 + bx + c = 0$, where a, b, c are integers with $d = b^2 - 4ac > 0$. We shall consider the binary form

$$f(x, y) = ax^2 + bxy + cy^2.$$

The substitution

$$x = p_n x' + p_{n-1} y', \qquad y = q_n x' + q_{n-1} y',$$

where $p_n/q_n (n = 1, 2, \ldots)$ denote the convergents to θ, has determinant

$$p_n q_{n-1} - p_{n-1} q_n = (-1)^{n-1},$$

and so, as in Section 5.1, we see that it takes f into a binary form

$$f_n(x, y) = a_n x^2 + b_n xy + c_n y^2$$

with the same discriminant d as f. Further, we have $a_n = f(p_n, q_n)$ and $c_n = a_{n-1}$. Now $f(\theta, 1) = 0$ and so

$$a_n/q_n^2 = f(p_n/q_n, 1) - f(\theta, 1)$$
$$= a((p_n/q_n)^2 - \theta^2) + b((p_n/q_n) - \theta).$$

By Section 6.2 we have $|\theta - p_n/q_n| < 1/q_n^2$, whence

$$|\theta^2 - (p_n/q_n)^2| < |\theta + p_n/q_n|/q_n^2 < (2|\theta| + 1)/q_n^2.$$

Thus we see that

$$|a_n| < (2|\theta| + 1)|a| + |b|,$$

that is, a_n is bounded independently of n. Since $c_n = a_{n-1}$ and $b_n^2 - 4a_n c_n = d$, it follows that b_n and c_n are likewise bounded. But, for $n \geq 1$, we have

$$\theta = \frac{p_n \theta_{n+1} + p_{n-1}}{q_n \theta_{n+1} + q_{n-1}},$$

where $\theta_1, \theta_2, \ldots$ denote the complete quotients of θ, and so $f_n(\theta_{n+1}, 1) = 0$. This implies that there are only finitely many possibilities for $\theta_1, \theta_2, \ldots$, whence $\theta_{l+m} = \theta_l$ for some positive l, m. Hence the continued fraction for θ is ultimately periodic, as required.

The continued fraction of a quadratic irrational θ is said to be purely periodic if $k = 0$ in the expression indicated above. It is easy to show that this occurs if and only if $\theta > 1$ and the conjugate θ' of θ, that is, the other root of the quadratic equation defining θ, satisfies $-1 < \theta' < 0$. Indeed if $\theta > 1$ and $-1 < \theta' < 0$ then it is readily verified by induction that the conjugates θ_n' of the complete quotients θ_n ($n = 1, 2, \ldots$) of θ likewise satisfy $-1 < \theta_n' < 0$; one needs to refer only to the relation $\theta_n' = a_n + 1/\theta_{n+1}'$, where $\theta = [a_0, a_1, \ldots]$, together with the fact that $a_n \geq 1$ for all n including $n = 0$. The inequality $-1 < \theta_n' < 0$ shows that $a_n = [-1/\theta_{n+1}']$. Now since θ is a quadratic irrational we have $\theta_m = \theta_n$ for some distinct m, n; but this gives $1/\theta_m' = 1/\theta_n'$, whence $a_{m-1} = a_{n-1}$. It follows that $\theta_{m-1} = \theta_{n-1}$, and repetition of this conclusion yields $\theta = \theta_{n-m}$, assuming that $n > m$. Hence θ is purely periodic. Conversely, if θ is purely periodic, then $\theta > a_0 \geq 1$. Further, for some $n \geq 1$, we have

$$\theta = \frac{p_n \theta + p_{n-1}}{q_n \theta + q_{n-1}},$$

where p_n/q_n ($n = 1, 2, \ldots$) denote the convergents to θ, and thus θ satisfies the equation

$$q_n x^2 + (q_{n-1} - p_n)x - p_{n-1} = 0.$$

Now the quadratic on the left has the value $-p_{n-1} < 0$ for $x = 0$, and it has the value $p_n + q_n - (p_{n-1} + q_{n-1}) > 0$ for $x = -1$. Hence the conjugate θ' of θ satisfies $-1 < \theta' < 0$, as required.

As an immediate corollary we see that the continued fractions of $\sqrt{d} + [\sqrt{d}]$ and $1/(\sqrt{d} - [\sqrt{d}])$ are purely periodic, where d is any positive integer, not a perfect square. Moreover this implies that the continued fraction of $\sqrt{d}$ is almost purely periodic in the sense that, here, $k = 1$; in other words, only the initial partial quotient a_0 precedes the repeated block. The convergents to $\sqrt{d}$, incidentally, are closely related to the solutions of the Pell equation, about which we shall speak in Chapter 8.

6.5 Liouville's theorem

The work of Section 6.4 shows that every quadratic irrational θ has bounded partial quotients. It follows from the results of Section 6.3 that there exists a number $c = c(\theta) > 0$ such that the inequality $|\theta - p/q| > c/q^2$ holds for all rationals p/q $(q > 0)$. Liouville proved in 1844 that a theorem of the latter kind is valid more generally for any algebraic irrational, and his discovery led to the first demonstration of the existence of transcendental numbers.

A real or complex number is said to be algebraic if it is a zero of a polynomial

$$P(x) = a_0 x^n + a_1 x^{n-1} + \cdots + a_n,$$

where $a_0, a_1, \ldots, a_n$ denote integers, not all 0. For each algebraic number α there is a polynomial P as above, with least degree, such that $P(\alpha) = 0$, and P is unique if one assumes that $a_0 > 0$ and that $a_0, a_1, \ldots, a_n$ are relatively prime; obviously P is irreducible over the rationals, and it is called the minimal polynomial for α. The degree of α is defined as the degree of P.

Liouville's theorem states that for any algebraic number α with degree $n > 1$ there exists a number $c = c(\alpha) > 0$ such that the inequality $|\alpha - p/q| > c/q^n$ holds for all rationals $p/q(q > 0)$. For the proof, we shall assume, as clearly we may, that α is real, and we shall apply the mean-value theorem to P, the minimal polynomial for α. We have, for any rational p/q $(q > 0)$,

$$P(\alpha) - P(p/q) = (\alpha - p/q) P'(\xi),$$

where $P'(x)$ denotes the derivative of P, and ξ lies between α and p/q. Now we have $P(\alpha) = 0$ and, since P is irreducible, we have also $P(p/q) \neq 0$. But $q^n P(p/q)$ is an integer and so $|P(p/q)| \geq 1/q^n$. We can suppose that $|\alpha - p/q| < 1$, for otherwise the theorem certainly holds; then we have $|\xi| < |\alpha| + 1$ and so $|P'(\xi)| < C$ for some $C = C(\alpha)$. This gives $|\alpha - p/q| > c/q^n$, where $c = 1/C$, as required.

The proof here enables one to furnish an explicit value for c in terms of the degree of P and its coefficients. Let us use this observation to confirm the assertion made in Section 6.3 concerning $\alpha = \frac{1}{2}(1 + \sqrt{5})$. In this case we have $P(x) = x^2 - x - 1$ and so $P'(x) = 2x - 1$. Let $p/q(q > 0)$ be any rational and let $\delta = |\alpha - p/q|$. Then $|P(p/q)| \leq \delta |P'(\xi)|$ for some ξ between α and p/q. Now clearly $|\xi| \leq \alpha + \delta$ and so

$$|P'(\xi)| \leq 2(\alpha + \delta) - 1 = 2\delta + \sqrt{5}.$$

But $|P(p/q)| \geq 1/q^2$, whence $\delta(2\delta + \sqrt{5}) \geq 1/q^2$. This implies that for any c' with $c' < 1/\sqrt{5}$ and for all sufficiently large q we have $\delta > c'/q^2$. Hence Hurwitz's theorem (see Section 6.3) is best possible.

A real or complex number that is not algebraic is said to be transcendental. It is now easy to give an example; consider, in fact, the series

$$\theta = 2^{-1!} + 2^{-2!} + 2^{-3!} + \cdots.$$

If we put

$$p_j = 2^{j!}(2^{-1!} + 2^{-2!} + \cdots + 2^{-j!}),$$
$$q_j = 2^{j!} \qquad (j = 1, 2, \ldots),$$

then p_j, q_j are integers, and we have

$$|\theta - p_j/q_j| = 2^{-(j+1)!} + 2^{-(j+2)!} + \cdots.$$

But the sum on the right is at most

$$2^{-(j+1)!}(1 + 2^{-1} + 2^{-2} + \cdots) = 2^{-(j+1)!+1} < q_j^{-j},$$

and it follows readily from Liouville's theorem that θ is transcendental. Indeed any real number θ for which there exists an infinite sequence of distinct rationals p_j/q_j satisfying $|\theta - p_j/q_j| < 1/q_j^{\omega_j}$, where $\omega_j \to \infty$ as $j \to \infty$, will be transcendental. For instance, this will hold for any infinite decimal in which there occur sufficiently long blocks of zeros or any continued fraction in which the partial quotients increase sufficiently rapidly.

There have been some remarkable improvements on Liouville's theorem, beginning with a famous work of Thue in 1909. He showed that for any algebraic number α with degree $n > 1$ and for any $\kappa > \frac{1}{2}n + 1$ there exists $c = c(\alpha, \kappa) > 0$ such that $|\alpha - p/q| > c/q^\kappa$ for all rationals $p/q(q > 0)$. The condition on κ was relaxed by Siegel in 1921 to $\kappa > 2\sqrt{n}$ and it was further relaxed by Dyson and Gelfond, independently, in 1947 to $\kappa > \sqrt{(2n)}$. Finally Roth proved in 1955 that it is enough to take $\kappa > 2$, and this is plainly best possible. There is an intimate connection between such results and the theory of Diophantine equations (see Chapter 8). In this context it is important to know whether the numbers $c(\alpha, \kappa)$ can be evaluated explicitly, that is, whether the results are effective. In fact all the improvements on Liouville's theorem referred to above are, in that sense, ineffective; for they involve a hypothetical assumption, made at the outset, that the inequalities in question have at least one large solution. Nevertheless effective results have been successfully obtained for particular algebraic numbers; for instance, Baker proved in 1964 from properties of hypergeometric functions that, for all rationals $p/q(q > 0)$, we have

$$|\sqrt[3]{2} - p/q| > 10^{-6}/q^{2.955}.$$

Moreover, a small but general effective improvement on Liouville's theorem, that is, valid for any algebraic α, has been established by way of the theory of linear forms in logarithms, referred to in the next section.

6.6 Transcendental numbers

In 1873 Hermite began a new era in number theory when he succeeded in proving that e, the natural base for logarithms, is transcendental. It had earlier been established that e was neither rational nor quadratic irrational; indeed the continued fraction for e was known, namely

$$e = [2, 1, 2, 1, 1, 4, 1, 1, 6, 1, 1, 8, \ldots].$$

But Hermite's work rested on quite different ideas concerning the approximation of analytic functions by rational functions. In 1882 Lindemann found a generalization of Hermite's argument and he obtained thereby his famous proof of the transcendence of π. This sufficed to solve the ancient Greek problem of constructing, with ruler and compasses only, a square with area equal to that of a given circle. In fact, given a unit length, all the points in the plane that are capable of construction are given by the intersection of lines and circles, whence their coordinates in a suitable frame of reference are algebraic numbers. Hence the transcendence of π implies that the length $\sqrt{\pi}$ cannot be classically constructed and so the quadrature of the circle is impossible. Lindemann actually proved that for any distinct algebraic numbers $\alpha_1, \ldots, \alpha_n$ and any non-zero algebraic numbers $\beta_1, \ldots, \beta_n$ we have

$$\beta_1 e^{\alpha_1} + \cdots + \beta_n e^{\alpha_n} \neq 0.$$

The transcendence of π follows in view of Euler's identity $e^{i\pi} = -1$; and the result plainly includes also the transcendence of e, of $\log \alpha$ for algebraic α not 0 or 1, and of the trigonometrical functions $\cos \alpha$, $\sin \alpha$ and $\tan \alpha$ for all non-zero algebraic α.

In the sense of Lebesgue measure, 'almost all' numbers are transcendental; in fact as Cantor observed in 1874, the set of all algebraic numbers is countable. However, it has proved notoriously difficult to demonstrate the transcendence of particular numbers; for instance, Euler's constant γ has resisted any attack, and the same applies to the values $\zeta(2n+1)(n=1, 2, \ldots)$ of the Riemann zeta-function, though Apéry demonstrated in 1978 that $\zeta(3)$ is irrational. In 1900, Hilbert raised, as the seventh of his famous list of 23 problems, the question of proving the transcendence of $2^{\sqrt{2}}$ and, more generally, that of α^β for algebraic α not 0 or 1 and algebraic irrational β. Hilbert expressed the opinion that a solution lay farther in the future than the Riemann hypothesis or Fermat's last theorem. But remarkably, in 1929, following studies on integral integer-valued functions, Gelfond succeeded in verifying the special case that $e^{\pi} = (-1)^{-i}$ is transcendental, and a complete solution to Hilbert's seventh problem was established by Gelfond and Schneider independently in 1934. A generalization of the Gelfond–Schneider theorem was obtained by Baker in 1966; this furnished, for instance, the transcendence of $e^{\beta_0}\alpha_1{}^{\beta_1}\cdots\alpha_n{}^{\beta_n}$, and indeed that of any non-vanishing linear form

$$\beta_1 \log \alpha_1 + \cdots + \beta_n \log \alpha_n,$$

where the αs and βs denote non-zero algebraic numbers. The work enabled quantitative versions of the results to be established, giving positive lower bounds for linear forms in logarithms, and these have played a crucial role in the effective solution of a wide variety of Diophantine problems. We have already referred to one such application at the end of Section 6.5; we shall mention some others later.

Several classical functions, apart from e^z, have been shown to assume transcendental values at non-zero algebraic values of the argument; these include the Weierstrass elliptic function $\wp(z)$, the Bessel function $J_0(z)$ and the elliptic modular function $j(z)$, where, in the latter case, z is necessarily neither real nor imaginary quadratic. In fact there is now a rich and fertile theory relating to the transcendence and algebraic independence of values assumed by analytic functions, and we refer to Section 6.8 for an introduction to the literature.

To illustrate a few of the basic techniques of the theory, we give now a short proof of the transcendence of e; the argument can be extended quite easily to furnish the transcendence of π and indeed the general Lindemann theorem. The proof depends on properties of the integral

$$I(t) = \int_0^t e^{t-x} f(x)dx,$$

defined for $t \geq 0$, where f is a real polynomial with degree m. By integration by parts we have

$$I(t) = e^t \sum_{j=0}^m f^{(j)}(0) - \sum_{j=0}^m f^{(j)}(t),$$

where $f^{(j)}(x)$ denotes the jth derivative of $f(x)$. Further, we observe that, if $\bar{f}$ denotes the polynomial obtained from f by replacing each coefficient with its absolute value, then

$$|I(t)| \leq \int_0^t |e^{t-x} f(x)|dx \leq t\, e^t\, \bar{f}(t).$$

Suppose now that e is algebraic, so that

$$a_0 + a_1 e + \cdots + a_n e^n = 0$$

for some integers $a_0, a_1, \ldots, a_n$ with $a_0 \neq 0$. We put

$$f(x) = x^{p-1}(x-1)^p \cdots (x-n)^p,$$

where p is a large prime; then the degree m of f is $(n+1)p - 1$. We shall compare estimates for

$$J = a_0 I(0) + a_1 I(1) + \cdots + a_n I(n).$$

By the above equations we see that

$$J = -\sum_{j=0}^{m} \sum_{k=0}^{n} a_k f^{(j)}(k).$$

Now, when $1 \leq k \leq n$, we have $f^{(j)}(k) = 0$ for $j < p$, and

$$f^{(j)}(k) = \binom{j}{p} p! g^{(j-p)}(k)$$

for $j \geq p$, where $g(x) = f(x)/(x-k)^p$. Thus, for all j, $f^{(j)}(k)$ is an integer divisible by $p!$. Further, we have $f^{(j)}(0) = 0$ for $j < p-1$, and

$$f^{(j)}(0) = \binom{j}{p-1}(p-1)! \, h^{(j-p+1)}(0)$$

for $j \geq p-1$, where $h(x) = f(x)/x^{p-1}$. Clearly $h^{(j)}(0)$ is an integer divisible by p for $j > 0$, and $h(0) = (-1)^{np}(n!)^p$. Thus, for $j \neq p-1$, $f^{(j)}(0)$ is an integer divisible by $p!$, and $f^{(p-1)}(0)$ is an integer divisible by $(p-1)!$ but not by p for $p > n$. It follows that J is a non-zero integer divisible by $(p-1)!$, whence $|J| \geq (p-1)!$. On the other hand, the trivial estimates $\bar{f}(k) \leq (2n)^m$ and $m \leq 2np$ give

$$|J| \leq |a_1| e \bar{f}(1) + \cdots + |a_n| n \, e^n \bar{f}(n) \leq c^p$$

for some c independent of p. The inequalities are inconsistent for p sufficiently large, and the contradiction shows that e is transcendental, as required.

6.7 Minkowski's theorem

Practically intuitive deductions relating to the geometry of figures in the plane, or, more generally, in Euclidean n-space, can sometimes yield results of great importance in number theory. It was Minkowski who first systematically exploited this observation, and he called the resulting study the Geometry of Numbers. The most famous theorem in this context is the convex-body theorem that Minkowski obtained in 1896. By a convex body we mean a bounded, open set of points in Euclidean n-space that contains $\lambda x + (1 - \lambda)y$ for all λ with $0 < \lambda < 1$ whenever it contains x and y. A set of points is said to be symmetric about the origin if it contains $-x$ whenever it contains x. The simplest form of Minkowski's theorem asserts that if a convex body $\mathcal{S}$, symmetric about the origin, has volume exceeding 2^n then it contains an integer point other than the origin. By an integer point we mean a point all of whose coordinates are integers.

For the proof, it will suffice to verify the following result due to Blichfeldt: any bounded region $\mathcal{R}$ with volume V exceeding 1 contains distinct points x, y such that $x - y$ is an integer point. Minkowski's theorem follows on taking $\mathcal{R} = \frac{1}{2}\mathcal{S}$, that is, the set of points $\frac{1}{2}x$ with x in $\mathcal{S}$, and noting that if x and y belong to $\mathcal{R}$ then $2x$ and $-2y$ belong to $\mathcal{S}$, whence $x - y = \frac{1}{2}(2x - 2y)$ also belongs to $\mathcal{S}$. To prove Blichfeldt's result, we note that $\mathcal{R}$ is the union of disjoint subsets $\mathcal{R}_u$, where $u = (u_1, \ldots, u_n)$ runs through all integer points and $\mathcal{R}_u$ denotes the part of $\mathcal{R}$ that lies in the interval $u_j \le x_j < u_j + 1$ $(1 \le j \le n)$. Thus $V = \sum V_u$, where V_u denotes the volume of $\mathcal{R}_u$, and, by hypothesis, we obtain $\sum V_u > 1$. It follows that if each of the regions $\mathcal{R}_u$ is translated by $-u$ so as to lie in the interval $0 \le x_j < 1$ $(1 \le j \le n)$, then at least two of the translated regions, say the translates of $\mathcal{R}_u$ and $\mathcal{R}_v$, must overlap. Hence there exist points x in $\mathcal{R}_u$ and y in $\mathcal{R}_v$ such that $x - u = y - v$, and so $x - y$ is an integer point, as required.

In order to state the more general form of Minkowski's theorem we need the concept of a lattice. First we recall that points $a_1, \ldots, a_n$ in Euclidean n-space are said to be linearly independent if the only real numbers $t_1, \ldots, t_n$ satisfying $t_1 a_1 + \cdots + t_n a_n = 0$ are $t_1 = \cdots = t_n = 0$; this is equivalent to the condition that $d = \det(a_{ij}) \ne 0$, where $a_j = (a_{1j}, \ldots, a_{nj})$. By a lattice Λ we mean a set of points of the form

$$x = u_1 a_1 + \cdots + u_n a_n,$$

where $a_1, \ldots, a_n$ are fixed linearly independent points, customarily referred to as generators or as a basis for the lattice, and $u_1, \ldots, u_n$ run through all the integers. The determinant of Λ is defined as $d(\Lambda) = |d|$. With this notation, the general Minkowski theorem asserts that if, for any lattice Λ, a convex body $\mathcal{S}$, symmetric about the origin, has volume exceeding $2^n d(\Lambda)$, then it contains a point of Λ other than the origin. The result can be established by simple modifications to the earlier arguments.

As an immediate application, let $\lambda_1, \ldots, \lambda_n$ be positive numbers and let S be the convex body $|x_j| < \lambda_j (1 \leq j \leq n)$; the volume of S is $2^n \lambda_1 \cdots \lambda_n$. Thus, on writing

$$L_j = u_1 a_{j1} + \cdots + u_n a_{jn} \qquad (1 \leq j \leq n),$$

we deduce that if $\lambda_1 \cdots \lambda_n > d(\Lambda)$ then there exist integers $u_1, \ldots, u_n$, not all 0, such that $|L_j| < \lambda_j (1 \leq j \leq n)$. This is referred to as Minkowski's linear forms theorem. It can be sharpened slightly to show that if $\lambda_1 \cdots \lambda_n = d(\Lambda)$ then there exist integers $u_1, \ldots, u_n$, not all 0, such that $|L_1| \leq \lambda_1$ and $|L_j| < \lambda_j$ $(2 \leq j \leq n)$. In fact, for each $m = 1, 2, \ldots$ there exists a non-zero integer point $\boldsymbol{u}_m$ for which $|L_1| < \lambda_1 + 1/m$ and $|L_j| < \lambda_j$ $(2 \leq j \leq n)$; but the $\boldsymbol{u}_m$ are bounded, and so $\boldsymbol{u}_m = \boldsymbol{u}$ for some fixed $\boldsymbol{u}$ and infinitely many m, whence $\boldsymbol{u} = (u_1, \ldots, u_n)$ has the required properties.

Minkowski's linear forms theorem implies that if $\theta_1, \ldots, \theta_n$ are any real numbers and if $Q > 0$ then there exist integers $p, q_1, \ldots, q_n$, not all 0, such that $|q_j| < Q(1 \leq j \leq n)$ and

$$|q_1 \theta_1 + \cdots + q_n \theta_n - p| \leq Q^{-n}.$$

Similarly we see that there exist integers $p_1, \ldots, p_n, q$, not all 0, such that $|q| \leq Q^n$ and $|q\theta_j - p_j| < 1/Q(1 \leq j \leq n)$. It follows that, if one at least of $\theta_1, \ldots, \theta_n$ is irrational, then

$$|\theta_j - p_j/q| < q^{-1-1/n} \qquad (1 \leq j \leq n)$$

for infinitely many rationals $p_j/q(q > 0)$. These results generalize Dirichlet's theorem discussed in Section 6.1. In the opposite direction, it is easy to extend the observation on quadratic irrationals made in Section 6.5 to show that, when θ is an algebraic number with degree $n + 1$, there exists $c = c(\theta) > 0$ such that

$$|q_1 \theta + \cdots + q_n \theta^n - p| > cq^{-n}$$

for all integers $p, q_1, \ldots, q_n$ with $q = \max |q_j| > 0$. This implies, by a classical transference principle, that the exponent $-1 - 1/n$ above is best possible. It is known from transcendence theory that, for any $\varepsilon > 0$, there exists $c > 0$ such that

$$|q_1 e + \cdots + q_n e^n - p| > cq^{-n-\varepsilon}$$

for all integers $p, q_1, \ldots, q_n$ with $q = \max |q_j| > 0$. Moreover, some deep work of Schmidt, generalizing the Thue–Siegel–Roth theorem, shows that the same holds when $e, \ldots, e^n$ are replaced by algebraic numbers $\theta_1, \ldots, \theta_n$ with $1, \theta_1, \ldots, \theta_n$ linearly independent over the rationals; in analogy with lattice points, we say that real numbers $\phi_1, \ldots, \phi_m$ are linearly independent over the rationals if the only rationals $t_1, \ldots, t_m$ satisfying $t_1\phi_1 + \cdots + t_m\phi_m = 0$ are $t_1 = \cdots = t_m = 0$.

Minkowski conjectured that if $L_1, \ldots, L_n$ are linear forms as above and if $\theta_1, \ldots, \theta_n$ are any real numbers then there exist integers $u_1, \ldots, u_n$ such that

$$|L_1 - \theta_1| \cdots |L_n - \theta_n| \leq 2^{-n} d(\Lambda).$$

At present the conjecture remains open, but it is trivial in the case $n = 1$ and Minkowski himself proved that it is valid in the case $n = 2$. It has subsequently been verified for $n = 3$, 4 and 5, and Tchebotarev showed that it holds for all n if 2^{-n} is replaced by $2^{-(1/2)n}$. Minkowski's work furnished a result to the effect that if θ is irrational and θ' is not of the form $r\theta + s$ for integers r, s, then there are infinitely many integers $q \neq 0$ such that, for some integer p,

$$|q\theta - p - \theta'| < 1/(4|q|);$$

and here the constant $1/4$ is best possible. The result implies that the numbers $\{n\theta\}$, where $n = 1, 2, \ldots$, are dense in the unit interval, that is, for every θ' with $0 < \theta' < 1$, and for every $\varepsilon > 0$, we have $|\{n\theta\} - \theta'| < \varepsilon$ for some n. A famous theorem of Kronecker implies that, more generally, the points

$$(\{n\theta_1\}, \ldots, \{n\theta_m\}) \qquad (n = 1, 2, \ldots),$$

where $1, \theta_1, \ldots, \theta_m$ are linearly independent over the rationals, are dense in the unit cube in Euclidean m-space.

6.8 Further reading

The classic text on continued fractions is Perron's *Die Lehre von den Kettenbrüchen* (Teubner, 1913). There are, however, useful accounts in most introductory works on number theory; see, in particular, Cassels' *An Introduction to Diophantine Approximation* (Cambridge University Press, 1957), and the books of Niven, Zuckerman and Montgomery (Wiley, 1991) and of Hardy and Wright (Oxford University Press, 2008) cited earlier. A nice, short work is Khintchine's *Kettenbrüche* (Teubner, 1956).

Numerous references to the literature relating to Sections 6.5 and 6.6 are given in Baker's *Transcendental Number Theory* (Cambridge University Press, 1990). The area is further covered by *Logarithmic Forms and Diophantine Geometry* (Cambridge University Press, 2007) by Baker and Wüstholz; this describes, in particular, the much fruitful interplay that now exists between transcendence theory and arithmetical algebraic geometry.

For advanced work concerning rational approximations to algebraic numbers see W. M. Schmidt, *Diophantine Approximation* (Springer, 1980). The topics referred to in Section 6.7 are discussed fully in Cassels' *An Introduction to the Geometry of Numbers* (Springer, 1971).

6.9 Exercises

(i) Express the continued fraction $[1, 2, \overline{3, 4}]$ as a quadratic irrational and hence determine its minimal polynomial.

(ii) Evaluate the continued fraction $[1, 2, 3, \overline{1, 4}]$.

(iii) Assuming that π is given by $3.1415926\ldots$, correct to seven decimal places, prove that the first three convergents to π are $\frac{22}{7}$, $\frac{333}{106}$ and $\frac{355}{113}$. Verify that $\left|\pi - \frac{355}{113}\right| < 10^{-6}$.

(iv) Assuming that $e = 2.71828182845\ldots$ correct to 11 decimal places, verify that the first few partial quotients to e are as indicated in Section 6.6. Show that $\frac{2721}{1001}$ is a convergent and verify that it differs from e by less than 2×10^{-7}.

(v) Let θ, θ' be the roots of the equation $x^2 - ax - 1 = 0$, where a is a positive integer and $\theta > 0$. Show that the denominators in the convergents to θ are given by $q_{n-1} = (\theta^n - \theta'^n)/(\theta - \theta')$. Verify that the Fibonacci sequence $1, 1, 2, 3, 5, \ldots$ is given by $q_0, q_1, \ldots$ in a special case.

(vi) Prove that the denominators q_n in the convergents to any real θ satisfy $q_n \geq (\frac{1}{2}(1 + \sqrt{5}))^{n-1}$. Prove also that, if the partial quotients are bounded above by a constant A, then $q_n \leq (\frac{1}{2}(A + \sqrt{(A^2 + 4)}))^n$.

(vii) Assuming that the continued fraction for e is as quoted in Section 6.6, show that $|e - p/q| > c/(q^2 \log q)$ for all rationals p/q ($q > 1$), where c is a positive constant.

(viii) Prove that, if the partial quotients $a_0, a_1, a_2, \ldots$ in the continued fraction of a real number θ form an increasing sequence, then the denominators q_n in the convergents to θ satisfy $q_n \leq (a_n + 1)^n$. Hence verify that, if $a_{n+1} > (a_n + 1)^{n^2}$ for all n, then θ is transcendental.

(ix) Assuming the Thue–Siegel–Roth theorem, show that the sum $a^{-b} + a^{-b^2} + a^{-b^3} + \ldots$ is transcendental for any integers $a \geq 2$, $b \geq 3$.

(x) Let α, β, γ, δ be real numbers with $\Delta = \alpha\delta - \beta\gamma \neq 0$. Prove that there exist integers x, y, not both 0, such that $|L| + |M| \leq \sqrt{(2|\Delta|)}$, where $L = \alpha x + \beta y$ and $M = \gamma x + \delta y$. Deduce that the inequality $|LM| \leq \frac{1}{2}|\Delta|$ is soluble non-trivially.

(xi) With the same notation, prove that the inequality $L^2 + M^2 \leq (4/\pi)|\Delta|$ is soluble in integers x, y, not both 0. Verify that the constant $4/\pi$ cannot be replaced by a number smaller than $\sqrt{(4/3)}$.

(xii) Assuming Kronecker's theorem and the transcendence of e^π, show that, for any primes $p_1, \ldots, p_m$, there exists an integer $n > 0$ such that

$$\cos(\log p_j^n) \leq -\tfrac{1}{2} \quad (j = 1, 2, \ldots, m).$$

7

Quadratic fields

7.1 Algebraic number fields

Although we shall be concerned principally in this chapter only with quadratic fields, we shall nevertheless begin with a short discussion of the more general concept of an algebraic number field. The theory relating to such fields arose from attempts to solve Fermat's last theorem and it is one of the most beautiful and profound in mathematics.

Let α be an algebraic number with degree n and let P be the minimal polynomial for α (see Section 6.5). By the conjugates of α we mean the zeros $\alpha_1, \ldots, \alpha_n$ of P. The algebraic number field k generated by α over the rationals $\mathbb{Q}$ is defined as the set of numbers $Q(\alpha)$, where $Q(x)$ is any polynomial with rational coefficients; the set can be regarded as being embedded in the complex number field $\mathbb{C}$ and thus its elements are subject to the usual operations of addition and multiplication. To verify that k is indeed a field we have to show that every non-zero element $Q(\alpha)$ has an inverse. Now, if P is the minimal polynomial for α as above, then P, Q are relatively prime and so there exist polynomials R, S such that $PS + QR = 1$ identically, that is, for all x. On putting $x = \alpha$ this gives $R(\alpha) = 1/Q(\alpha)$, as required. The field k is said to have degree n over $\mathbb{Q}$, and one writes $[k : \mathbb{Q}] = n$.

The construction can be continued analogously to furnish, for every algebraic number field k and every algebraic number β, a field $K = k(\beta)$ with elements given by polynomials in β with coefficients in k. The degree $[K : k]$ of K over k is defined in the obvious way as the degree of β over k. Now K is in fact an algebraic number field over $\mathbb{Q}$, for it can be shown that $K = \mathbb{Q}(\gamma)$, where $\gamma = u\alpha + v\beta$ for some rationals u, v; thus we have

$$[K : k][k : \mathbb{Q}] = [K : \mathbb{Q}].$$

An algebraic number is said to be an algebraic integer if the coefficient of the highest power of x in the minimal polynomial P is 1. The algebraic integers in

61

an algebraic number field k form a ring R. The ring has an integral basis, that is, there exist elements $\omega_1, \ldots, \omega_n$ in R such that every element in R can be expressed uniquely in the form $u_1\omega_1 + \cdots + u_n\omega_n$ for some rational integers $u_1, \ldots, u_n$. We can write $\omega_i = p_i(\alpha)$, where p_i denotes a polynomial over $\mathbb{Q}$, and it is then readily verified that the number $(\det p_i(\alpha_j))^2$ is a rational integer independent of the choice of basis; it is called the discriminant of k and it turns out to be an important invariant.

An algebraic integer α is said to be divisible by an algebraic integer β if α/β is an algebraic integer. An algebraic integer ε is said to be a unit if $1/\varepsilon$ is an algebraic integer. Suppose now that R is the ring of algebraic integers in a number field k. Two elements α, β of R are said to be associates if $\alpha = \varepsilon\beta$ for some unit ε, and this is an equivalence relation on R. An element α of R is said to be irreducible if every divisor of α in R is either an associate or a unit. One calls R a unique factorization domain if every element of R can be expressed essentially uniquely as a product of irreducible elements. The fundamental theorem of arithmetic asserts that the ring of integers in $k = \mathbb{Q}$ has this property; but it does not hold for every k. Nevertheless, it is known from pioneering studies of Kummer and Dedekind that a unique factorization property can be restored by the introduction of ideals, and this forms the central theme of algebraic number theory. The work on Fermat's last theorem that motivated much of the subject related to the particular case of the cyclotomic field $\mathbb{Q}(\zeta)$ where ζ is a root of unity.

7.2 The quadratic field

Let d be a square-free integer, positive or negative, but not 1. The quadratic field $\mathbb{Q}(\sqrt{d})$ is the set of all numbers of the form $u + v\sqrt{d}$ with rational u, v, subject to the usual operations of addition and multiplication. For any element $\alpha = u + v\sqrt{d}$ in $\mathbb{Q}(\sqrt{d})$ one defines the norm of α as the rational number $N(\alpha) = u^2 - dv^2$. Clearly $N(\alpha) = \alpha\bar{\alpha}$, where $\bar{\alpha} = u - v\sqrt{d}$; $\bar{\alpha}$ is called the conjugate of α. Now for any elements α, β in $\mathbb{Q}(\sqrt{d})$ we see that $\overline{\alpha\beta} = \bar{\alpha}\bar{\beta}$ and thus we have the important formula $N(\alpha)N(\beta) = N(\alpha\beta)$. It is readily verified that $\mathbb{Q}(\sqrt{d})$ is indeed a field; in particular, the inverse of any non-zero element α is $\bar{\alpha}/N(\alpha)$. The special field $\mathbb{Q}(\sqrt{(-1)})$ is called the Gaussian field and it is customary to express its elements in the form $u + iv$; in this case we have $N(\alpha) = u^2 + v^2$ and so the product formula is precisely the identity referred to in Section 5.4.

We proceed now to determine the algebraic integers in $\mathbb{Q}(\sqrt{d})$. Suppose that $\alpha = u + v\sqrt{d}$ is such an integer and let $a = 2u, b = 2v$. Then α is a zero of the polynomial $P(x) = x^2 - ax + c$, where $c = N(\alpha)$, and so the rational numbers a, c must in fact be integers. We have $4c = a^2 - b^2d$ and, since d is square-free, it follows that also b is a rational integer. Now if $d \equiv 2$ or 3 (mod 4) then, since a square is congruent to 0 or 1 (mod 4), we see that a, b are even and thus u, v are rational integers; hence, in this case, an integral basis for $\mathbb{Q}(\sqrt{d})$ is given by 1, $\sqrt{d}$. If $d \equiv 1$ (mod 4), which is the only other possibility, then $a \equiv b$ (mod 2) and thus $u - v$ is a rational integer; recalling that $v = \frac{1}{2}b$, we conclude that, in this case, an integral basis for $\mathbb{Q}(\sqrt{d})$ is given by 1, $\frac{1}{2}(1 + \sqrt{d})$. The discriminant D of $\mathbb{Q}(\sqrt{d})$, as defined in Section 7.1, is therefore $4d$ when $d \equiv 2$ or 3 (mod 4) and it is d when $d \equiv 1$ (mod 4).

There is a close analogy between the theory of quadratic fields and the theory of binary quadratic forms as described in Chapter 5. In particular, the discriminant D of $\mathbb{Q}(\sqrt{d})$ is congruent to 0 or 1 (mod 4) and so D is also the discriminant of a binary quadratic form. Now if α is any algebraic integer in $\mathbb{Q}(\sqrt{d})$ then, for some rational integers x, y, we have $\alpha = x + y\sqrt{d}$ when $d \equiv 2$ or 3 (mod 4) and $\alpha = x + \frac{1}{2}y(1 + \sqrt{d})$ when $d \equiv 1$ (mod 4). Thus we see that $N(\alpha) = F(x, y)$, where F denotes the principal form with discriminant D, that is, $x^2 - dy^2$ when $D \equiv 0$ (mod 4) and $(x + \frac{1}{2}y)^2 - \frac{1}{4}dy^2$ when $D \equiv 1$ (mod 4).

7.3 Units

By a unit in $\mathbb{Q}(\sqrt{d})$ we mean an algebraic integer ε in $\mathbb{Q}(\sqrt{d})$ such that $1/\varepsilon$ is an algebraic integer. Plainly if ε is a unit then $N(\varepsilon)$ and $N(1/\varepsilon)$ are rational integers and, since $N(\varepsilon)N(1/\varepsilon) = 1$, we see that $N(\varepsilon) = \pm 1$. Conversely, if $N(\varepsilon) = \pm 1$, then $\varepsilon\bar{\varepsilon} = \pm 1$ and so ε is a unit. Thus, by the above remarks, the units in $\mathbb{Q}(\sqrt{d})$ are determined by the integer solutions x, y of the equation $F(x, y) = \pm 1$.

We shall distinguish two cases according as $d < 0$ or $d > 0$; in the first case the quadratic field is said to be imaginary and in the second it is said to be real. Now in an imaginary quadratic field there are only finitely many units. In fact if $D < -4$ then, as is readily verified, the equation $F(x, y) = \pm 1$ has only the solutions $x = \pm 1$, $y = 0$ and so the only units in $\mathbb{Q}(\sqrt{d})$ are ± 1. For $d = -1$, that is, for the Gaussian field, we have $F(x, y) = x^2 + y^2$ and there are therefore four units ± 1, $\pm i$. For $d = -3$ we have $F(x, y) = x^2 + xy + y^2$ and the equation $F(x, y) = \pm 1$ has six solutions, namely $(\pm 1, 0)$, $(0, \pm 1)$, $(1, -1)$ and $(-1, 1)$; thus the units of $\mathbb{Q}(\sqrt{(-3)})$ are ± 1, $\frac{1}{2}(\pm 1 \pm \sqrt{(-3)})$. It follows that the units in an imaginary quadratic field are all roots of unity; they are given by the zeros of $x^2 - 1$ when $D < -4$, by those of $x^4 - 1$ when $D = -4$

and by those of $x^6 - 1$ when $D = -3$. Hence the number of units is the same as the number w for forms with discriminant D indicated in Section 5.3.

We turn now to real quadratic fields; in this case there are infinitely many units. To establish the result it suffices to show that, when $d > 0$, there is a unit η in $\mathbb{Q}(\sqrt{d})$ other than ± 1; for then η^m is a unit for all integers m, and, since the only roots of unity in $\mathbb{Q}(\sqrt{d})$ are ± 1, we see that different m give distinct units. We shall use Dirichlet's theorem on Diophantine approximation (see Section 6.1); the theorem implies that, for any integer $Q > 1$, there exist rational integers p, q, with $0 < q < Q$, such that $|\alpha| \le 1/Q$, where $\alpha = p - q\sqrt{d}$. Now the conjugate $\bar{\alpha} = \alpha + 2q\sqrt{d}$ satisfies $|\bar{\alpha}| \le 3Q\sqrt{d}$ and thus we have $|N(\alpha)| \le 3\sqrt{d}$. Further, since $\sqrt{d}$ is irrational, we obtain, as $Q \to \infty$, infinitely many α with this property. But $N(\alpha)$ is a rational integer bounded independently of Q, and thus, for infinitely many α, it takes some fixed value, say N. Moreover we can select two distinct elements from the infinite set, say $\alpha = p - q\sqrt{d}$ and $\alpha' = p' - q'\sqrt{d}$, such that $p \equiv p' \pmod{N}$ and $q \equiv q' \pmod{N}$. We now put $\eta = \alpha/\alpha'$. Then $N(\eta) = N(\alpha)/N(\alpha') = 1$. Further, η is clearly not 1, and it is also not -1 since $\sqrt{d}$ is irrational and q, q' are positive. Furthermore we have $\eta = x + y\sqrt{d}$, where $x = (pp' - dqq')/N$ and $y = (pq' - p'q)/N$, and the congruences above imply that x, y are rational integers. Thus η is a non-trivial unit in $\mathbb{Q}(\sqrt{d})$, as required. The argument here shows, incidentally, that the Pell equation $x^2 - dy^2 = 1$ has a non-trivial solution; we shall discuss the equation more fully in Chapter 8.

We can now give a simple expression for all the units in a real quadratic field. In fact consider the set of all units in the field that exceed 1. The set is not empty, for if η is the unit obtained above then one of the numbers $\pm\eta$ or $\pm 1/\eta$ is a member. Further, each element of the set has the form $u + v\sqrt{d}$, where u, v are either integers, or, in the case $d \equiv 1 \pmod 4$, possibly halves of odd integers. Furthermore u and v are positive, for $u + v\sqrt{d}$ is greater than its conjugate $u - v\sqrt{d}$, which lies between -1 and 1. It follows that there is a smallest element in the set, say ε. Now if ε' is any positive unit in the field then there is a unique integer m such that $\varepsilon^m \le \varepsilon' < \varepsilon^{m+1}$; this gives $1 \le \varepsilon'/\varepsilon^m < \varepsilon$. But $\varepsilon'/\varepsilon^m$ is also a unit in the field and thus, from the definition of ε, we conclude that $\varepsilon' = \varepsilon^m$. This shows that all the units in the field are given by $\pm\varepsilon^m$, where $m = 0, \pm 1, \pm 2, \ldots$.

The results established here for quadratic fields are special cases of a famous theorem of Dirichlet concerning units in an arbitrary algebraic number field. Suppose that the field k is generated by an algebraic number α with degree n and that precisely s of the conjugates $\alpha_1, \ldots, \alpha_n$ of α are real; then $n = s + 2t$, where t is the number of complex conjugate pairs. Dirichlet's theorem asserts that there exist $r = s + t - 1$ fundamental units $\varepsilon_1, \ldots, \varepsilon_r$ in k such that every unit in k can be expressed uniquely in the form $\rho\varepsilon_1{}^{m_1}\cdots\varepsilon_r{}^{m_r}$, where $m_1,\ldots,m_r$ are rational integers and ρ is a root of unity in k.

7.4 Primes and factorization

Let R be the ring of algebraic integers in a quadratic field $\mathbb{Q}(\sqrt{d})$. By a prime π in R we mean an element of R that is neither 0 nor a unit and which has the property that if π divides $\alpha\beta$, where α, β are elements of R, then either π divides α or π divides β. It will be noted at once that a prime π is irreducible in the sense indicated in Section 7.1; for if $\pi = \alpha\beta$ then either α/π or β/π is an element of R, whence either β or α is a unit. However, an irreducible element need not be a prime. Consider, for example, the number 2 in the quadratic field $\mathbb{Q}(\sqrt{(-5)})$. It is certainly irreducible, for if $2 = \alpha\beta$ then $4 = N(\alpha)N(\beta)$; but $N(\alpha)$ and $N(\beta)$ have the form $x^2 + 5y^2$ for some integers x, y, and, since the equation $x^2 + 5y^2 = \pm 2$ has no integer solutions, it follows that either $N(\alpha) = \pm 1$ or $N(\beta) = \pm 1$ and thus either α or β is a unit. On the other hand, 2 is not a prime in $\mathbb{Q}(\sqrt{(-5)})$, for it divides

$$(1 + \sqrt{(-5)})(1 - \sqrt{(-5)}) = 6,$$

but it does not divide either $1 + \sqrt{(-5)}$ or $1 - \sqrt{(-5)}$; indeed, by taking norms, it is readily verified that each of the latter is irreducible.

Now every element α of R that is neither 0 nor a unit can be factorized into a finite product of irreducible elements. For if α is not itself irreducible than $\alpha = \beta\gamma$ for some β, γ in R, neither of which is a unit. If β were not irreducible then it could be factorized likewise, and the same holds for γ. The process must terminate, for if $\alpha = \beta_1 \cdots \beta_n$, where none of the βs is a unit, then, since $|N(\beta_j)| \geq 2$, we see that $|N(\alpha)| \geq 2^n$. The ring R is said to be a unique factorization domain if the expression for α as a finite product of irreducible elements is essentially unique, that is, unique except for the order of the factors and the possible replacement of irreducible elements by their associates. A fundamental problem in number theory is to determine which domains have unique factorization, and here the definition of a prime plays a crucial role. In fact we have the basic theorem that R is a unique factorization domain if and only if every irreducible element of R is also a prime in R. To verify the assertion, note that, if factorization in R is unique and if π is an irreducible element such that π divides $\alpha\beta$ for some α, β in R, then π must be an associate of one of the irreducible factors of α or β and so π divides α or β, as required. Conversely, if every irreducible element is also a prime then we can argue as in the demonstration of the fundamental theorem of arithmetic given in Section

1.5; thus if $\alpha = \pi_1 \cdots \pi_k$ as a product of irreducible elements, and if π' is an irreducible element occurring in another factorization, then π' must divide π_j for some j, whence π' and π_j are associates, and assuming by induction that the result holds for α/π', the required uniqueness of factorization follows.

All the imaginary quadratic fields $\mathbb{Q}(\sqrt{d})$ which have the unique factorization property are known; they are given by $d = -1, -2, -3, -7, -11, -19, -43, -67$ and -163. The theorem has a long history, dating back to Gauss, and it was finally proved by Baker and Stark, independently, in 1966; the methods of proof were quite different, one depending on transcendence theory (cf. Section 6.6) and the other on the study of elliptic modular functions. The theorem shows, incidentally, that the nine discriminants d indicated in Exercise (i) of Section 5.7 are the only values for which $h(d) = 1$. The problem of finding all the real quadratic fields $\mathbb{Q}(\sqrt{d})$ with unique factorization remains open; it is generally conjectured that there are infinitely many such fields but even this has not been proved. Nevertheless all such fields with d relatively small, for instance with $d < 100$, are known; we shall discuss some particular cases in the next section.

7.5 Euclidean fields

A quadratic field $\mathbb{Q}(\sqrt{d})$ is said to be Euclidean if its ring of integers R has the property that, for any elements α, β of R with $\beta \neq 0$, there exist elements γ, δ of R such that $\alpha = \beta\gamma + \delta$ and $|N(\delta)| < |N(\beta)|$. For such fields there exists a Euclidean algorithm analogous to that described in Chapter 1. In fact we can generate the sequence of equations $\delta_{j-2} = \delta_{j-1}\gamma_j + \delta_j$ ($j = 1, 2, \ldots$), where $\delta_{-1} = \alpha$, $\delta_0 = \beta$, $\delta_1 = \delta$, $\gamma_1 = \gamma$ and $|N(\delta_j)| < |N(\delta_{j-1})|$; the sequence terminates when $\delta_{k+1} = 0$ for some k and then δ_k has the properties of a greatest common divisor, that is, δ_k divides α and β, and every common divisor of α, β divides δ_k. Moreover we have $\delta_k = \alpha\lambda + \beta\mu$ for some λ, μ in R. This can be verified either by successive substitution or by observing that $|N(\delta_k)|$ is the least member of the set of positive integers of the form $|N(\alpha\lambda + \beta\mu)|$, where λ, μ run through the elements of R. In fact the set certainly has a least member $|N(\delta')|$, say, where $\delta' = \alpha\lambda + \beta\mu$ for some λ, μ in R; thus every common divisor of α, β divides δ'. Further, δ' divides α, since from $\alpha = \delta'\gamma + \delta''$, with $|N(\delta'')| < |N(\delta')|$, we see that $\delta'' = \alpha\lambda' + \beta\mu'$ for some λ', μ' in R, whence $N(\delta'') = 0$ and so $\delta'' = 0$; similarly δ' divides β. Hence we have $\delta' = \delta_k$. It is clear that if δ_k is a unit then, by division, we obtain elements λ, μ in R with $\alpha\lambda + \beta\mu = 1$.

We proceed now to prove that a Euclidean field has unique factorization. It suffices, in view of Section 7.4, to show that every irreducible element π

in R is a prime; accordingly suppose that π divides $\alpha\beta$ but that π does not divide α. Then, by the remarks above, there exist integers λ, μ in R such that $\alpha\lambda + \pi\mu = 1$. This gives $\alpha\beta\lambda + \pi\beta\mu = \beta$, whence π divides β. Thus π is a prime and the desired result follows.

It was proved by Chatland and Davenport in 1950, and independently by Inkeri at about the same time, that there are precisely 21 Euclidean fields $\mathbb{Q}(\sqrt{d})$; the values of d are given by -11, -7, -3, -2, -1, 2, 3, 5, 6, 7, 11, 13, 17, 19, 21, 29, 33, 37, 41, 57 and 73. It had been proved earlier by Heilbronn that the list must be finite, and it had been verified as a consequence of works by Dickson, Perron, Oppenheimer, Remak and Rédei that the fields listed here are indeed Euclidean; we shall confirm the assertion for the first eight fields in a moment. It is easy to see that there can be no other Euclidean fields with $d < 0$. In fact if $d \equiv 2$ or 3 (mod 4) and $d \leq -5$ then we cannot have $\sqrt{d} = 2\gamma + \delta$ with $|N(\delta)| < 4$; for we can express γ, δ as $x + y\sqrt{d}$ and $x' + y'\sqrt{d}$ respectively, where x, y and x', y' are rational integers, and since $N(\delta) \geq x'^2 + 5y'^2$ we would obtain $y' = 0$, contrary to $2y + y' = 1$. Similarly if $d \equiv 1$ (mod 4) and $d \leq -15$, then we cannot have $\frac{1}{2}(1 + \sqrt{d}) = 2\gamma + \delta$ with $|N(\delta)| < 4$. The most difficult part of the theorem is the proof that there are no other Euclidean fields with $d > 0$. In this connection, Davenport showed by an ingenious algorithm derived from studies on Diophantine approximation that if $d > 2^{14}$ then $\mathbb{Q}(\sqrt{d})$ is not Euclidean; this reduced the problem to a finite checking of cases. Incidentally, Rédei claimed originally that the field $\mathbb{Q}(\sqrt{97})$ was Euclidean but Barnes and Swinnerton-Dyer proved this to be erroneous.

We shall show now that if $d = -2$, -1, 2 or 3 then $\mathbb{Q}(\sqrt{d})$ is Euclidean. Accordingly let α, β be any algebraic integers in $\mathbb{Q}(\sqrt{d})$ with $\beta \neq 0$. Then $\alpha/\beta = u + v\sqrt{d}$ for some rationals u, v. We select integers x, y as close as possible to u, v and put $r = u - x$, $s = v - y$; then $|r| \leq \frac{1}{2}$ and $|s| \leq \frac{1}{2}$. On writing $\gamma = x + y\sqrt{d}$ we obtain $\alpha = \beta\gamma + \delta$, where $\delta = \beta(r + s\sqrt{d})$. This gives $N(\delta) = N(\beta)(r^2 - ds^2)$. But for $|d| \leq 2$ we have $|r^2 - ds^2| \leq r^2 + 2s^2 \leq \frac{3}{4}$, and for $d = 3$ we have $|r^2 - ds^2| \leq \max(r^2, ds^2) \leq \frac{3}{4}$. Hence $|N(\delta)| < |N(\beta)|$, as required.

Finally we prove that $\mathbb{Q}(\sqrt{d})$ is Euclidean when $d = -11$, -7, -3, 5 and 13. In these cases we have $d \equiv 1$ (mod 4) and so 1, $\frac{1}{2}(1 + \sqrt{d})$ is an integral basis for $\mathbb{Q}(\sqrt{d})$. Again let α, β be any algebraic integers in $\mathbb{Q}(\sqrt{d})$, with $\beta \neq 0$, and let $\alpha/\beta = u + v\sqrt{d}$ with u, v rational. We select an integer y as close as possible to $2v$ and put $s = v - \frac{1}{2}y$; then $|s| \leq \frac{1}{4}$. Further, we select an integer x as close as possible to $u - \frac{1}{2}y$ and put $r = u - x - \frac{1}{2}y$; then $|r| \leq \frac{1}{2}$. On writing $\gamma = x + \frac{1}{2}y(1 + \sqrt{d})$ we see that $\alpha = \beta\gamma + \delta$, where $\delta = \beta(r + s\sqrt{d})$. Now, for $|d| \leq 11$, we have $|r^2 - ds^2| \leq \frac{1}{4} + \frac{11}{16} < 1$, and, for $d = 13$, we have $|r^2 - ds^2| \leq \frac{13}{16}$. The result follows.

7.6 The Gaussian field

To conclude this chapter we shall describe the principal properties of the most fundamental quadratic field, namely the Gaussian field $\mathbb{Q}(\sqrt{(-1)})$ or $\mathbb{Q}(i)$. We have already seen that the integers in the field, that is, the Gaussian integers, have the form $x + iy$ with x, y rational integers. Thus the norm of a Gaussian integer has the form $x^2 + y^2$, and, in particular, it is non-negative. It was noted in Section 7.3 that there are just four units ± 1 and $\pm i$. Moreover we proved in Section 7.5 that the field is Euclidean and so has unique factorization. Hence there is no need to distinguish between irreducible elements and primes, and we shall use the latter terminology in preference; in fact we shall refer to the elements as Gaussian primes.

Our purpose now is to determine all the Gaussian primes. We begin with two preliminary observations which actually apply analogously to all quadratic fields with unique factorization. First, if α is any Gaussian integer and if $N(\alpha)$ is a rational prime then α is a Gaussian prime; for plainly if $\alpha = \beta\gamma$ for some Gaussian integers β, γ then $N(\alpha) = N(\beta)N(\gamma)$ and so either $N(\beta) = 1$ or $N(\gamma) = 1$, whence either β or γ is a unit. Secondly we observe that every Gaussian prime π divides just one rational prime p. For π certainly divides $N(\pi)$ and so there is a least positive rational integer p such that π divides p; and p is a rational prime, for if $p = mn$, where m, n are rational integers, then, since π is a Gaussian prime, we have either π divides m or π divides n, whence, by the minimal property of p, either m or n is 1. The prime p is unique, for if p' is any other rational prime then there exist rational integers a, a' such that $ap + a'p' = 1$; thus if π were to divide both p and p' then it would divide 1 and so be a unit contrary to definition.

We note next that a rational prime p is either itself a Gaussian prime or is the product $\pi\pi'$ of two Gaussian primes, where π, π' are conjugates. Indeed p is divisible by some Gaussian prime π and thus we have $p = \pi\lambda$ for some Gaussian integer λ; this gives $N(\pi)N(\lambda) = p^2$ and the two cases correspond to the possibilities $N(\lambda) = 1$, implying that λ is a unit and that p is an associate of π, and $N(\lambda) = p$, implying that $N(\pi) = p$. Now the first case applies when $p \equiv 3 \pmod 4$ and the second when $p \equiv 1 \pmod 4$. For $N(\pi)$ has the form $x^2 + y^2$ and a square is congruent to 0 or 1 (mod 4). Further, if $p \equiv 1 \pmod 4$, then -1 is a quadratic residue (mod p), whence p divides $x^2 + 1 = (x + i)(x - i)$ for some rational integer x; but if p were a Gaussian prime then it would divide either $x + i$ or $x - i$, contrary to the fact that neither $x/p + i/p$ nor $x/p - i/p$ is a Gaussian integer. With regard to the prime 2, we have $2 = (1 + i)(1 - i)$ and here $1 + i$ and $1 - i$ are Gaussian primes and, moreover, associates. Combining our results, we find therefore that the totality of Gaussian primes are given by the rational primes $p \equiv 3$ (mod 4), by the factors π, π' in the expression $p = \pi \pi'$ appertaining to primes $p \equiv 1$ (mod 4), and by $1 + i$, together with all their associates formed by multiplying with ± 1 and $\pm i$. The argument here furnishes, incidentally, another proof of the result that every prime $p \equiv 1$ (mod 4) can be expressed as a sum of two squares (see Section 5.4).

Many of the definitions and theorems discussed earlier for the rational field possess natural analogues in the Gaussian field. Thus, for example, one can specify greatest common divisors and congruences in an obvious way, and there is an analogue of Fermat's theorem to the effect that if π is a Gaussian prime and α is a Gaussian integer, with $(\alpha, \pi) = 1$, then $\alpha^{N(\pi)-1} \equiv 1$ (mod π). There is also, for instance, an analogue of the prime number theorem to the effect that the number of non-associated Gaussian primes π with $N(\pi) \leq x$ is asymptotic to $x / \log x$ as $x \to \infty$.

7.7 Further reading

The structure of quadratic fields can be properly appreciated only in the wider context of algebraic number theory and with reference especially to the theory of ideals. This will be discussed in later chapters. As an initial guide, we mention here that the classic text in this connection is that of Hecke. It was originally published in German in 1923 and it appeared in English translation under the title *Lectures on the Theory of Algebraic Numbers* (Springer, 1981); it remains one of the best works on the subject. There are several newer expositions, however. In particular, the book *Algebraic Number Theory and Fermat's Last Theorem* by I. Stewart and D. Tall (A. K. Peters, 2002) is relatively elementary and easy to read, and other generally accessible works are those by S. Alaca and K. S. Williams, *Introductory Algebraic Number Theory* (Cambridge University Press, 2004), and by W. Narkiewicz, *Elementary and Analytic Theory of Algebraic Numbers* (Springer, 2004). Further literature on the subject is mentioned in Section 10.9.

An account of the solution to the problem of determining all imaginary quadratic fields with unique factorization, referred to in Section 7.4, can be found in Chapter 5 of Baker's *Transcendental Number Theory* (Cambridge University Press, 1990). The work, referred to in Section 7.5, of Chatland and Davenport on Euclidean fields appeared in the *Canadian J. Math.* **2** (1950), 289–296; the article is reprinted in *The Collected Works of Harold Davenport*, Vol. I (eds B. J. Birch *et al.*, Academic Press, 1977), pp. 366–373. For a proof of the result on Gaussian primes cited at the end of Section 7.6 see

E. Landau's *Einführung in die Elementare und Analytische Theorie der Algebraischen Zahlen und der Ideale* (Teubner, 1918).

7.8 Exercises

(i) Show that the units in $\mathbb{Q}(\sqrt{2})$ are given by $\pm(1 + \sqrt{2})^n$, where $n = 0$, $\pm 1, \pm 2, \ldots$. Find the units in $\mathbb{Q}(\sqrt{3})$.

(ii) Determine the integers n and d for which $(1 + n\sqrt{d})/(1 - n\sqrt{d})$ is a unit in $\mathbb{Q}(\sqrt{d})$.

(iii) By considering products of norms, or otherwise, prove that there are infinitely many irreducible elements in the integral domain of any quadratic field.

(iv) Explain why the equation $2.11 = (5 + \sqrt{3})(5 - \sqrt{3})$ is not inconsistent with the fact that $\mathbb{Q}(\sqrt{3})$ has unique factorization.

(v) Prove that the equation $2.3 = (\sqrt{(-6)})(-\sqrt{(-6)})$ implies that $\mathbb{Q}(\sqrt{(-6)})$ does not have unique factorization.

(vi) Show that $1 + \sqrt{(-17)}$ is irreducible in $\mathbb{Q}(\sqrt{(-17)})$. Verify that $\mathbb{Q}(\sqrt{(-17)})$ does not have unique factorization.

(vii) Find equations to show that $\mathbb{Q}(\sqrt{d})$ does not have unique factorization for $d = -10, -13, -14$ and -15.

(viii) By considering congruences (mod 5), show that there are no algebraic integers in $\mathbb{Q}(\sqrt{10})$ with norm ± 2 and ± 3. Prove that $4 + \sqrt{10}$ is irreducible in $\mathbb{Q}(\sqrt{10})$. Hence verify that $\mathbb{Q}(\sqrt{10})$ does not have unique factorization.

(ix) Use the fact that $\mathbb{Q}(\sqrt{3})$ is Euclidean to determine algebraic integers α, β in $\mathbb{Q}(\sqrt{3})$ such that $(1 + 2\sqrt{3})\alpha + (5 + 4\sqrt{3})\beta = 1$.

(x) Prove that the primes in $\mathbb{Q}(\sqrt{2})$ are given by the rational primes $p \equiv \pm 3$ (mod 8), the factors π, π' in the expression $p = \pi\pi'$ appertaining to primes $p \equiv \pm 1$ (mod 8), and by $\sqrt{2}$, together with all their associates.

(xi) Show that if π is a Gaussian prime then the numbers $1, 2, \ldots, N(\pi)$ form a complete set of residues (mod π); that is, show that none of the differences is divisible by π, but that for any Gaussian integer α there is a rational integer a with $1 \leq a \leq N(\pi)$, such that π divides $\alpha - a$. Apply this result to establish the analogue of Fermat's theorem quoted at the end of Section 7.6.

8

Diophantine equations

8.1 The Pell equation

Diophantine analysis has its genesis in the fertile mind of Fermat. He had studied Bachet's edition, published in 1621, of the first six books that then remained of the famous *Arithmetica*; this was a treatise, originally consisting of 13 books, written by the Greek mathematician Diophantus of Alexandria at about the third century AD. The *Arithmetica* was concerned only with the determination of particular rational or integer solutions of algebraic equations, but it inspired Fermat to initiate researches into the nature of all such solutions, and herewith the modern theories began.

An especially notorious Diophantine equation, in fact the issue of a celebrated challenge from Fermat to the English mathematicians of his time, is the equation

$$x^2 - dy^2 = 1,$$

where d is a positive integer other than a perfect square. It is usually referred to as the Pell equation but the nomenclature, due to Euler, has no historical justification since Pell apparently made no contribution to the topic. Fermat conjectured that there is at least one non-trivial solution in integers x, y, that is, a solution other than $x = \pm 1$, $y = 0$; the conjecture was proved by Lagrange in 1768. In fact we have already established the result in Section 7.3; it was assumed there that d is square-free but the argument plainly holds for any d that is not a perfect square. Now there is a unique solution to the Pell equation in which the integers x, y have their smallest positive values; it is called the fundamental solution. Let x', y' be this solution and put $\varepsilon = x' + y' \sqrt{d}$. Then, by the arguments of Section 7.3, we see that all solutions are given by $x + y\sqrt{d} = \pm\varepsilon^n$, where $n = 0, \pm 1, \pm 2, \ldots$. In particular, the equation has infinitely many solutions.

71

More insight into the character of the solutions is provided by the continued-fraction algorithm. First we observe that any solution in positive integers x, y satisfies $x - y\sqrt{d} = 1/(x + y\sqrt{d})$, whence $x > y\sqrt{d}$ and $x - y\sqrt{d} < 1/(2y\sqrt{d})$. This gives $|\sqrt{d} - x/y| < 1/(2y^2)$, and it follows from Section 6.3 that x/y is a convergent to $\sqrt{d}$. Now it was noted in Section 6.4 that the continued fraction for $\sqrt{d}$ has the form

$$[a_0, \overline{a_1, \ldots, a_m}];$$

the number m of repeated partial quotients is called the period of $\sqrt{d}$. Let $p_n/q_n (n = 1, 2, \ldots)$ be the convergents to $\sqrt{d}$ and let $\theta_n (n = 1, 2, \ldots)$ be the complete quotients. We have $x = p_n$, $y = q_n$ for some n, that is, $p_n{}^2 - dq_n{}^2 = 1$. Here n must be odd for, by Section 6.3,

$$\sqrt{d} = \frac{p_n\theta_{n+1} + p_{n-1}}{q_n\theta_{n+1} + q_{n-1}},$$

whence, on recalling that $p_{n-1}q_n - p_nq_{n-1} = (-1)^n$, we obtain

$$q_n\sqrt{d} - p_n = (-1)^n/(q_n\theta_{n+1} + q_{n-1}),$$

and so, for even n, $q_n\sqrt{d} > p_n$. In fact n must have the form $lm - 1$, where $l = 1, 2, 3, \ldots$ when m is even and $l = 2, 4, 6, \ldots$ when m is odd. For the expression for $\sqrt{d}$ above gives

$$(p_n - q_n\sqrt{d})\theta_{n+1} = q_{n-1}\sqrt{d} - p_{n-1},$$

and thus

$$(p_n{}^2 - dq_n{}^2)\theta_{n+1} = (q_{n-1}\sqrt{d} - p_{n-1})(q_n\sqrt{d} + p_n)$$
$$= (-1)^{n-1}\sqrt{d} + c,$$

where c is an integer. But $p_n{}^2 - dq_n{}^2 = 1$ and n is odd; hence $\theta_{n+1} = \sqrt{d} + c$. Now $\sqrt{d} = a_0 + 1/\theta_1$, where θ_1 is purely periodic, and we have $\theta_{n+1} = a_{n+1} + 1/\theta_{n+2}$. Since $\theta_1 > 1$, $\theta_{n+2} > 1$, we obtain $a_{n+1} = a_0 + c$ and $\theta_1 = \theta_{n+2}$; it follows that $n + 1$ is divisible by m and so n has the form $lm - 1$, as asserted.

We have therefore shown that the only possible positive solutions x, y to the Pell equation are given by $x = p_n$, $y = q_n$, where p_n/q_n is a convergent to $\sqrt{d}$ with n of the form $lm - 1$ as above. In fact all of these p_n, q_n satisfy the equation and thus they comprise the full set of positive solutions. For, in view of the periodicity of $\sqrt{d}$, we have $\theta_1 = \theta_{n+2}$ for all $n = lm - 1$ as above, and hence

$$\sqrt{d} = \frac{p_{n+1}\theta_1 + p_n}{q_{n+1}\theta_1 + q_n}.$$

But $\sqrt{d} = a_0 + 1/\theta_1$, and on substituting for θ_1 and using the fact that $\sqrt{d}$ is irrational, we obtain

$$p_n = q_{n+1} - a_0 q_n, \quad p_{n+1} - a_0 p_n = q_n d.$$

On eliminating a_0 we see that

$$p_n^2 - dq_n^2 = p_n q_{n+1} - p_{n+1} q_n,$$

and, since n is odd, it follows that $p_n^2 - dq_n^2 = 1$, as required.

A similar analysis applies to the equation $x^2 - dy^2 = -1$. In this case there is no solution when the period m of $\sqrt{d}$ is even. When m is odd, all positive solutions are given by $x = p_n$, $y = q_n$, where p_n/q_n is the nth convergent to $\sqrt{d}$ and $n = lm - 1$ with $l = 1, 3, 5, \ldots$. Further, when the equation is soluble, the solution of smallest value, sometimes called the fundamental solution, is $x' = p_{m-1}$, $y' = q_{m-1}$. Then, on writing $\eta = x' + y'\sqrt{d}$, one deduces that all solutions are given by $x + y\sqrt{d} = \pm\eta^n$, where $n = \pm 1, \pm 3, \pm 5, \ldots$. The result is in fact easily obtained on noting that the fundamental solution to $x^2 - dy^2 = 1$ is given by η^2. Thus, for instance, the equation $x^2 - 2y^2 = -1$ is soluble and all solutions are given by $x + y\sqrt{2} = \pm(1 + \sqrt{2})^n$ with $n > 0$ and odd; the solutions of $x^2 - 2y^2 = 1$ are given similarly with n even and for $j = 0, \pm 1, \pm 2, \ldots$ we have

$$2x = \pm((3 + 2\sqrt{2})^j + (3 - 2\sqrt{2})^j), \ 2\sqrt{2}y = \pm((3 + 2\sqrt{2})^j - (3 - 2\sqrt{2})^j).$$

In summary we can say that the solutions of the equation $x^2 - dy^2 = \pm 1$ are given by $x + y\sqrt{d} = \pm\eta^n$ for $n = 0, \pm 1, \pm 2, \ldots$. If $N\eta = 1$ then they all satisfy $x^2 - dy^2 = 1$; otherwise we have $N\eta = -1$ and they satisfy $x^2 - dy^2 = 1$ for even n and $x^2 - dy^2 = -1$ for odd n. Plainly η is a unit in $K = \mathbb{Q}(\sqrt{d})$ and if $d \equiv 2, 3 \pmod 4$ then it is the fundamental unit. An analogous result holds for the more general equation $x^2 - dy^2 = k$, where k is a non-zero integer. Here, when the equation is soluble, one can specify a finite set of solutions x', y' such that, on writing $\zeta = x' + y'\sqrt{d}$, all solutions are given by $x + y\sqrt{d} = \pm\zeta\varepsilon^n$ with ε the fundamental unit in K and $n = 0, \pm 1, \pm 2, \ldots$.

As an example, consider the equation $x^2 - 131y^2 = 1$. The continued fraction for $\sqrt{131}$ is $[11, \overline{2, 4, 11, 4, 2, 22}]$ and thus the period m of $\sqrt{131}$ is 6. Since m is even, the solutions are given by the convergents p_n/q_n with $n = lm - 1$ and $l = 1, 2, 3, \ldots$; the smallest solution is $x = p_5$, $y = q_5$. Now, by the recurrence relations, the first five convergents are $\frac{23}{2}, \frac{103}{9}, \frac{1156}{101}, \frac{4727}{413}$ and $\frac{10610}{927}$. Hence we obtain $x = 10610$, $y = 927$.

As another example, consider the equation $x^2 - 97y^2 = -1$. The continued fraction for $\sqrt{97}$ is

$$[9, \overline{1, 5, 1, 1, 1, 1, 1, 1, 5, 1, 18}].$$

Thus the period m of $\sqrt{97}$ is 11 and, since m is odd, the equation is soluble. Indeed the fundamental solution is given by $x = p_{10}$, $y = q_{10}$, where p_n/q_n ($n = 1, 2, \ldots$) denote the convergents to $\sqrt{97}$. Now the first ten convergents to $\sqrt{97}$ are 10, $\frac{59}{6}$, $\frac{69}{7}$, $\frac{128}{13}$, $\frac{197}{20}$, $\frac{325}{33}$, $\frac{522}{53}$, $\frac{847}{86}$, $\frac{4757}{483}$ and $\frac{5604}{569}$. Hence the fundamental solution to $x^2 - 97y^2 = -1$ is $x = 5604$, $y = 569$. Further, if we write $\eta = 5604 + 569\sqrt{97}$ then $\varepsilon = \eta^2$ gives the fundamental solution to $x^2 - 97y^2 = 1$; the solution is in fact $x = 62\,809\,633$, $y = 6\,377\,352$.

Incidentally, the continued fraction for $\sqrt{d}$ always has the form

$$[a_0, \overline{a_1, a_2, a_3, \ldots, a_3, a_2, a_1, 2a_0}],$$

as for $\sqrt{131}$ and $\sqrt{97}$ above, and moreover the period m of $\sqrt{d}$ is always odd when d is a prime $p \equiv 1 \pmod 4$. In fact, for such p, the equation $x^2 - py^2 = -1$ is always soluble. For if x', y' is the fundamental solution to $x^2 - py^2 = 1$ then x' is odd and so $(x' + 1, x' - 1) = 2$; this gives either $x' + 1 = 2u^2$, $x' - 1 = 2pv^2$ or $x' - 1 = 2u^2$, $x' + 1 = 2pv^2$ for some positive integers u, v with $y' = 2uv$, whence $u^2 - pv^2 = \pm 1$, and here the minus sign must hold since $v < y'$.

8.2 The Thue equation

A multitude of special techniques have been devised through the centuries for solving particular Diophantine equations. The scholarly treatise by Dickson on the history of the theory of numbers (see Section 7.6) contains numerous references to early works in the field. Most of these were of an *ad hoc* nature, the arguments involved being specifically related to the example under consideration, and there was little evidence of a coherent theory. In 1900, as the tenth of his famous list of 23 problems, Hilbert asked for a universal algorithm for deciding whether or not an equation of the form $f(x_1, \ldots, x_n) = 0$, where f denotes a polynomial with integer coefficients, is soluble in integers $x_1, \ldots, x_n$. The problem was resolved in the negative by Matiyasevich, developing ideas of Davis, Robinson and Putnam on recursively enumerable sets. The proof has subsequently been refined to show that an algorithm of the kind sought by Hilbert does not exist even if one limits attention to polynomials in just nine variables, and it seems to me quite likely that it does not in fact exist for polynomials in only three variables. For polynomials in two variables, however, the situation would appear to be quite different.

In 1909, a new technique based on Diophantine approximation was introduced by the Norwegian mathematician Axel Thue. He considered the equation $F(x, y) = m$, where F denotes an irreducible binary form with integer coefficients and degree at least 3, and m is any integer. The equation can be expressed as

$$a_0 x^n + a_1 x^{n-1} y + \cdots + a_n y^n = m,$$

and this can be written in the form

$$a_0(x - \alpha_1 y) \cdots (x - \alpha_n y) = m,$$

where $\alpha_1, \ldots, \alpha_n$ signify a complete set of conjugate algebraic numbers. Thus if the equation is soluble in positive integers x, y then the nearest of the numbers $\alpha_1, \ldots, \alpha_n$ to x/y, say α, satisfies $|x - \alpha y| \ll 1$. Here we are using Vinogradov's notation; by $a \ll b$ we mean $a < bc$ for some constant c, that is, in this case, a number independent of x and y, and similarly by $a \gg b$ we shall mean $b < ac$ for some such c. Now, for y sufficiently large and for $\alpha \neq \alpha_j$, we have

$$|x - \alpha_j y| = |(x - \alpha y) + (\alpha - \alpha_j)y| \gg y;$$

this gives $|x - \alpha y| \ll 1/y^{n-1}$, whence $|\alpha - x/y| \ll 1/y^n$. But by Thue's improvement on Liouville's theorem mentioned in Section 6.5, we have $|\alpha - x/y| \gg 1/y^\kappa$ for any $\kappa > \frac{1}{2}n + 1$. It follows that y is bounded above and so there are only finitely many possibilities for x and y. The argument obviously extends to integers x, y of arbitrary sign, and hence we obtain the remarkable result that the Thue equation has only finitely many solutions in integers. Plainly the condition $n \geq 3$ is necessary here, for, as we have shown, the Pell equation has infinitely many solutions.

The demonstration of Thue just described has a major limitation. Although it yields an estimate for the number of solutions of $F(x, y) = m$, it does not enable one to furnish the complete list of solutions in a given instance or indeed to determine whether or not the equation is soluble. This is a consequence of the ineffective nature of the original Thue inequality on which the proof depends. Some effective cases of the inequality have been derived and, in these instances, one can easily solve the related Thue equation even for quite large values of m; for example, from the result on $\sqrt[3]{2}$ mentioned in Section 6.5 we obtain the bound $(10^6 |m|)^{23}$ for all solutions of $x^3 - 2y^3 = m$. But still the basic limitation of Thue's argument remains. Another approach was initiated by Delaunay and Nagell in the 1920s. It involved factorization in algebraic number fields and it enabled certain equations of Thue type with small degree to be solved completely. In particular, the method applied to the equation $x^3 - dy^3 = 1$, where d is a cube-free integer, and it yielded the result that there is at most one solution in non-zero integers x, y. The method was developed by Skolem using analysis in the p-adic domain, and he furnished thereby a new proof of Thue's theorem in the case when not all of the zeros of $F(x, 1)$ are real. The work depended on the compactness property of the p-adic integers and so was generally ineffective, but Ljunggren succeeded in applying the technique to deal with several striking examples. For instance, he showed that the only integer solutions (x, y) of the equation $x^3 - 3xy^2 - y^3 = 1$ are $(1, 0), (0, -1), (-1, 1), (1, -3), (-3, 2)$ and $(2, 1)$.

An entirely different demonstration of Thue's theorem was given by Baker in 1968. It involved the theory of linear forms in logarithms (see Section 6.6) and it led to explicit bounds for the sizes of all the integer solutions x, y of $F(x, y) = m$; in fact the method yielded bounds of the form $c|m|^{c'}$, where c, c' are numbers depending only on F. Thus, in principle, the complete list of solutions can be determined in any particular instance by a finite amount of computation. In practice the bounds that arise in Baker's method are large, typically of order $10^{10^{500}}$, but it has been shown that they can usually be reduced to manageable figures by simple observations from Diophantine approximation.

8.3 The Mordell equation

Some profound results relating to the equation $y^2 = x^3 + k$, where k is a non-zero integer, were discovered by Mordell in 1922, and the equation continued to be one of Mordell's major interests throughout his life. The theorems that he initiated divide naturally according as one is dealing with integer solutions x, y or rational solutions. Let us begin with a few words about the latter.

The equation $y^2 = x^3 + k$ represents an elliptic curve in the real projective plane. By a rational point on the curve we shall mean either a pair (x, y) of rational numbers satisfying the equation, or the point at infinity on the curve; in other words, the rational points are given in homogeneous coordinates by (x, y, z), where $\lambda x, \lambda y, \lambda z$ are rational for some λ. It had been noted, at least by the time of Bachet, that the chord joining any two rational points on the curve intersects the curve again at a rational point, and similarly that the tangent at a rational point intersects again at a rational point. Thus, Fermat remarked, if there is a rational point on the curve other than the point at infinity, then, by taking chords and tangents, one would expect, in general, to obtain an infinity of rational points; a precise result of this kind was established by Fueter in 1930. It was also well known that the set of all rational points on the curve form a group under the chord and tangent process (see Fig. 8.1); the result is in fact an

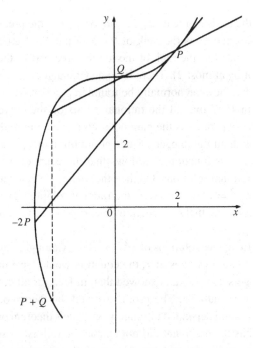

Fig. 8.1   Illustration of the group law on $y^2 = x^3 + 17$. The points P, Q and $P + Q$ are $(2, 5)$, $(\frac{1}{4}, \frac{33}{8})$ and $(-2, -3)$ respectively. The tangent at P meets the curve again at $-2P = (-\frac{64}{25}, -\frac{59}{125})$.

immediate consequence of the addition formulae for the Weierstrass functions $x = \wp(u)$, $y = \frac{1}{2}\wp'(u)$ that parameterize the curve. Indeed, with this notation, the group law becomes simply the addition of parameters. Mordell proved that the group has a finite basis, that is, there is a finite set of parameters $u_1, \ldots, u_n$ such that all rational points on the curve are given by $u = m_1 u_1 + \cdots + m_n u_n$, where $m_1, \ldots, m_n$ run through all rational integers. This is equivalent to the assertion that there is a finite set of rational points on the curve such that, on starting from the set and taking all possible chords and tangents, one obtains the totality of rational points on the curve. The demonstration involved an ingenious technique, usually attributed to Fermat, known as the method of infinite descent; we shall refer to the method again in Section 8.4. The work applied more generally to the equation $y^2 = x^3 + ax + b$ with a, b rational, and so, by birational transformation, to any curve of genus 1. Weil extended the theory to curves of higher genus and the subject of the Mordell–Weil theorem, as the result became known, subsequently gained great notoriety and stimulated much further research. The latter has been directed especially to the problem of determining the basis elements $u_1, \ldots, u_n$ or at least the precise value of n; this is usually referred to as the rank of the Mordell–Weil group (the rank r of the curve is defined as the rank of the torsion-free part of the group and it can differ from n by at most 2). There is no general algorithm for determining these quantities but they can normally be found in practice. Thus, for instance, Billing proved in 1937 that all the rational points on the curve $y^2 = x^3 - 2$ are given by mu_1, where u_1 is the parameter corresponding to the point $(3, 5)$ and m runs through all the integers. Since no multiple of u_1 is a period of the associated Weierstrass function, it follows that the equation $y^2 = x^3 - 2$ has infinitely many rational solutions. On the other hand, it is known, for instance, that the equation $y^2 = x^3 + 1$ has only the rational solutions given by $(0, \pm 1)$, $(-1, 0)$ and $(2, \pm 3)$, and that the equation $y^2 = x^3 - 5$ has no rational solutions whatever.

We turn now to integer solutions of $y^2 = x^3 + k$. Although, initially, Mordell believed that, for certain values of k, the equation would have infinitely many solutions in integers x, y, he later showed that, in fact, for all k, there are only finitely many such solutions. The proof involved the theory of reduction of binary cubic forms and depended ultimately on Thue's theorem on the equation $F(x, y) = m$. Thus the argument did not enable the full list of solutions to be determined in any particular instance. However, the situation was changed by Baker's work referred to in Section 8.2. This gave an effective demonstration of Thue's theorem, and, as a consequence, it furnished, for all solutions of $y^2 = x^3 + k$, a bound for $|x|$ and $|y|$ of the form $\exp(c|k|^{c'})$, where c, c' are absolute constants; Stark later showed that c' could be taken as $1 + \varepsilon$ for any $\varepsilon > 0$, provided that c was allowed to depend on ε. Thus, in principle, the complete list of solutions can now be determined for any particular value of k by a finite amount of computation. In practice, the bounds that arise are too large to enable one to check the finitely many remaining possibilities for x and y directly; but, as for the Thue equation, this can usually be accomplished by some supplementary analysis. In this way it has been shown, for instance, that all integer solutions of the equation $y^2 = x^3 - 28$ are given by $(4, \pm 6)$, $(8, \pm 22)$ and $(37, \pm 225)$.

Nevertheless, in many cases, much readier methods of solution are available. In particular, it frequently suffices to appeal to simple congruences. Consider, for example, the equation $y^2 = x^3 + 11$. Since $y^2 \equiv 0$ or $1 \pmod 4$, we see that, if there is a solution, then x must be odd and in fact $x \equiv 1 \pmod 4$. Now we have

$$x^3 + 11 = (x + 3)(x^2 - 3x + 9) - 16,$$

and $x^2 - 3x + 9$ is positive and congruent to 3 (mod 4). Hence there is a prime $p \equiv 3$ (mod 4) that divides $y^2 + 16$. But this gives $y^2 \equiv -16$ (mod p), whence $(yz)^2 \equiv -1$ (mod p), where $z = \frac{1}{4}(p+1)$. Thus -1 is a quadratic residue (mod p), contrary to Section 4.2. We conclude therefore that the equation $y^2 = x^3 + 11$ has no solution in integers x, y. Several more examples of this kind are given in Mordell's book (see Section 8.7).

Another typical method of solution is by factorization in quadratic fields. Consider the equation $y^2 = x^3 - 11$. Since $y^2 \equiv 0$, 1 or 4 (mod 8) we see that, if there is a solution, then x must be odd. We shall use the result established in Section 8.5, that the field $\mathbb{Q}(\sqrt{(-11)})$ is Euclidean and so has unique factorization. We have

$$(y + \sqrt{(-11)})(y - \sqrt{(-11)}) = x^3,$$

and the factors on the left are relatively prime; for any common divisor would divide $2\sqrt{(-11)}$, contrary to the fact that neither 2 nor 11 divides x. Thus, on recalling that the units in $\mathbb{Q}(\sqrt{(-11)})$ are ± 1, we obtain $y + \sqrt{(-11)} = \pm\omega^3$ and $x = N(\omega)$ for some algebraic integer ω in the field. Actually we can omit the minus sign since -1 can be incorporated in the cube. Now, since $-11 \equiv 1$ (mod 4), we have $\omega = a + \frac{1}{2}b(1 + \sqrt{(-11)})$ for some rational integers a, b. Hence, on equating coefficients of $\sqrt{(-11)}$, we see that

$$1 = 3(a + \tfrac{1}{2}b)^2(\tfrac{1}{2}b) - 11(\tfrac{1}{2}b)^3,$$

that is, $(3a^2 + 3ab - 2b^2)b = 2$. This gives $b = \pm 1$ or ± 2, and so the solutions (a, b) are $(0, -1)$, $(1, -1)$, $(1, 2)$ and $(-3, 2)$. But we have $x = a^2 + ab + 3b^2$. Thus we conclude that the integer solutions of the equation $y^2 = x^3 - 11$ are $(3, \pm4)$ and $(15, \pm58)$. A similar analysis can be carried out for the equation $y^2 = x^3 + k$ whenever $\mathbb{Q}(\sqrt{k})$ has unique factorization and $k \equiv 2$, 3, 5, 6 or 7 (mod 8).

Soon after establishing his theorem on the finiteness of the number of solutions of $y^2 = x^3 + k$, Mordell extended the result to the equation

$$y^2 = ax^3 + bx^2 + cx + d,$$

where the cubic on the right has distinct zeros; the work again rested ultimately on Thue's theorem but utilized the reduction of quartic forms rather than cubic. In a letter to Mordell, an extract from which was published in 1926 under the pseudonym X, Siegel described an alternative argument that applied more generally to the hyperelliptic equation $y^2 = f(x)$, where f denotes a polynomial with integer coefficients and with at least three simple zeros; indeed it applied to the superelliptic equation $y^m = f(x)$, where m is any integer ≥ 2. The theory was still further extended by Siegel in 1929; in a major work combining his refinement of Thue's inequality, referred to in Section 6.5, together with the Mordell–Weil theorem, Siegel succeeded in giving a simple condition for the equation $f(x, y) = 0$, where f is any polynomial with integer coefficients, to possess only finitely many solutions in integers x, y. In particular he showed that it suffices if the curve represented by the equation has genus at least 1. The result was employed by Schinzel, in conjunction with an old method of Runge concerning algebraic functions, to furnish a striking extension of Thue's theorem; this asserts that the equation $F(x, y) = G(x, y)$ has only finitely many solutions in integers x, y, where F is a binary form as in Section 8.2, and G is any polynomial with degree less than that of F.

8.4 The Fermat equation

In the margin of his well-worn copy of Bachet's edition of the works of Diophantus, Fermat wrote 'It is impossible to write a cube as the sum of two cubes, a fourth power as the sum of two fourth powers, and, in general, any power beyond the second as the sum of two similar powers. For this I have discovered a truly wonderful proof but the margin is too small to contain it'. As is well known, after efforts of numerous mathematicians over several centuries, Fermat's conjecture was finally established by Wiles in 1995. The work rested on deep theories on the modularity of elliptic curves; it went well beyond anything available at the time of Fermat and there is now considerable doubt as to whether Fermat really had a proof.

Many special cases of Fermat's conjecture were verified, mainly as a consequence of the work of Kummer in the nineteenth century. Indeed, as mentioned in Chapter 7, it was Kummer's remarkable researches that led to the foundation of the theory of algebraic numbers. Kummer showed in fact that the Fermat problem is closely related to questions concerning cyclotomic fields. The latter arise by writing the Fermat equation $x^n + y^n = z^n$ in the form

$$(x + y)(x + \zeta y) \cdots (x + \zeta^{n-1} y) = z^n,$$

where ζ is a root of unity. As we shall see in a moment, the case $n = 4$ can be readily treated; thus it suffices to prove that the equation has no solution in positive integers x, y, z when n is an odd prime p. The factors on the left are algebraic integers in the cyclotomic field $\mathbb{Q}(\zeta)$ and, when $p \leq 19$, the field has unique factorization; it is then relatively easy to establish the result. Kummer derived various more general criteria. In particular, he introduced the concept of a regular prime p and proved that Fermat's conjecture holds for all such p;

a prime is said to be regular if it does not divide any of the numerators of the first $\frac{1}{2}(p-3)$ Bernoulli numbers, that is, the coefficients B_j in the equation

$$t/(e^t-1)=1-\tfrac{1}{2}t+\sum_{j=1}^{\infty}(-1)^{j-1}B_jt^{2j}/(2j)!.$$

Kummer also established the result for certain classes of irregular primes. Thus, in particular, he covered all $p<100$; there are only three irregular primes in the range and these he was able to deal with separately. The most notable result arising from this approach was obtained by Wagstaff in 1978; by extensive computations he succeeded in establishing Fermat's conjecture for all $p<125\,000$.

Before the work of Kummer, Fermat's equation had already been solved for several small values of n. The special case $x^2+y^2=z^2$ dates back to the Greeks and the solutions (x, y, z) in positive integers are called Pythagorean triples. It suffices to determine all such triples with x, y, z relatively prime and with y even; for if x and y were both odd, we would have $z^2\equiv 2\pmod 4$ which is impossible. On writing the equation in the form $(z+x)(z-x)=y^2$, and noting that $(z+x,z-x)=2$, we obtain $z+x=2a^2$, $z-x=2b^2$ and $y=2ab$ for some positive integers a, b with $(a,b)=1$. This gives

$$x=a^2-b^2, \qquad y=2ab, \qquad z=a^2+b^2.$$

Moreover, since z is odd, we see that a and b have opposite parity. Conversely, it is readily verified that if a, b are positive integers with $(a,b)=1$ and of opposite parity then x, y, z above furnish a Pythagorean triple with $(x,y,z)=1$ and with y even. Thus we have found the most general solution of $x^2+y^2=z^2$. The first four Pythagorean triples, that is, with smallest values of z, are $(3,4,5)$, $(5, 12, 13)$, $(15, 8, 17)$ and $(7, 24, 25)$.

The next simplest case of Fermat's equation is $x^4+y^4=z^4$. This was solved by Fermat himself, using the method of infinite descent. He considered in fact the equation $x^4+y^4=z^2$. If there is a solution in positive integers, then it can be assumed that x, y, z are relatively prime and that y is even. Now (x^2, y^2, z) is a Pythagorean triple and there exist integers a, b as above such that $x^2=a^2-b^2$, $y^2=2ab$ and $z=a^2+b^2$. Further, b must be even, for otherwise we would have a even and b odd, and so $x^2\equiv-1\pmod 4$, which is impossible. Furthermore, (x, b, a) is a Pythagorean triple. Hence we obtain $x=c^2-d^2$, $b=2cd$ and $a=c^2+d^2$ for some positive integers c, d with $(c,d)=1$. This gives $y^2=2ab=4cd(c^2+d^2)$. But c, d and c^2+d^2 are coprime in pairs, whence $c=e^2$, $d=f^2$ and $c^2+d^2=g^2$ for some positive integers e, f, g. Thus we have $e^4+f^4=g^2$ and $g\le g^2=a\le a^2<z$. It follows, on supposing

that z is chosen minimally at the outset, that the equation $x^4 + y^4 = z^2$ has no solution in positive integers x, y, z.

The first apparent proof that the equation $x^3 + y^3 = z^3$ has no non-trivial solution was published by Euler in 1770, but the argument depended on properties of integers of the form $a^2 + 3b^2$ and there has long been some doubt as to its complete validity. An uncontroversial demonstration was given later by Gauss using properties of the quadratic field $\mathbb{Q}(\sqrt{(-3)})$. The proof is another illustration of the method of infinite descent. By considering congruences (mod λ^4), where λ is the prime $\frac{1}{2}(3 - \sqrt{(-3)})$ in $\mathbb{Q}(\sqrt{(-3)})$, it is readily verified that if the equation $x^3 + y^3 = z^3$ has a solution in positive integers, then one at least of x, y, z is divisible by λ. Hence, for some integer $n \geq 2$, the equation $\alpha^3 + \beta^3 + \eta\lambda^{3n}\gamma^3 = 0$ has a solution with η a unit in $\mathbb{Q}(\sqrt{(-3)})$ and with α, β, γ non-zero algebraic integers in the field. It is now easily deduced, by factorizing $\alpha^3 + \beta^3$, that the same equation, with n replaced by $n - 1$, has a solution as above, and the desired result follows. The equation $x^5 + y^5 = z^5$ was solved by Legendre and Dirichlet about 1825, and the equation $x^7 + y^7 = z^7$ by Lamé in 1839. By then, however, the *ad hoc* arguments were becoming quite complicated and it was not until the fundamental work of Kummer that Fermat's conjecture was established for equations with higher prime exponents.

Numerous results were obtained concerning special classes of solutions. For instance, Sophie Germain proved in 1823 that if p is an odd prime such that $2p + 1$ is also a prime then the 'first case' of Fermat's conjecture holds for p, that is, the equation $x^p + y^p = z^p$ has no solution in positive integers with xyz not divisible by p. Further, Wieferich proved in 1909 that the same conclusion is valid for any p that does not satisfy the congruence $2^{p-1} \equiv 1 \pmod{p^2}$. These results were greeted with great admiration at the times of their discovery. The latter condition is not in fact very stringent; there are only two primes up to $3 \cdot 10^9$ that satisfy the congruence, namely 1093 and 3511.

In another direction, Faltings succeeded in 1983 in proving a long-standing conjecture of Mordell to the effect that every curve defined over the rationals with genus at least 2 has only finitely many rational points. The argument involved deep theories in algebraic geometry and was a very important advance. As a particular case it follows at once that, for any given $n \geq 4$, there are only finitely many solutions to the Fermat equation in relatively prime integers x, y, z.

The work of Wiles referred to at the beginning rested on a study of the elliptic curve $y^2 = x(x - a)(x + b)$. Motivated by earlier work of Hellegouarch, it was suggested by Frey in 1985 that if there is a solution to the Fermat equation $u^p + v^p = w^p$ with a prime $p > 3$ and integers u, v, w then the elliptic curve indicated above with $a = u^p$ and $b = v^p$, now often referred to as the Frey curve, is not modular, that is, it cannot be paramaterized by modular functions. The suggestion was shown to be true by Ribet in 1990 and this opened the way to the eventually successful attack on Fermat's last theorem. For there was a well-known conjecture of Taniyama–Shimura dating back to 1955 asserting that every elliptic curve defined over the rationals is modular, and the Fermat problem was now reduced to verifying the latter. Wiles succeeded in proving the Taniyama–Shimura conjecture for a large class of elliptic curves including the case relevant to Fermat and this established the result. Actually Wiles' first version of the proof, which he announced in 1993, turned out to be incomplete but the exposition was put right a little later through some joint work of Taylor and Wiles. The full Taniyama–Shimura conjecture, based on the same sphere of ideas, was verified in 2001 by Breuil, Conrad, Diamond and Taylor.

The Wolfskehl Prize, offered by the Academy of Sciences in Göttingen in 1908 for the first demonstration of Fermat's last theorem, was conferred on Andrew Wiles in 1997; it amounted then to DM75,000.

8.5 The Catalan equation

In 1844, Catalan conjectured that the only solution of the equation $x^p - y^q = 1$ in integers x, y, p, q, all > 1, is given by $3^2 - 2^3 = 1$. After the endeavours of many mathematicians, as we shall describe below, the conjecture was finally established by Mihăilescu in 2004. He used new ideas from the theory of cyclotomic fields and his work was a remarkable achievement.

Previously, the most notable advance towards a demonstration was made by Tijdeman in 1976. He proved, by means of the theory of linear forms in logarithms (see Section 6.6), that the Catalan equation has only finitely many integer solutions and all of these can be effectively bounded. Thus, in principle, Tijdeman's work reduced the problem simply to the checking of finitely many cases. However, despite much effort, the bounds furnished by the theory have turned out, so far at least, to be too large to make the computation practical.

Nonetheless it is of interest to illustrate the approach. Accordingly let us consider the simpler equation $ax^n - by^n = c$, where a, b and $c \neq 0$ are given integers, and we seek to bound all the solutions in positive integers x, y and $n \geq 3$. We can assume, without loss of generality, that a and b are positive, and it will suffice to treat the case $y \geq x$. The equation can be written in the form $e^\Lambda - 1 = c/(by^n)$, where

$$\Lambda = \log(a/b) + n \log(x/y).$$

Now we can suppose that $y^n > (2c)^2$, for otherwise the solutions can obviously be bounded; then we have $|e^\Lambda - 1| < 1/(2y^{(1/2)n})$. But, for any real number u, the inequality $|e^u - 1| < \frac{1}{2}$ implies that $|u| \le 2|e^u - 1|$. Hence we obtain $|\Lambda| < y^{-(1/2)n}$, that is, $\log|\Lambda| < -\frac{1}{2}n \log y$. On the other hand, by the theory of linear forms in logarithms, we have $\log|\Lambda| \gg -\log n \log y$, where the implied constant depends only on a and b. Thus we see that $\frac{1}{2}n \ll \log n$, whence n is bounded in terms of a and b. The required bounds for x and y now follow from the effective result of Baker on the Thue equation referred to at the end of Section 8.2.

The work of Tijdeman on the equation $x^p - y^q = 1$ runs on the same lines. One can assume that p, q are odd primes and then, by elementary factorization, one obtains $x = kX^q + 1$, $y = lY^p - 1$ for some integers X, Y, where k is 1 or $1/p$ and l is 1 or $1/q$. Plainly we have $|p \log x - q \log y| \ll y^{-q}$, and substituting for x and y on the left yields a linear form,

$$\Lambda = p \ \log k - q \ \log l + pq \ \log(X/Y),$$

for which $|\Lambda|$ is small; similar forms arise by substituting for just one of x and y. The theory of linear forms in logarithms now furnishes the desired bounds for p and q, and those for x and y then follow from an effective version of the result on the superelliptic equation referred to in Section 8.3.

Several instances of Catalan's equation were solved long before the advent of the theories alluded to above. Indeed, in the Middle Ages, Leo Hebraeus had already dealt with the case $x = 3$, $y = 2$ and, in 1738, Euler had solved the case $p = 2$, $q = 3$. The case $q = 2$ was treated by V. A. Lebesgue in 1850, the cases $p = 3$ and $q = 3$ by Nagell in 1921, the case $p = 4$ by S. Selberg in 1932 and the case $p = 2$, which includes the result for $p = 4$, by Chao Ko in 1967. Moreover Cassels proved in 1960 that if p, q are primes, as one can assume, then p divides y and q divides x. Let us convey a little of the flavour of these works by proving that the equation $x^5 - y^2 = 1$ has no solution in integers except $x = 1$, $y = 0$. We shall use the unique factorization property of the Gaussian field. Clearly, since $y^2 \equiv 0$ or 1 (mod 4), we have x odd and y even. The equation can be written in the form $x^5 = (1 + iy)(1 - iy)$ and, since x is odd, the factors on the right are relatively prime. Thus we have $1 + iy = \varepsilon \omega^5$, where ε is a Gaussian unit and ω is a Gaussian integer. Now $\varepsilon = \pm 1$ or $\pm i$, and so $\varepsilon = \varepsilon^5$. Hence, on writing $\varepsilon \omega = u + iv$, where u, v are rational integers, we have $1 + iy = (u + iv)^5$. This gives $u^5 - 10u^3v^2 + 5uv^4 = 1$, whence $u = \pm 1$ and $1 - 10v^2 + 5v^4 = \pm 1$. It follows easily that $u = 1$, $v = 0$ and so $x = 1$, $y = 0$, as required. The argument can readily be extended to establish Lebesgue's more general theorem that, for any odd prime p, the equation $x^p - y^2 = 1$ has no solution in positive integers.

Mihăilescu, in his studies referred to at the beginning, showed first, following on work of Inkeri and Mignotte, that any primes p, q with $x^p - y^q = 1$ for some positive integers x, y must satisfy a double Wieferich criterion, namely

$$p^{q-1} \equiv 1 \ (\mathrm{mod}\ q^2), \quad q^{p-1} \equiv 1 \ (\mathrm{mod}\ p^2).$$

By combining this criterion with new results from the theory of cyclotomic fields, especially a deep theorem of Thaine on cyclotomic units, Mihăilescu succeeded in eliminating the case $p \not\equiv 1 \ (\mathrm{mod}\ q)$. There remained the case $p \equiv 1 \ (\mathrm{mod}\ q)$ and, initially, Mihăilescu excluded this through the theory of linear forms in logarithms. Later, however, he showed that, here too, one could give a purely algebraic argument. In any event, the celebrated conjecture of Catalan was finally solved.

A particularly striking result on an exponential Diophantine equation rather like those of Fermat and Catalan was obtained by Erdős and Selfridge in 1975; they proved that a product of consecutive integers cannot be a perfect power, that is, the equation

$$y^n = x(x+1)\cdots(x+m-1)$$

has no solution in integers x, y, m, n, all > 1.

8.6 The *abc*-conjecture

The *abc*-conjecture is a simple statement about integers. It was first formulated by Oesterlé and subsequently refined by Masser and it encapsulates many important results. Indeed it has come to be seen as of one of the key problems for the future direction of mathematics.

Let a, b, c be relatively prime non-zero integers satisfying

$$a + b + c = 0$$

and let N denote the 'radical' or 'conductor' of abc, that is, the product of all the distinct prime factors of abc. Then the *abc*-conjecture asserts that, for any $\varepsilon > 0$, we have

$$\max(|a|, |b|, |c|) \ll N^{1+\varepsilon},$$

where the implied constant depends only on ε (for the $\ll$ notation see Section 8.2).

The conjecture implies at once that the Fermat equation $x^n + y^n = z^n$ with $n > 3$ has only finitely many solutions in relatively prime positive integers

x, y, z. Indeed, on taking $a = x^n, b = y^n, c = -z^n$, we see that the radical N of abc is the same as that of xyz and so $N \le xyz \le z^3$; hence the abc-conjecture gives $z^n \ll z^{3(1+\varepsilon)}$ and, on assuming $\varepsilon < \frac{1}{3}$, it follows that z is bounded. In the same way we deduce that the Catalan equation $x^m - y^n = 1$ has only finitely many solutions in integers x, y, m, n, all > 1, provided we exclude $m = n = 2$. For on taking $a = x^m, b = -y^n, c = -1$ we see that $N \le xy$, whence the conjecture gives $\max(x^m, y^n) \ll (xy)^{1+\varepsilon}$. It follows that $xy \ll (xy)^{(1/m+1/n)(1+\varepsilon)}$ and so, since $1/m + 1/n < 1$, we conclude that x and y are bounded; then $\max(x^m, y^n)$ is bounded and thus also m and n. In fact the argument applies more generally to the Fermat–Catalan equation

$$ax^r + by^s + cz^t = 0,$$

where a, b, c are given non-zero integers, and shows that, if the conjecture holds, then the equation has only finitely many solutions in relatively prime integers x, y, z, all > 1, and positive exponents r, s, t satisfying $(1/r) + (1/s) + (1/t) < 1$.

The conjecture has many further applications and it has been widely discussed in the literature. The only significant approach to date with regard to a verification is due to Stewart and Kunrui Yu; they have shown, through deep studies concerning linear forms in logarithms, that

$$\log \max(|a|, |b|, |c|) \ll N^{1/3} (\log N)^3,$$

where the implied constant is absolute, that is, it can be given a numerical value independent of any parameters.

The work of Oesterlé was motivated by an earlier conjecture of Szpiro on elliptic curves, and Masser was influenced by a theorem which Mason had obtained through researches on linear forms in logarithms over function fields and which we now recognize as the analogue of the abc-conjecture in that setting. Stothers, as it turned out, had come upon the same result independently through studies on Riemann surfaces; his work had been published before Mason's but its significance was not realized at the time and it was only much later, when Mason's theorem had long been in the public domain, that it came to light.

Let k denote an algebraically closed field with characteristic 0. Further, let $a(x), b(x), c(x)$ be non-zero polynomials in $k[x]$ with no common factor and not all constant. Suppose that the polynomials have degrees l, m, n and that the numbers of their distinct zeros are given by p, q, r respectively. The Mason–Stothers theorem, as it is now known, states that if $a(x) + b(x) + c(x) = 0$ then we have

$$\max(l, m, n) < p + q + r.$$

For the proof, let $w(x)$ be the monic polynomial with simple zeros consisting of the distinct zeros of $a(x)b(x)c(x)$. Then $w(x)$ has degree $p + q + r$. We have $w(x)a'(x) = a(x)A(x)$, where $a'(x)$ denotes the derivative of $a(x)$ with respect to x and $A(x)$ denotes a polynomial with degree $< p + q + r$. Similarly we have $w(x)b'(x) = b(x)B(x)$ and $w(x)c'(x) = c(x)C(x)$. Further, the $A(x)$, $B(x)$, $C(x)$ are distinct. For if, for instance, $A(x) = B(x)$ then $a(x)b'(x) = b(x)a'(x)$ and so $a(x)$ and $b(x)$ would divide $a'(x)$ and $b'(x)$ respectively; this is plainly impossible since, by hypothesis, at most one of $a(x), b(x), c(x)$ is constant and so $a'(x)$ and $b'(x)$ cannot both be zero.

Now since $a'(x) + b'(x) + c'(x) = 0$ we obtain

$$a(x)A(x) + b(x)B(x) + c(x)C(x) = 0.$$

Then from $a(x) + b(x) + c(x) = 0$ it follows that

$$a(x)(A(x) - C(x)) + b(x)(B(x) - C(x)) = 0,$$

whence $a(x)$ divides $B(x) - C(x)$ and $b(x)$ divides $A(x) - C(x)$. Thus we get $\max(l, m) < p + q + r$ and, since clearly $n \leq \max(l, m)$, this proves the theorem.

8.7 Further reading

As remarked in Section 8.2, an excellent source for early results on Diophantine equations is Dickson's *History of the Theory of Numbers* (Washington, 1920). Another good reference work is Skolem's *Diophantische Gleichungen* (Springer, 1938).

For general reading, Mordell's *Diophantine Equations* (Academic Press, 1969) is to be highly recommended; the author was one of the great contributors to the subject and, as one would expect, he covers a broad range of material with clarity and considerable skill. Several other books on number theory contain valuable sections on Diophantine equations; this applies especially to Nagell's *Introduction to Number Theory* (Wiley, 1951).

A good survey of results on the effective solution of Diophantine equations is given by Győry in his article in *A Panorama of Number Theory or the View from Baker's Garden* (ed. G. Wüstholz, Cambridge University Press, 2002), pp. 38–72. For further accounts see the books by Baker and by Baker and Wüstholz cited in Section 6.8. The latter includes, in particular, references to the literature on the *abc*-conjecture.

In connection with Section 8.4, Ribenboim's *13 Lectures on Fermat's Last Theorem* (Springer, 1979) deals well with the classical material, and there is also a good discussion of the classical aspects relating to the Fermat equation in Borevich and Shafarevich's *Number Theory* (Academic Press, 1966). For an introduction to Wiles' work see Stewart and Tall's *Algebraic Number Theory and Fermat's Last Theorem* (A. K. Peters, 2002). As regards Mihăilescu's work referred to in Section 8.5, see Schoof's *Catalan's Conjecture* (Springer, 2008).

The theorem of Schinzel referred to at the end of Section 8.3 appeared in *Comment. Pontificia Acad. Sci.* **2** (1969), no. 20, 1–9. The theorem of Wiles referred to in Section 8.4 appeared in *Ann. Math.* **141** (1995), 443–551. The theorem of Faltings referred to in Section 8.4 appeared in *Invent. Math.* **73** (1983), 349–366. The theorem of Mihăilescu referred to in Section 8.5 appeared in *J. Reine Angew. Math.* **572** (2004), 123–144. The theorem of Erdős and Selfridge referred to at the end of Section 8.5 appeared in *Illinois J. Math.* **19** (1975), 292–301.

8.8 Exercises

(i) Prove that, if (x_n, y_n), with $n = 1, 2, \ldots$, is the sequence of positive solutions of the Pell equation $x^2 - dy^2 = 1$, written according to increasing values of x or y, then x_n and y_n satisfy a recurrence relation $u_{n+2} - 2au_{n+1} + u_n = 0$, where a is a positive integer. Find a when $d = 7$.

(ii) Determine whether or not the equation $x^2 - 31y^2 = -1$ is soluble in integers x, y.

(iii) Find the minimal solution in positive integers x, y of the equation $x^2 - dy^2 = 1$ when $d = 39, 41, 55$ and 1003.

(iv) Show that if p, q are primes $\equiv 3 \pmod{4}$ then at least one of the equations $px^2 - qy^2 = \pm 1$ is soluble in integers x, y.

(v) Prove, by congruences, that if a, c are integers with $a > 1$, $c > 1$ and $a + c \le 16$, then the equation $x^4 - ay^4 = c$ has no solution in rationals x, y.

(vi) Show that the equation $x^3 + 2y^3 = 7(z^3 + 2w^3)$ has no solution in relatively prime integers x, y, z, w.

(vii) By considering the intersection of the quartic surface $x^4 + y^4 + z^4 = 2$ with the line $y = z - x = 1 - tx$, where t is a parameter, show that the equation $x^4 + y^4 + z^4 = 2w^4$ has infinitely many solutions in relatively prime integers x, y, z, w.

(viii) Solve the equation $y^2 = x^3 - 17$ in integers x, y by considering the factors of $x^3 + 8$.

(ix) Solve the equation $y^2 = x^3 - 2$ in integers x, y by factorization in $\mathbb{Q}(\sqrt{(-2)})$.

(x) Prove, by the method of infinite descent, that the equation $x^4 - y^4 = z^2$ has no solution in positive integers x, y, z.

(xi) Prove that the equation $x^4 - 3y^4 = z^2$ has no solution in positive integers x, y, z. Deduce that the equation $x^4 + y^4 = z^3$ has no such solution with $(x, y) = 1$. (For the first part see Pocklington, *Proc. Cambridge Phil. Soc.* **17** (1914), 108–21.)

(xii) By considering $(x + 1)^3 + (x - 1)^3$, show that every integer divisible by 6 can be represented as a sum of four integer cubes. Show further that every integer can be represented as a sum of five integer cubes.

(xiii) Prove, by factorization in $\mathbb{Q}(\sqrt{(-7)})$, that the equation $x^2 + x + 2 = y^3$ has no solution in integers x, y except $x = 2$, $y = 2$ and $x = -3$, $y = 2$. Verify that the equation $x^2 + 7 = 2^{3k+2}$ has no solution in integers k, x with $k > 1$. (This is a special case of a conjecture of Ramanujan to the effect that the equation $x^2 + 7 = 2^n$ has only the integer solutions given by $n = 3, 4, 5, 7$ and 15. The conjecture was proved by Nagell; for a demonstration see page 205 of Mordell's book, referred to in Section 8.7).

9

Factorization and primality testing

9.1 Fermat pseudoprimes

By a primality test we mean a criterion which, if it is not satisfied, guarantees that a natural number n is composite. If the number passes several of these tests – that is, if it satisfies the criterion in each case – then it is likely, though in general not certain, that it is a prime. It turns out that in cryptography it is often enough to know that a number is 'probably' a prime and this is where the concept originates. A method which definitely establishes that a number is a prime is called deterministic; otherwise it is called probabilistic. The simplest example of a deterministic method is based on the criterion that n be not divisible by any integer between 2 and $\sqrt{n}$; if n passes the test for each possible divisor then it is, without doubt, a prime. But verifying in a particular instance is a very time-consuming process.

Suppose now that n is composite and odd. If there exists an integer b, with $(b, n) = 1$, such that $b^{n-1} \equiv 1 \pmod{n}$ then n is called a pseudoprime (or Fermat pseudoprime) to the base b. Thus a pseudoprime is a number that has a property analogous to that in Fermat's theorem (see Section 3.3) but is not a prime. For example, since the order of 2 $(\mod 21)$ is 6, we obtain $8^{20} \equiv 1 \pmod{21}$ and so 21 is a pseudoprime to the base 8; but $2^{20} \equiv 4 \pmod{21}$, whence 21 is not a pseudoprime to the base 2.

To explain the way the concept is used in practice, let n be a large odd integer. For any integer b with $0 < b < n$ we can determine $d = (b, n)$ by Euclid's algorithm. If $d > 1$ then certainly n is composite; otherwise we calculate the least positive residue of $b^{n-1} \pmod{n}$ using, say, the repeated-squaring method (see Section 9.7) and if it turns out to be 1 then there is a chance that n is a prime. If, on repeating the process for several further b, we find that in each case n is either a prime or a pseudoprime then there is a high probability that n is in fact a prime; indeed it is not difficult to show (see Koblitz's book referred

90

to in Section 9.8) that the chance that it is composite after k tests is at most $1/2^k$ except if it happens to have the special property that it is a pseudoprime for all possible b.

The question now arises as to whether the latter can actually occur. The answer is 'yes': the composite odd integers n such that $b^{n-1} \equiv 1 \pmod{n}$ for all b with $(b, n) = 1$ are called Carmichael numbers. The smallest example is $561 = 3 \times 11 \times 17$. From Fermat's theorem, one sees that, if n is square-free and $p - 1$ divides $n - 1$ for each prime factor p of n (as in the case of 561), then n is a Carmichael number. And the converse holds. For if n is square-free and p divides n then, by the Chinese remainder theorem, there is an integer b such that $b \equiv g \pmod{p}$ and $b \equiv 1 \pmod{n/p}$, where g is a primitive root $\pmod{p}$; then $(b, n) = 1$, whence, if n is a Carmichael number, we have $g^{n-1} \equiv b^{n-1} \equiv 1 \pmod{p}$ and thus $p - 1$ divides $n - 1$. The proof that a Carmichael number is square-free follows on similar lines, taking g as a primitive root $\pmod{p^2}$, and we refer again to Koblitz's book. As a corollary, a Carmichael number n must be the product of at least three primes; for if $n = pq$ with p, q primes and $p < q$ then $q - 1$ divides $n - 1 = p(q - 1) + (p - 1)$ which is impossible since $p - 1 < q - 1$. A result of Alford, Granville and Pomerance of 1992 (see Section 9.8) gives the existence of infinitely many Carmichael numbers.

9.2 Euler pseudoprimes

Let n be composite and odd and let b be an integer such that $(b, n) = 1$. We call n an Euler pseudoprime to the base b if, with the Jacobi symbol, we have

$$b^{\frac{1}{2}(n-1)} \equiv \left(\frac{b}{n}\right) \pmod{n}.$$

Thus an Euler pseudoprime has a property analogous to that in Euler's criterion and yet it is not a prime. If n is an Euler pseudoprime to the base b then it is also a Fermat pseudoprime to that base; this is obvious on squaring both sides of the above congruence. Further, there is no analogue of a Carmichael number: that is, every composite n is not an Euler pseudoprime for some b'. In fact, if n is square-free, we can take b' as a solution to the congruences $x \equiv a \pmod{p}$ and $x \equiv 1 \pmod{n/p}$, where p is a prime divisor of n and a is some quadratic non-residue $\pmod{p}$; if n is divisible by p^2 for some prime p, we can take $b' = 1 + n/p$. Now if n is an Euler pseudoprime to a base b then, by the multiplicative property of the Jacobi symbol, we see that it is not an Euler pseudoprime for bb'. Thus at most half of the b with $0 < b < n$ and $(b, n) = 1$ can be bases. It follows that if an integer n, after k trials with different random

b, satisfies the above congruence in each case then there is a probability of at most $1/2^k$ that it is composite; this gives an efficient probabilistic method, due to Solovay and Strassen, of testing for primality.

Every n is trivially an Euler pseudoprime to the bases ± 1. A non-trivial example is given by $703 = 19 \times 37$; since 3 has order 18 mod 19 and mod 37, we have $3^{351} \equiv 3^9 \equiv -1 \pmod{703}$, and since also

$$\left(\frac{3}{703}\right) = -\left(\frac{703}{3}\right) = -\left(\frac{1}{3}\right) = -1,$$

it follows that 703 is an Euler pseudoprime to the base 3.

There is a refinement of the Solovay–Strassen test due to Miller and Rabin based on the concept of a strong pseudoprime; it derives from the observation that the only solutions to the congruence $x^2 \equiv 1 \pmod{p}$ for a prime p are $x \equiv \pm 1$. Let n and b be as at the beginning of this section and let $n - 1 = 2^s m$, where s is a positive integer and m is odd. Then n is called a strong pseudoprime with respect to b if either $b^m \equiv 1 \pmod{n}$ or $b^{lm} \equiv -1 \pmod{n}$ for some $l = 2^r$ where $0 \le r < s$. It is easily seen that a composite integer $n \equiv 3 \pmod 4$ is a strong pseudoprime to the base b if and only if it is an Euler pseudoprime to that base. For in this case we have $s = 1, l = 1$ and $m = \frac{1}{2}(n-1)$, whence the criterion for a strong pseudoprime becomes $b^m \equiv \pm 1 \pmod{n}$; this clearly holds if n is an Euler pseudoprime and the converse is also valid, for if the congruence is satisfied then from $b^m = b(b^2)^{\frac{1}{4}(n-3)}$ and $n \equiv 3 \pmod 4$ we obtain

$$\left(\frac{b}{n}\right) = \left(\frac{b^m}{n}\right) = \left(\frac{\pm 1}{n}\right) = \pm 1$$

with linked signs. An example of a strong pseudoprime is now 703 as above. There is a theorem to the effect that a strong pseudoprime to a base b is also an Euler pseudoprime to that base but the proof is not simple and we refer again to Koblitz's book for details.

In the practical application of the Miller–Rabin test, one determines, for a random b, first whether $b^m \equiv \pm 1 \pmod{n}$ and then, by sequential squaring, whether $b^{lm} \equiv -1 \pmod{n}$ for $l = 2, 4, \ldots, 2^{s-1}$. If none of the congruences is satisfied then n must be composite since, for a prime p, the first element in the sequence $b^{p-1}, b^{\frac{1}{2}(p-1)}, b^{\frac{1}{4}(p-1)}, \ldots$ which is not congruent to 1, if any, must be congruent to $-1 \pmod{p}$. The probability that n is composite after passing k trials with different random b is at most $1/4^k$, a useful improvement on $1/2^k$ as in the earlier tests.

9.3 Fermat factorization

As we shall see from Section 9.7, it is of interest to have an efficient technique for the factorization of large numbers. One of the simplest is the method of Fermat factorization, based on the fact that if n is odd and $n = ab$ with $a \geq b > 0$ then $n = r^2 - s^2$, where $r = \frac{1}{2}(a+b)$, $s = \frac{1}{2}(a-b)$, and, conversely, if $n = r^2 - s^2$ then $n = ab$ with $a = r+s$, $b = r-s$. Thus if a and b happen to be close together so that s is small then r will be only slightly bigger than $\sqrt{n}$ and hence, on checking whether $r^2 - n$ is a perfect square for $r = [\sqrt{n}]+1$, $[\sqrt{n}]+2, \ldots$, we shall quickly come upon an instance when it is so and this will yield a and b.

As an example, consider $n = 644\,773$; here $[\sqrt{n}] = 802$ and $803^2 - 644\,773 = 36 = 6^2$, whence we immediately get $n = (803+6)(803-6) = 809 \times 797$. As a further example consider $n = 1\,485\,151$; here $[\sqrt{n}] = 1218$ and $1219^2 - n = 810$ is not a square but $1220^2 - n = 3249 = 57^2$, whence we get $n = (1220 + 57)(1220 - 57) = 1277 \times 1163$.

A modification of the method works when b is close to a small odd multiple of a rather than a itself. Thus consider $n = 5\,933\,299$. Here the technique above does not readily yield the desired factorization but let us study instead $m = 3n$. We have $[\sqrt{m}] = 4218$ and $4219^2 - m = 64$, whence we obtain $m = (4219 + 8)(4219 - 8) = 4227 \times 4211$ and so $n = 1409 \times 4211$.

Fermat factorization, in its simplest form, will succeed eventually in any particular instance but it may be impractical: that is, it may use up too much time.

9.4 Fermat bases

There is another, more efficient, generalization of Fermat factorization based on the fact that, whenever we find integers r, s such that $r^2 \equiv s^2 \pmod{n}$, the number $(r - s, n)$ is a proper divisor of n unless $r \equiv s \pmod{n}$, and similarly for $(r + s, n)$ unless $r \equiv -s \pmod{n}$. The search for suitable r and s leads to the following notion.

A factor base B is a set $\{p_1, \ldots, p_k\}$ of distinct primes with the proviso that p_1 is allowed to be -1. For a given odd n, the square of an integer b is called a B-number if the numerically least residue b' of $b^2 \pmod{n}$, that is, the residue satisfying $-\frac{1}{2}n < b' < \frac{1}{2}n$, can be expressed as a product of powers of elements of B. It is customary to associate to a B-number the k-tuple $(\varepsilon_1, \ldots, \varepsilon_k)$, where $\varepsilon_j = 0$ if p_j occurs to an even power in the canonical factorization of b' and $\varepsilon_j = 1$ otherwise. Thus, for $n = 233$, since $50^2 \equiv -63 \pmod{233}$, we see that 50^2 is a B-number, where $B = \{-1, 3, 7\}$, and its associated k-tuple is $(1, 0, 1)$.

Now to factorize a large odd integer n we seek a factor base B and a set of B-numbers b_m^2 such that the product of their numerically least residues b'_m is a perfect square, say s^2. In other words, we seek the B-numbers such that every element p_j of B occurs to an even power in the canonical factorization of the product, and this will be so if $\sum_m \varepsilon_{mj}$ is even for each $j = 1, 2, \ldots, k$, where $(\varepsilon_{m1}, \ldots, \varepsilon_{mk})$ is the k-tuple associated to b_m^2 as above. On taking r as the product of the b_m, we have $r^2 \equiv s^2 \pmod{n}$ and, provided that r is not congruent to $\pm s \pmod{n}$, we obtain a proper factor $(r \pm s, n)$ of n. If n is composite then, since r is a random square-root of s^2, the probability that $r \equiv \pm s \pmod{n}$ is at most $\frac{1}{2}$; repeated trials with different bases B and sets of B-numbers can be used to reduce the probability. Moreover, by linear algebra over the mod 2 field $\mathbb{F}_2$, one knows that as soon as one has found $k + 1$ distinct B-numbers b_m^2 then some subset can be determined that will have the desired property.

As an example, let us attempt to factorize $n = 70751$. It seems reasonable to hope that taking integers b_m close to $[\sqrt{(ln)}]$ for $l = 1, 2, 3, \ldots$ will yield small residues b'_m of $b_m^2 \pmod{n}$ and that sufficiently many of these will be products of small primes. On testing $[\sqrt{n}], [\sqrt{n}] + 1, \ldots, [\sqrt{n}] + 11$, and then $[\sqrt{(2n)}], [\sqrt{(3n)}], \ldots, [\sqrt{(9n)}]$, we find that

$$
\begin{aligned}
&[\sqrt{n}] + 11 = 276, && 276^2 \equiv 5425 \pmod{n}, && 5425 = 5^2 \times 7 \times 31; \\
&[\sqrt{(8n)}] = 752, && 752^2 \equiv -504 \pmod{n}, && -504 = -2^3 \times 3^2 \times 7; \\
&[\sqrt{(9n)}] = 797, && 797^2 \equiv -1550 \pmod{n}, && -1550 = -2 \times 5^2 \times 31.
\end{aligned}
$$

Hence we have a factor base $\{-1, 2, 3, 5, 7, 31\}$ and B-numbers 276^2, 752^2, 797^2 with the desired property and thus we can take $r = 276 \times 752 \times 797$ and $s = -2^2 \times 3 \times 5^2 \times 7 \times 31$. Then $r \equiv 3106 \pmod{n}$ and $s \equiv 5651 \pmod{n}$, whence $r + s \equiv 8757 \pmod{n}$ and $r - s \equiv -2545 \pmod{n}$. Now it is readily seen that $(8757, 70751) = 139$ and $(2545, 70751) = 509$ and so finally $n = 139 \times 509$.

9.5 The continued-fraction method

A way of finding a factor base B and B-numbers with the property described above is by the use of continued fractions.

The method depends on the fact that if $\theta > 1$ is a real number and p_m/q_m ($m = 0, 1, 2, \ldots$) are the convergents to θ then we have $|p_m^2 - \theta^2 q_m^2| < 2\theta$. For the proof, we recall that θ occurs in the interval between p_m/q_m and p_{m+1}/q_{m+1} and that the length of this interval is $1/(q_m q_{m+1})$. Thus we have $|\theta - p_m/q_m| <$

$1/(q_m q_{m+1})$, whence $|p_m - \theta q_m| < 1/q_{m+1}$ which gives $p_m < \theta q_m + 1/q_{m+1}$. Hence we obtain

$$|p_m^2 - \theta^2 q_m^2| = |p_m + \theta q_m||p_m - \theta q_m| < (2\theta q_m + 1/q_{m+1})(1/q_{m+1})$$
$$< 2\theta(q_m + 1)/q_{m+1} < 2\theta$$

as required. On applying the result with $\theta = \sqrt{n}$, where n is a positive integer, not a perfect square, we see that $|p_m^2 - n q_m^2| < 2\sqrt{n}$ for each convergent p_m/q_m to $\sqrt{n}$. This shows that, if $n > 16$ so that $2\sqrt{n} < \frac{1}{2}n$, then $p_m^2 - n q_m^2$ is the numerically least residue of p_m^2 (mod n) and that the latter is less than $2\sqrt{n}$.

Now, to factor n, one generates as far as necessary the partial quotients in the continued fraction for $\sqrt{n}$ and then the numerators p_m ($m = 0, 1, 2, \ldots$) of the corresponding convergents reduced, say, to their least positive residues (mod n). This enables one to compute the quantities $b_m' = p_m^2 - n q_m^2 \equiv p_m^2$ (mod n); one knows, by the result obtained above, that all of these are less than $2\sqrt{n}$ and so there is a reasonable chance that, with respect to some factor base B, a subset of the p_m^2 will be B-numbers with the desired property and that moreover $(r \pm s, n)$ will be a proper divisor of n as before. Clearly, when p_m, q_m are solutions to the Pell equation $x^2 - n y^2 = 1$, we have $b_m' = 1$ and p_m^2 is a B-number to the base $\{-1\}$. Thus it is enough simply to determine whether the corresponding divisors $(p_m \pm 1, n)$ are non-trivial and this can always be done unless the period of the continued fraction for $\sqrt{n}$ is unduly large.

As an example, consider $n = 27\,323$. The continued fraction for $\sqrt{n}$ is given by $[165, \overline{3, 2, 1, 2, 3, 330}]$ and it has period 6. The numerators of the first 6 convergents are $165, 496, 1157, 1653, 4463, 15\,042$ and the numerically least residues of their squares (mod n) are $-98, 109, -178, 109, -98, 1$. Thus 496^2 and 1653^2 are B-numbers with respect to the base $\{109\}$ and since $496 \times 1653 \equiv 198$ (mod n) we can take $r = 198$ and $s = 109$. This gives $r + s = 307$ and $r - s = 89$, whence $n = 307 \times 89$. The pattern of residues here is typical and, though we chose to use the second and fourth of them, we could just as well have used the sixth as indicated above.

As another example, consider $n = 12\,403$. The continued fraction for $\sqrt{n}$ is $[111, \overline{2, 1, 2, 2, 7, 1, 4, 1, 4, 1, 7, 2, 2, 1, 2, 222}]$ and it has period 16. The numerators of the first nine convergents, reduced (mod n), are $111, 223, 334, 891, 2116, 3300, 5416, 158, 5574$ and the residues of their squares are $-82, 117, -71, 89, -27, 166, -39, 158, -39$. Thus 5416^2 and 5574^2 are B-numbers with respect to the base $\{-1, 3, 13\}$ and since $5416 \times 5574 \equiv -118$ (mod n) we can take $r = 118$ and $s = 39$. This gives $r + s = 157$ and $r - s = 79$, whence

$n = 157 \times 79$. It will be seen that we could have simply used the congruence $158^2 \equiv 158 \pmod{n}$ which comes from the eighth residue and that 223^2 is also a B-number to the above base.

As a final example, consider $n = 36\,581$. The continued fraction for $\sqrt{n}$ begins $[191, 3, 1, 4, 1, 1, 1, 3, 7, 12, 4, 1, \ldots]$ and it turns out to have period 66. The numerators of the first 12 convergents, reduced $\pmod{n}$, are 191, 574, 765, 3634, 4399, 8033, 12432, 8748, 506, 14820, 23205, 1444 and the residues of their squares are -100, 247, -71, 215, -148, 205, -101, 52, -31, 76, -295, 19. Now we have $247 \times 52 \times 76 = (2^2 \times 13 \times 19)^2$ but this yields only $r \equiv s \pmod{n}$. However, we have also $76 \times 19 = (2 \times 19)^2$ and since $14820 \times 1444 \equiv 195 \pmod{n}$ we can take $r = 195$ and $s = 38$. This gives $n = 157 \times 233$.

9.6 Pollard's method

Let n be a composite positive integer. If some prime factor p of n has the property that $p - 1$ has no large prime divisor then there is a method, due to Pollard, customarily referred to as Pollard's $p - 1$ method, that is almost certain to find p. It is based on the fact that if a positive integer k is divisible by all integers up to a certain bound K and if $p - 1$ is divisible only by prime powers less than K then $p - 1$ divides k, whence, by Fermat's theorem, we have $a^k \equiv 1 \pmod{p}$ for all integers $a > 0$ with $(a, p) = 1$. Hence $(a^k - 1, n)$ is divisible by p and it is a proper divisor of n unless $a^k \equiv 1 \pmod{n}$.

To apply Pollard's method, we choose a bound K, an integer, and we take k to be $K!$ or, say, the lowest common multiple of the integers not exceeding K. Then, for an integer a with $1 < a < n - 1$, we find $a^k \pmod{n}$ by repeated squaring (see Section 9.7) and we compute $(a^k - 1, n)$ by Euclid's algorithm. If the latter is non-trivial we are through; otherwise we try again with different pairs a, k.

As an example, consider $n = 212\,899$. We choose $K = 7$ and we take k to be the lowest common multiple of the integers up to 7, that is, $k = 420$. Then, on taking $a = 2$ and noting that $420 = 2^2 + 2^5 + 2^7 + 2^8$, we obtain $a^k = 2^{420} \equiv 54\,861 \pmod{n}$ and $(54\,860, n) = 211$. Thus we get $n = 211 \times 1009$.

To apply Pollard's method when $n = 732\,661$, for instance, we would need a bound K as big as 359; for we have the prime factorizations $n = 719 \times 1019$, $718 = 2 \times 359$ and $1018 = 2 \times 509$. Lenstra introduced a technique analogous to Pollard's based on the theory of elliptic curves which has much greater flexibility. It incorporates elements of interest to both number theorists and cryptographers; see the books mentioned in Section 9.8.

9.7 Cryptography

A public key cryptosystem based on Euler's theorem was introduced by Rivest, Shamir and Adleman in 1978. The RSA algorithm, as it is called, begins with each user choosing two large primes p, q and a natural number m with $m < \phi(n)$ and $(m, \phi(n)) = 1$, where $n = pq$. From his knowledge of p, q the user can compute $\phi(n) = (p-1)(q-1) = n + 1 - p - q$ and moreover he can find the least positive solution l to the congruence $lm \equiv 1 \pmod{\phi(n)}$: that is, the multiplicative inverse of m. The pair m, n is called the enciphering key for the particular user and it is made public; the pair l, n is called the deciphering key and it is concealed.

Now to send a plaintext message P, in digital form, to a recipient with enciphering key m, n, the user transmits as ciphertext C a residue of $P^m \pmod n$, say the least positive residue; the recipient retrieves the original message by computing the least positive residue of $C^l \pmod n$. Since $C \equiv P^m$ we have, by Euler's theorem, $C^l \equiv P^{lm} \equiv P \pmod n$ and so the recipient does indeed obtain P assuming, as one may, that $0 < P \leq n$. (Note here that, since n is square-free, $P^{lm} \equiv P \pmod n$ holds for all P including the instance when $(P, n) > 1$. Indeed $\phi(n)$ is divisible by $p - 1$, whence, if $(P, p) = 1$, we have, by Fermat's theorem, $P^{\phi(n)} \equiv 1 \pmod p$ and thus $P^{lm} \equiv P \pmod p$; the latter is obvious when $(P, p) > 1$ and it holds similarly for q.) The success of the cryptosystem depends on the fact that finding large primes is much easier than factorizing large integers even when one knows that there are just two prime factors.

As an example, consider sending the message NO to a user with enciphering key $5, 703$. Let us make the message digital by taking digraphs in the usual 26-letter alphabet: then the numerical equivalent of NO is $P = 13.26 + 14 = 342$. Now $342^5 \equiv 589 \pmod{703}$ and therefore $C = 589$ (that is, $22.26 + 17$ or just WR in equivalent digraphs). To decipher, we observe that $703 = 19.37$, whence $\phi(703) = 648$. Then, to determine l, we use Euclid's algorithm to solve $5l \equiv 1 \pmod{648}$; we have $648 = 129.5 + 3$, $5 = 1.3 + 2$, $3 = 1.2 + 1$ and so $1 = 2(648 - 129.5) - 5 = 2.648 - 259.5$. This gives $l \equiv -259 \pmod{648}$ and hence $l = 389$. Now $389 = 1 + 2^2 + 2^7 + 2^8$ and, mod 703, we have $589^4 \equiv 266$, $589^{128} \equiv 342$, $589^{256} \equiv 266$. Thus $589^{389} \equiv 342$ as required. The technique used here for calculating the latter residues is called the repeated-squaring method.

9.8 Further reading

The book most closely associated with our text and which we particularly recommend is *A Course in Number Theory and Cryptography* (Springer, 1994) by N. Koblitz. This contains, in particular, references to the original papers on the RSA algorithm. Another relevant work is H. Riesel's *Prime Numbers and Computer Methods for Factorization* (Birkhäuser, 1994). For developments in relation to elliptic curves, see L. C. Washington's *Elliptic Curves: Number Theory and Cryptography* (Chapman & Hall/CRC, 2008) and the book by Blake, Seroussi and Smart, *Elliptic Curves in Cryptography* (Cambridge University Press, 2000). The result of Alford, Granville and Pomerance mentioned at the end of Section 9.1 appeared in *Ann. Math.* **139** (1994), 703–722.

9.9 Exercises

(i) Show that, if p is a prime, then p^2 is a pseudoprime with respect to a base b if and only if $b^{p-1} \equiv 1 \pmod{p^2}$. What is the analogous result for p^3?

(ii) Verify that 341 is a pseudoprime to the base 2. Show that, if n is a pseudoprime to the base 2, then so is $N = 2^n - 1$. Deduce that there are infinitely many pseudoprimes to the base 2.

(iii) With the same hypothesis, show that N is both an Euler pseudoprime and a strong pseudoprime to the base 2.

(iv) Prove that, if n is a strong pseudoprime with respect to a base b, then it is also a strong pseudoprime with respect to the base b^k for any positive integer k.

(v) Find all Carmichael numbers of the form $91p$, where p is a prime.

(vi) Determine the divisors of 1 324 703 and 7 009 529 by the Fermat factorization method.

(vii) Determine the divisors of 10 349 and 30 523 by the continued-fraction method.

(viii) Determine the divisors of 61 549 and 219 341 by Pollard's method.

(ix) In the RSA algorithm, what is the deciphering key corresponding to the enciphering key 7, 1027? Find also the deciphering keys corresponding to 7, 1661 and 7, 4661.

(x) Use the repeated-squaring method to find the least positive residue of $5^{35} \pmod{1019}$.

10

Number fields

10.1 Introduction

The subject of number fields was originated by Kummer, Dedekind, Weber and others during the nineteenth century. It is closely related to the theory of Diophantine equations – in fact it was motivated to a large extent by attempts to solve Fermat's last theorem – and it now impinges on most branches of number theory. We have already given a short discussion of the topic, mainly with respect to quadratic fields, in Chapter 7; here and in the next three chapters we shall develop the subject in more detail.

As prerequisites we shall assume only the elementary properties of rings, fields and vector spaces. Some knowledge of Galois theory is useful but, as we shall present the material, not essential. However, to avoid the theory, we shall need to appeal in a few places to the classical symmetric function theorem; it asserts that if R is any ring then every symmetric polynomial in $R[x_1, \ldots, x_n]$ is expressible as a polynomial over R in the elementary symmetric functions $s_1, \ldots, s_n$, where

$$(t + x_1) \cdots (t + x_n) = t^n + s_1 t^{n-1} + \cdots + s_n,$$

so that $s_1 = x_1 + \cdots + x_n$, $s_2 = x_1 x_2 + \cdots + x_{n-1} x_n$, $\ldots$, $s_n = x_1 \cdots x_n$. In other words, the symmetric polynomials form a polynomial ring $R[s_1, \ldots, s_n]$. A proof is given, for example, on page 178 of Vol. 1 of P. M. Cohn's *Algebra* (Wiley, 1982).

We shall frequently mention the division algorithm; this, as we recall from Section 1.2, asserts that if a, b are in $\mathbb{Z}$ and if $b \neq 0$ then there exist q, r in $\mathbb{Z}$ such that $a = bq + r$ with $0 \leq r < |b|$. We shall need an analogue for polynomials to the effect that if $a(x), b(x)$ are in the polynomial ring $\mathbb{Q}[x]$ (more generally $R[x]$, where R is a field) and if $b(x)$ is not constant then there exist $q(x), r(x)$ in $\mathbb{Q}[x]$ such that $a(x) = b(x)q(x) + r(x)$ and degree $r(x) <$ degree $b(x)$. For the proof, consider the polynomials of the form $a(x) - b(x)q(x)$ with $q(x)$ in

$\mathbb{Q}[x]$; one of these, say $r(x)$, has least degree and if this were $\geq$ the degree of $b(x)$ then we could diminish it by subtracting a multiple of $b(x)$ with an element of $\mathbb{Q}[x]$ of the form λx^j. In fact $q(x), r(x)$ are unique as one readily verifies.

10.2 Algebraic numbers

An algebraic number α is a zero of a polynomial $P(x)$ with rational coefficients. The minimum polynomial for α is the P as above that is monic and has least degree; a polynomial $P(x) = a_0 x^d + a_1 x^{d-1} + \cdots + a_d$ is said to be monic if a_0, the leading coefficient, is 1.[†] The degree of α is defined as the degree d of the minimum polynomial P for α. The conjugates of α are defined as the zeros $\alpha_1, \ldots, \alpha_d$ of P; thus α is one of $\alpha_1, \ldots, \alpha_d$ and we can write $P(x) = (x - \alpha_1) \ldots (x - \alpha_d)$. Note further that

 (i) P is the minimum polynomial for each of the conjugates $\alpha_1, \ldots, \alpha_d$ of α, for, by the division algorithm, the minimum polynomial P_j for α_j divides P and so α would be a zero of P_j or P/P_j, a contradiction unless $P = P_j$.
 (ii) All the α_j are distinct. For let P' denote the derivative of P (that is, with the above notation, $P'(x) = a_0 d x^{d-1} + \cdots + a_{d-1}$). Then P, P' are relatively prime by (i) and hence P has no squared factor.
(iii) If Q is a polynomial with rational coefficients such that $Q(\alpha_j) = 0$ for some j then P divides Q and thus $Q(\alpha_j) = 0$ for all j.

The totality of algebraic numbers form a field Ω. This is readily verified from the axioms for a field. For example we observe that $\alpha + \beta$ is a zero of the polynomial $\prod_{i=1}^m \prod_{j=1}^n (x - (\alpha_i + \beta_j))$, where $\alpha_1, \ldots, \alpha_m$ and $\beta_1, \ldots, \beta_n$ are the conjugates of α, β; this polynomial has rational coefficients by the symmetric function theorem. Also, for example, $1/\beta$, for $\beta \neq 0$, is a zero of $x^n P(1/x)$, where P is the minimum polynomial for β.

10.3 Algebraic number fields

The algebraic number field $\mathbb{Q}(\alpha)$ generated by the algebraic number α over the rationals $\mathbb{Q}$ is defined as the set $p(\alpha)$ where p is a polynomial with rational coefficients. The set can be regarded as being embedded in the complex numbers and so we have the usual operations of addition and multiplication. Then the set is a field; this is obvious $(p(\alpha) + q(\alpha) = (p+q)(\alpha),\ p(\alpha)q(\alpha) = (pq)(\alpha), \ldots)$

[†] The minimum polynomial defined here is the minimal polynomial of Section 6.5 divided by a_0; we argue now in terms of polynomials with rational coefficients rather than integral.

except for the existence of the inverse. But if $p(\alpha) \neq 0$ then p and the minimum polynomial P for α are relatively prime by (iii) above, whence there exist polynomials q, Q in $\mathbb{Q}[x]$ with $pq + PQ = 1$ identically and so $q(\alpha) = 1/p(\alpha)$.

Now let $K = \mathbb{Q}(\alpha)$ be an algebraic number field. The degree of K is defined as the degree of α, say n. It is clear from the division algorithm that every element in K has the form $p(\alpha)$ where $p(x)$ is a polynomial in $\mathbb{Q}[x]$ with degree at most $n - 1$. Thus K is a vector space over $\mathbb{Q}$ with basis $1, \alpha, \ldots, \alpha^{n-1}$ and its dimension, say $[K : \mathbb{Q}]$, is n. Further, if β is any other generator of K, that is, $K = \mathbb{Q}(\beta)$, then β must have degree n.

The conjugate fields $K_1, \ldots, K_n$ of K are defined as the embeddings of K into $\mathbb{C}$ given by the monomorphisms $\sigma_1, \ldots, \sigma_n$, where $\sigma_j(\alpha) = \alpha_j$ and $\alpha_1, \ldots, \alpha_n$ are the conjugates of α. Although, by (ii), the $\sigma_1, \ldots, \sigma_n$ are distinct, the conjugate fields are not necessarily distinct; for instance $\mathbb{Q}(\sqrt{2}) = \mathbb{Q}(-\sqrt{2})$. If $\theta = p(\alpha)$ is any element of K then the images $\sigma_j(\theta)(= p(\alpha_j))$ of θ are called the field conjugates of θ. Further, we say that $f(x) = \prod(x - \sigma_j(\theta))$ is the field polynomial for θ. We have $f(x) = (g(x))^m$ for some positive integer m, where g is the minimum polynomial for θ; for we can write $f(x) = (g(x))^m h(x)$ with $(g, h) = 1$ and $m \geq 1$ since certainly g divides f. Now, if $h(x) \neq 1$, then, since f, g are monic, we would have $h(\sigma_j(\theta)) = 0$ for some j, that is, $h(p(\alpha_j)) = 0$, and thus, by (iii), $h(p(\alpha)) = h(\theta) = 0$; this contradicts $(g, h) = 1$. It follows that the field conjugates of θ are just the conjugates of θ repeated m times.

10.4 Dimension theorem

Let k be an algebraic number field and let $K = k(\alpha)$, that is, the set of elements $p(\alpha)$ where $p(x) \in k[x]$. Then plainly K is a field. In fact it is a number field over $\mathbb{Q}$ in the sense defined in Section 10.3. For if $k = \mathbb{Q}(\beta)$ then the following result holds.

Theorem 10.1 *We have $K = k(\alpha) = \mathbb{Q}(\alpha, \beta) = \mathbb{Q}(\gamma)$, where $\gamma = u\alpha + v\beta$ for some integers u, v.*

Proof Let $\alpha_1, \ldots, \alpha_n$ and $\beta_1, \ldots, \beta_m$ be the conjugates of α, β respectively. Take u, v so that the mn numbers $\gamma_{ij} = u\alpha_i + v\beta_j$ are distinct. We put $\gamma = u\alpha + v\beta$ and assume that $\alpha = \alpha_1$, $\beta = \beta_1$ so that $\gamma = \gamma_{1,1}$. We have to show that $K = \mathbb{Q}(\gamma)$.

Let $q(x) = \prod_{i,j}(x - \gamma_{ij})$ and put

$$s(x) = \sum_{i=1}^{n} \sum_{j=1}^{m} \frac{\alpha_i q(x)}{x - \gamma_{ij}}.$$

By the symmetric function theorem we have $q(x)$ and $s(x) \in \mathbb{Q}[x]$. On taking $x = \gamma$ and recalling that the γ_{ij} are distinct we obtain $s(\gamma) = \alpha q'(\gamma)$ (for clearly $q(x)/(x - \gamma_{ij})$ is 0 for $x = \gamma$ unless $i = j = 1$ and then it is $q'(\gamma)$). Hence α is a rational function of γ. Similarly for β and the result follows.   □

In analogy with the rational case, we define the degree $[K : k]$ of $K = k(\alpha)$ over k as the degree of α over k, that is, the degree of the minimum polynomial for α in $k[x]$; since k contains $\mathbb{Q}$, there certainly exists such a minimum polynomial. Then K is a vector space over k with dimension $[K : k]$. But, by the theorem, K is also a vector space over $\mathbb{Q}$ with dimension $[K : \mathbb{Q}]$ and, since k has dimension $[k : \mathbb{Q}]$ as a vector space over $\mathbb{Q}$, we obtain the equation

$$[K : k][k : \mathbb{Q}] = [K : \mathbb{Q}].$$

10.5 Norm and trace

For any algebraic number α with conjugates $\alpha_1, \ldots, \alpha_d$ we define the (absolute) norm and trace of α by $N\alpha = \alpha_1 \ldots \alpha_d$ and $T\alpha = \alpha_1 + \cdots + \alpha_d$.

Now suppose that α is an element of a number field K with degree n and let $\sigma_1, \ldots, \sigma_n$ be the monomorphisms of K indicated in Section 10.3. Then the field norm and trace of α are defined by

$$N_{K/\mathbb{Q}}(\alpha) = \sigma_1(\alpha) \ldots \sigma_n(\alpha), \quad T_{K/\mathbb{Q}}(\alpha) = \sigma_1(\alpha) + \cdots + \sigma_n(\alpha).$$

By the result at the end of Section 10.3 we have $N_{K/\mathbb{Q}}(\alpha) = (N\alpha)^m$ and $T_{K/\mathbb{Q}}(\alpha) = mT\alpha$ for some positive integer m. Further, we have, for any elements α, β in K,

$$N_{K/\mathbb{Q}}(\alpha\beta) = N_{K/\mathbb{Q}}(\alpha)N_{K/\mathbb{Q}}(\beta),$$
$$T_{K/\mathbb{Q}}(\alpha + \beta) = T_{K/\mathbb{Q}}(\alpha) + T_{K/\mathbb{Q}}(\beta).$$

Finally suppose that $K = k(\theta)$ where k is an algebraic number field over $\mathbb{Q}$ with degree s. Then $r = [K : k]$ is the degree of θ over k and we have $rs = n$. Further, there are r distinct monomorphisms from K to $\mathbb{C}$ that fix k, namely the mappings $\tau_1, \ldots, \tau_r$ given by $\tau_j(\theta) = \theta_j$, where $\theta_1, \ldots, \theta_r$ are the zeros of the minimum polynomial for θ over k. We define the relative norm and trace by

$$N_{K/k}(\alpha) = \tau_1(\alpha) \ldots \tau_r(\alpha), \quad T_{K/k}(\alpha) = \tau_1(\alpha) + \cdots + \tau_r(\alpha),$$

where α is any element in K. Then $N_{K/k}(\alpha)$ is in k and we have

$$N_{K/k}(\alpha\beta) = N_{K/k}(\alpha)N_{K/k}(\beta)$$

and similarly for T. Note that the definition is consistent with the case $k = \mathbb{Q}$ given earlier. It is an easy exercise to show that $N_{K/\mathbb{Q}} = N_{k/\mathbb{Q}}(N_{K/k})$ and that this holds also for T.

10.6 Algebraic integers

An algebraic number α is an algebraic integer if the coefficients in the minimum polynomial for α are (rational) integers. Plainly the conjugates of an algebraic integer are themselves algebraic integers. Further, by Gauss' lemma, we see that the zeros of any monic polynomial with rational integer coefficients are algebraic integers. Furthermore, the totality of algebraic integers form a ring $\mathcal{O}$ (verification as for Ω) and the totality of algebraic integers in an algebraic number field K (that is, $K \cap \mathcal{O}$) form a ring $\mathcal{O}_K$.

Now suppose that α is any algebraic number with minimum polynomial p. We define the denominator of α as the least integer $a > 0$ such that ap has (relatively prime) integer coefficients; in other words, a is the lowest common multiple (l.c.m.) of the denominators of the coefficients of p. Then the following holds.

Lemma 10.1 *We have* $a\alpha \in \mathcal{O}$, *that is,* $a\alpha$ *is an algebraic integer. More generally,* $a\alpha_1 \ldots \alpha_m$ *is an algebraic integer for any distinct conjugates* $\alpha_1, \ldots, \alpha_m$ *of* α.

Proof If $p(x) = x^n + a_{n-1}x^{n-1} + \cdots + a_0$ then $a^n p(x) = q(ax)$ where $q(x) = x^n + a a_{n-1}x^{n-1} + \cdots + a^n a_0$. Thus $q(x)$ is the minimum polynomial for $a\alpha$ and the first assertion follows.

We show now that if $f(x) = \beta_r x^r + \cdots + \beta_0$ ($\beta_r \neq 0$) is in $\mathcal{O}[x]$ and if $f(\gamma) = 0$ then $f(x)/(x - \gamma) \in \mathcal{O}[x]$. This will suffice to prove the second assertion; for on taking $f(x) = ap(x)$ we deduce that $f(x)/\prod_l (x - \alpha_l) \in \mathcal{O}[x]$, where α_l runs through all the conjugates of α other than $\alpha_1, \ldots, \alpha_m$, that is, $a(x - \alpha_1) \ldots (x - \alpha_m)$ is in $\mathcal{O}[x]$, whence $a\alpha_1 \ldots \alpha_m$ is in $\mathcal{O}$ as required.

The assertion holds for $r = 1$ and we get the result in general by induction on r. For consider $\phi(x) = f(x) - \beta_r x^{r-1}(x - \gamma)$. This is a polynomial with degree $\leq r - 1$ and with $\phi(\gamma) = 0$. Further, we have $\beta_r \gamma \in \mathcal{O}$ and so $\phi(x) \in \mathcal{O}[x]$; indeed the argument furnishing the first assertion is easily extended to give $\beta_r^{r-1} f(x) = g(\beta_r x)$ with a monic g in $\mathcal{O}[x]$ and it is also easily seen that the zeros of any such g are in $\mathcal{O}$. Hence, by induction, $\phi(x)/(x - \gamma) \in \mathcal{O}[x]$ and thus $f(x)/(x - \gamma) \in \mathcal{O}[x]$ as required. $\qquad\square$

10.7 Basis and discriminant

Let K be an algebraic number field. Further, let $\theta_1, \ldots, \theta_n$ be a basis for K as a vector space over $\mathbb{Q}$. We define the discriminant of the basis as

$$\Delta(\theta_1, \ldots, \theta_n) = (\det S)^2,$$

where S is the matrix

$$\begin{pmatrix} \sigma_1(\theta_1) & \sigma_2(\theta_1) & \cdots & \sigma_n(\theta_1) \\ \sigma_1(\theta_2) & \sigma_2(\theta_2) & \cdots & \sigma_n(\theta_2) \\ \vdots & \vdots & \ddots & \vdots \\ \sigma_1(\theta_n) & \sigma_2(\theta_1) & \cdots & \sigma_n(\theta_n) \end{pmatrix}.$$

Note that then

$$\Delta(\theta_1, \ldots, \theta_n) = \det(SS') = \det(T_{K/\mathbb{Q}}(\theta_i\theta_j)),$$

where S' denotes the transpose of S.

If we have another basis, say $\phi_1, \ldots, \phi_n$, then $\phi_i = \sum_{j=1}^n f_{ij}\theta_j$ for some rationals f_{ij}. Let $F = \det(f_{ij})$. Then $F \neq 0$ and

$$\Delta(\phi_1, \ldots, \phi_n) = F^2 \Delta(\theta_1, \ldots, \theta_n). \tag{$*$}$$

Now if θ is a generator of K then $\Delta(1, \theta, \ldots, \theta^{n-1})$ is the square of a Vandermonde determinant and thus

$$\Delta(1, \theta, \ldots, \theta^{n-1}) = \prod_{1 \leq i < j \leq n} (\sigma_i(\theta) - \sigma_j(\theta))^2.$$

Hence, on taking $\phi_j = \theta^{j-1}$, the equation $(*)$ shows that $\Delta(\theta_1, \ldots, \theta_n) \neq 0$ for all bases of K.

Consider next the ring of integers $\mathcal{O}_K$ of K. A basis for $\mathcal{O}_K$ over $\mathbb{Z}$ is called an integral basis for K. Thus elements $\omega_1, \ldots, \omega_n$ of $\mathcal{O}_K$ form an integral basis for K if and only if every element α in $\mathcal{O}_K$ can be expressed in the form

$$\alpha = u_1\omega_1 + \cdots + u_n\omega_n$$

for some rational integers $u_1, \ldots, u_n$.

Theorem 10.2 *There exists an integral basis for K.*

Proof There certainly exists an $\mathcal{O}_K$-basis for K over $\mathbb{Q}$, that is, a basis for K over $\mathbb{Q}$ with elements in $\mathcal{O}_K$; for instance $1, a\theta, \ldots, (a\theta)^{n-1}$ where $K = \mathbb{Q}(\theta)$ and a is the denominator for θ. Now for any $\mathcal{O}_K$-basis $\omega_1, \ldots, \omega_n$ of K over $\mathbb{Q}$, the number $|\Delta(\omega_1, \ldots, \omega_n)|$ is a rational integer since, by symmetry, it is rational and $\omega_j \in \mathcal{O}_K$ implies $\Delta \in \mathcal{O}_K$. Thus there exist $\omega_1, \ldots, \omega_n$ such that $|\Delta(\omega_1, \ldots, \omega_n)|$ takes its smallest value. We proceed to prove that $\omega_1, \ldots, \omega_n$ is then an integral basis for K.

Let α be any element of $\mathcal{O}_K$. Then there exist rationals $u_1, \ldots, u_n$ such that $\alpha = u_1\omega_1 + \cdots + u_n\omega_n$ and we have to show that $u_1, \ldots, u_n$ must be integers. But if, say, $u_1 = u + v$ where u is an integer and $0 < v < 1$ then, on writing

$$\omega_1' = \alpha - u\omega_1 = v\omega_1 + u_2\omega_2 + \cdots + u_n\omega_n,$$

we would have $\omega_1', \omega_2, \ldots, \omega_n$ an $\mathcal{O}_K$-basis for K over $\mathbb{Q}$. Further, by (*), we see that

$$\Delta(\omega_1', \omega_2, \ldots, \omega_n) = F^2 \Delta(\omega_1, \ldots, \omega_n),$$

where F is the determinant of the matrix

$$\begin{pmatrix} v & u_2 & \cdots & u_n \\ 0 & 1 & \cdots & 0 \\ \vdots & \vdots & \ddots & \vdots \\ 0 & 0 & \cdots & 1 \end{pmatrix}$$

and thus $F = v$. Since we have $0 < v < 1$ this contradicts the minimal choice of $|\Delta(\omega_1, \ldots, \omega_n)|$ and the theorem follows. $\square$

It is clear that the determinant F from one integral basis to another is a rational integer, and the same holds for its inverse; hence it must be ± 1. Then $F^2 = 1$ and it follows from (*) that $\Delta(\omega_1, \ldots, \omega_n)$ is the same for all integral bases. Moreover, as already noted, it is not 0. The quantity is called the discriminant of K.

In view of the relations given at the beginning of the section, we can express the discriminant d of K alternatively as $\det(T_{K/\mathbb{Q}}(\omega_i\omega_j))$ and, since the traces here are rational integers, it follows that d is a rational integer. Further, we have Stickelberger's criterion, namely $d \equiv 0$ or 1 (mod 4). To verify this we define $\Omega = \det(\sigma_j(\omega_i))$, so that $d = \Delta(\omega_1, \ldots, \omega_n) = \Omega^2$, and we observe that Ω is given by a sum of $n!$ terms of the form $\sigma_{j_1}(\omega_1) \ldots \sigma_{j_n}(\omega_n)$ where the suffixes $j_1, \ldots, j_n$ run through the permutations of $1, \ldots, n$. Now $\Omega = a - b$, where a and b denote the sums of the terms over all even and odd permutations repectively, and by the symmetric function theorem we have $a + b$ and ab rational integers. But $d = (a - b)^2 = (a + b)^2 - 4ab$ and the result follows.

The concept of the discriminant of a field derives from an earlier and more primitive concept, namely that of the discriminant of a polynomial.

Definition 10.1 The discriminant of the polynomial

$$f(x) = a_0 x^n + a_1 x^{n-1} + \cdots + a_n \qquad (a_0 \neq 0)$$

with coefficients in $\mathbb{Q}$ and with zeros $\alpha_1, \ldots, \alpha_n$ is defined as

$$\operatorname{disc}(f) = a_0^{2n-2} \prod_{1 \leq i < j \leq n} (\alpha_i - \alpha_j)^2.$$

Now let $K = \mathbb{Q}(\theta)$ with $\theta \in \mathcal{O}_K$. If f is the minimum polynomial of θ then

$$\operatorname{disc}(f) = \Delta(1, \theta, \ldots, \theta^{n-1})$$

and, denoting by d the discriminant of K, we see that (*) gives $\operatorname{disc}(f) = m^2 d$ for an integer m. The latter is called the index of θ and in terms of modules we have $m = [\mathcal{O}_K : \mathbb{Z}(\theta)]$, that is, m is the number of elements in $\mathcal{O}_K / \mathbb{Z}(\theta)$ (cf. Section 11.5). Plainly $m = 1$, that is, $\operatorname{disc}(f) = d$, if and only if $1, \theta, \ldots, \theta^{n-1}$ is an integral basis for K; a basis of this form is sometimes referred to as a power integral basis. In particular, when $\operatorname{disc}(f)$ is square-free then $1, \theta, \ldots, \theta^{n-1}$ is an integral basis. More generally it follows from (*) that if, with the notation at the beginning of this section, $\theta_1, \ldots, \theta_n$ are in $\mathcal{O}_K$ and $\Delta(\theta_1, \ldots, \theta_n)$ is square-free then $\theta_1, \ldots, \theta_n$ is an integral basis for K.

Finally we observe that, with f as in the definition above, we have $f'(\alpha_i) = a_0 \prod_{j \neq i} (\alpha_i - \alpha_j)$, where $f'(x)$ denotes the derivative of $f(x)$, and hence

$$\operatorname{disc}(f) = (-1)^{\frac{1}{2}n(n-1)} a_0^{n-2} f'(\alpha_1) \ldots f'(\alpha_n).$$

Further, in the special case when $f(x) = x^n + px + q \ (q \neq 0)$, we have $f'(x) = nx^{n-1} + p$, whence $\alpha_i f'(\alpha_i) = -nq + (1-n)p\alpha_i$; this gives

$$q f'(\alpha_1) \ldots f'(\alpha_n) = (1-n)^n p^n f(nq/((1-n)p))$$

and thus

$$\operatorname{disc}(f) = (-1)^{\frac{1}{2}n(n-1)} \left(n^n q^{n-1} + (1-n)^{n-1} p^n \right).$$

10.8 Calculation of bases

The quadratic field $K = \mathbb{Q}(\sqrt{d})$, where d is a square-free integer, has already been discussed in Chapter 7. In particular it was shown in Section 7.2 that if $d \equiv 2$ or $3 \pmod 4$ then an integral basis for K is $1, \sqrt{d}$ and the discriminant of K is $4d$; if $d \equiv 1 \pmod 4$ then an integral basis for K is $1, \frac{1}{2}(1 + \sqrt{d})$ and the discriminant of K is d. We verify this again briefly with the current notation. The elements of K have the form $\alpha = u + v\sqrt{d}$ with u, v rational and, on putting $a = 2u, b = 2v$, the field polynomial (minimum if $v \neq 0$) for α is $x^2 - ax + c$ where $c = N\alpha = u^2 - dv^2$. Thus if $\alpha \in \mathcal{O}_K$ then $a, c \in \mathbb{Z}$ and from $4c = a^2 - b^2 d$ and the fact that d is square-free we see that $b \in \mathbb{Z}$. If $d \equiv 2$ or $3 \pmod 4$ then, since a square is $\equiv 0, 1 \pmod 4$, we have a, b even and u, v integers. If $d \equiv 1 \pmod 4$ then a, b have the same parity and $\alpha = \frac{1}{2}(a - b) + b(\frac{1}{2}(1 + \sqrt{d}))$ where $\frac{1}{2}(a - b)$ is an integer. This establishes the assertion about the bases and then the discriminants are given in the two cases respectively by

$$\begin{vmatrix} 1 & \sqrt{d} \\ 1 & -\sqrt{d} \end{vmatrix}^2 = 4d \quad \text{and} \quad \begin{vmatrix} 1 & \frac{1}{2}(1 + \sqrt{d}) \\ 1 & \frac{1}{2}(1 - \sqrt{d}) \end{vmatrix}^2 = d.$$

We now give some examples to show how one can find integral bases for fields of higher degree. We begin with a simple instance.

Example 10.1 Let α denote a zero of the polynomial $x^3 + px + q$, where p, q are integers and $q \neq 0$. By Section 10.7 we have disc$(f) = -(27q^2 + 4p^3)$ and if the latter is square-free then it is the discriminant of the field $\mathbb{Q}(\alpha)$ and $1, \alpha, \alpha^2$ is an integral basis. In particular this applies when $p = q = -1$ and $p = -2, q = -1$; the discriminants are then -23 and 5 respectively.

Example 10.2 Consider the cubic field $K = \mathbb{Q}(\alpha)$, where $\alpha = \sqrt[3]{10}$. Let $\beta = \frac{1}{3}(\alpha^2 + \alpha + 1)$. We show first that β is in $\mathcal{O}_K$. In fact $(\alpha - 1)\beta = \frac{1}{3}(\alpha^3 - 1) = 3$, whence $(1 + 3/\beta)^3 = 10$, that is, $(\beta + 3)^3 = 10\beta^3$. The expression on the left is $\beta^3 + 9\beta^2 + 27\beta + 27$ and so the field polynomial for β is $x^3 - x^2 - 3x - 3$. This has integer coefficients and so $\beta \in \mathcal{O}_K$ as asserted. We proceed now to prove that $1, \alpha, \beta$ is an integral basis for K. Since $1, \alpha, \alpha^2$ is a basis for K over $\mathbb{Q}$, the elements θ in $\mathcal{O}_K$ have the form $\theta = a + b\alpha + c\alpha^2$ with a, b, c rational. The conjugates of α are $\alpha, \omega\alpha, \omega^2\alpha$, where ω is a primitive cube root of unity, and so the traces of $\theta, \alpha\theta, \alpha^2\theta$ are $3a, 30c, 30b$ respectively. These must be rational integers, say u, w, v. The field polynomial of $v + w\alpha$ is $(x - v)^3 - 10w^3$, whence $N(v + w\alpha) = v^3 + 10w^3$. Since $N(\alpha) = 10$ this gives $N(3\alpha(b + c\alpha)) = \frac{1}{100}(v^3 + 10w^3)$ and, since $3(\theta - a)$ is in $\mathcal{O}_K$, it follows that the latter must be a rational integer; thus 10 divides v and w. Hence we have $\theta = \frac{1}{3}(r + s\alpha) + t\beta$ with integers $r = u - \frac{1}{10}w$, $s = \frac{1}{10}(v - w)$, $t = \frac{1}{10}w$. The field polynomial of $\frac{1}{3}(r + s\alpha)$ is $(x - \frac{1}{3}r)^3 - 10(\frac{1}{3}s)^3$ and, since $\theta - t\beta$ is in $\mathcal{O}_K$, this must have integer coefficients. The coefficient of x is $\frac{1}{3}r^2$, whence 3 divides r and from the constant coefficient we see that then 3 divides s. Hence $1, \alpha, \beta$ form an integral basis for K as asserted. On denoting the basis by $\omega_1, \omega_2, \omega_3$ we see that the discriminant of K is given by

$$\det(T_{K/\mathbb{Q}}(\omega_i\omega_j)) = \begin{vmatrix} 3 & 0 & 1 \\ 0 & 0 & 10 \\ 1 & 10 & 21 \end{vmatrix} = -300.$$

Example 10.3 Consider the quartic field $K = \mathbb{Q}(\alpha)$, where $\alpha = \sqrt[4]{2}$. We proceed to prove that $1, \alpha, \alpha^2, \alpha^3$ is an integral basis for K. Plainly K contains the subfield $k = \mathbb{Q}(\sqrt{2})$ and, for any element θ of K, we have $T_{K/k}(\theta) = \theta + \bar{\theta}$, where $\bar{\theta}$ is the conjugate of θ given by the mapping $\alpha \to -\alpha$; in fact the latter is the only element of the Galois group of K apart from the identity that fixes k. Now suppose that $\theta \in \mathcal{O}_K$ and let $\theta = a + b\alpha + c\alpha^2 + d\alpha^3$ with rational a, b, c, d. Then $T_{K/k}(\theta) = 2a + 2c\sqrt{2}$ and similarly $T_{K/k}(\alpha\theta) = 2b\sqrt{2} + 4d$. Since these are in $\mathcal{O}_K$ and since also $1, \sqrt{2}$ is an integral basis for k it follows that $2a, 2b, 2c$ and $4d$ are rational integers, say p, q, r, s. Further, we have $N_{K/k}(\theta) = \theta\bar{\theta} = (a + c\sqrt{2})^2 - \sqrt{2}(b + d\sqrt{2})^2$; the latter simplifies to $A + B\sqrt{2}$, where $A = a^2 + 2c^2 - 4bd$ and $B = 2ac - b^2 - 2d^2$ and these must be rational integers. Now $4A = p^2 + 2r^2 - 2qs$ and $8B = 4pr - 2q^2 - s^2$; this shows successively that p, s, r, q are even and furthermore that 8 divides s^2, whence 4 divides s. Hence a, b, c, d are rational integers and so $1, \alpha, \alpha^2, \alpha^3$ is an integral basis for K as asserted. We write the basis as $\omega_1, \omega_2, \omega_3, \omega_4$ and note that the conjugates of α are $\alpha, i\alpha, -\alpha, -i\alpha$; then the discriminant of K is given by

$$\det(T_{K/\mathbb{Q}}(\omega_i \omega_j)) = \begin{vmatrix} 4 & 0 & 0 & 0 \\ 0 & 0 & 0 & 8 \\ 0 & 0 & 8 & 0 \\ 0 & 8 & 0 & 0 \end{vmatrix} = -4 \times 8^3 = -2^{11}.$$

Example 10.4 Consider the biquadratic field $K = \mathbb{Q}(\sqrt{2}, \sqrt{7})$. This has Galois group $1, \alpha, \beta, \alpha\beta$ where α takes $\sqrt{2}$ to $-\sqrt{2}$ leaving $\sqrt{7}$ unchanged and β takes $\sqrt{7}$ to $-\sqrt{7}$ leaving $\sqrt{2}$ unchanged. Now suppose that $\theta \in \mathcal{O}_K$ and let $\theta = a + b\sqrt{7} + c\sqrt{2} + d\sqrt{14}$ with rational a, b, c, d. When $k = \mathbb{Q}(\sqrt{7})$ we have $T_{K/k}(\theta) = 2a + 2b\sqrt{7}$ and, since $1, \sqrt{7}$ is an integral basis for k, it follows that $2a$ and $2b$ are rational integers, say p and q. Similarly, from the subfields $\mathbb{Q}(\sqrt{2})$ and $\mathbb{Q}(\sqrt{14})$ we obtain $2c = r$ and $2d = s$ for some rational integers r and s. Also when $k = \mathbb{Q}(\sqrt{7})$ we have $N_{K/k}(\theta) = (a + b\sqrt{7})^2 - 2(c + d\sqrt{7})^2$ and this can be written as $A + B\sqrt{7}$, where $A = a^2 + 7b^2 - 2(c^2 + 7d^2)$ and $B = 2(ab - 2cd)$. Now A and B must be integers and we have $4A = p^2 + 7q^2 - 2(r^2 + 7s^2)$ and $4B = 2(pq - 2rs)$. Thus $p^2 + 7q^2$ and pq are even, whence p, q are even and a, b are integers. Further, $r^2 + 7s^2$ must be even and so r and s, that is, $2c$ and $2d$, have the same parity. It follows that an integral basis for K is $1, \sqrt{2}, \sqrt{7}, \frac{1}{2}(1 + \sqrt{7})\sqrt{2}$. For certainly the last element is in $\mathcal{O}_K$ since the square of it is $4 + \sqrt{7}$ and it has minimum polynomial $x^4 - 8x^2 + 9$. To calculate the discriminant of K we denote the integral basis just obtained by $\omega_1, \omega_2, \omega_3, \omega_4$. Then on recalling that the Galois group of K is $1, \alpha, \beta, \alpha\beta$ as above we see that the discriminant of K is given by

$$\det(T_{K/\mathbb{Q}}(\omega_i \omega_j)) = \begin{vmatrix} 4 & 0 & 0 & 0 \\ 0 & 8 & 0 & 4 \\ 0 & 0 & 28 & 0 \\ 0 & 4 & 0 & 16 \end{vmatrix} = 4^4 \begin{vmatrix} 1 & 0 & 0 & 0 \\ 0 & 2 & 0 & 1 \\ 0 & 0 & 7 & 0 \\ 0 & 1 & 0 & 4 \end{vmatrix} = 4^4 \times 49 = 2^8 7^2.$$

10.9 Further reading

An introduction to the literature was given in Section 7.7 and many of the works included under further reading in other chapters contain valuable expositions relating to the subject. Here we mention in addition the books by J. Esmonde and M. R. Murty, *Problems in Algebraic Number Theory* (Springer, 2004), and by A. Fröhlich and M. J. Taylor, *Algebraic Number Theory* (Cambridge University Press, 1991), both of which cover the topic well. The book *Algebraic Number Theory* by S. Lang (Springer, 1994) has been a standard reference for many years, and *Number Fields* (Springer, 1995) by D. A. Marcus is another accessible work. The volume *Basic Number Theory* (Springer, 1974) by A. Weil covers similar ground but it is written on a very sophisticated level. In connection with Section 10.8, the classic text is Berwick's *Integral Bases* (Cambridge University Press, 1927).

10.10 Exercises

(i) Find the minimum polynomials over $\mathbb{Q}$ of $(1+i)\sqrt{3}$, $i+\sqrt{3}$, $i+e^{i\pi/3}$.

(ii) Find the field polynomials of i and $\sqrt[3]{5}$ in $\mathbb{Q}(i+\sqrt[3]{5})$.

(iii) By the symmetric function theorem, or otherwise, prove that any zero of a monic polynomial with algebraic integer coefficients is an algebraic integer.

(iv) Which of the following are algebraic integers?

$$1/2, \ (\sqrt{3}+\sqrt{5})/2, \ (\sqrt{3}+\sqrt{7})/\sqrt{2}, \ (1+\sqrt[3]{19}+(\sqrt[3]{19})^2)/3.$$

(v) We know that the kernel of the map 'evaluation at i' given by $\mathbb{Z}[X] \to \mathbb{Z}[i]$, that is, $g(X) \mapsto g(i)$, is $\mathbb{Z}[X] \cap (X^2+1)\mathbb{Q}[X]$. Show that in fact the kernel is $(X^2+1)\mathbb{Z}[X]$ and deduce that the above map induces an isomorphism of rings $\mathbb{Z}[X]/(X^2+1) \to \mathbb{Z}[i]$.

(vii) Show that, for $a, b \in \mathbb{Q}^*$, where $\mathbb{Q}^*$ denotes the multiplicative group of non-zero elements of $\mathbb{Q}$, the degree of $\mathbb{Q}(\sqrt{a}, \sqrt{b})$ is equal to the order of the subgroup of $\mathbb{Q}^*/\mathbb{Q}^{*2}$ generated by a, b. Determine whether the field $\mathbb{Q}(\sqrt{(2+\sqrt{2})})$ is of the form $\mathbb{Q}(\sqrt{a}, \sqrt{b})$ with $a, b \in \mathbb{Q}$.

(viii) Show that if a number field K has degree $d = s + 2t$, where s is the number of real conjugate fields and $2t$ is the number of complex conjugate fields, then the discriminant D of K satisfies $(-1)^t D > 0$.

(ix) Let $K = \mathbb{Q}(\alpha)$ where $\alpha = \sqrt[3]{d}$ for a square-free integer $d \neq 0, \pm 1$. Show that the ring of algebraic integers $\mathcal{O}_K$ of K satisfies

$$\mathbb{Z}[\sqrt[3]{d}] \subseteq \mathcal{O}_K \subseteq \tfrac{1}{3}\mathbb{Z}[\sqrt[3]{d}].$$

Verify that the field polynomial of $a + b\alpha + c\alpha^2$ for rational a, b, c is

$$x^3 - 3ax + 3(a^2 - bcd)x - (a^3 + b^3 d + c^3 d^2 - 3abcd).$$

By considering the cases when $3a, 3b, 3c$ are $0, \pm 1$, prove that an integral basis for K is given by $1, \alpha, \alpha^2$ when $d \not\equiv \pm 1 \pmod 9$ and by $1, \alpha, \beta$ otherwise where $\beta = \tfrac{1}{3}(1 \pm \alpha + \alpha^2)$ with corresponding $\pm$ signs.

(x) Let $k \subset K$ be number fields. Show that, for $\alpha \in \mathcal{O}_K$, the trace $T_{K/k}(\alpha)$ and norm $N_{K/k}(\alpha)$ are in $\mathcal{O}_k$. Let now $K = \mathbb{Q}(\sqrt{3}, \sqrt{5})$. By computing traces and norms for the three quadratic subfields k of K, show that an integral basis for K is $1, \sqrt{3}, \tfrac{1}{2}(1 + \sqrt{5}), \tfrac{1}{2}(1 + \sqrt{5})\sqrt{3}$.

(xi) Prove that an integral basis for $\mathbb{Q}(i, \sqrt{2})$ is $1, i, \sqrt{2}, \tfrac{1}{2}(1 + i)\sqrt{2}$. Prove further that an integral basis for $\mathbb{Q}(\sqrt{2}, \sqrt{p})$, where p is a prime with $p \equiv 3 \pmod 4$, is $1, \sqrt{2}, \sqrt{p}, \tfrac{1}{2}(1 + \sqrt{p})\sqrt{2}$. Calculate the discriminants of the fields.

11

Ideals

11.1 Origins

The introduction of ideals was motivated by a desire to restore the property of unique factorization in number fields; certainly, as already observed in Section 7.4, the property is not universal. Consider, as there, the quadratic field $K = \mathbb{Q}(\sqrt{(-5)})$ with basis $1, \sqrt{(-5)}$. We have

$$21 = 3 \times 7 = (1 + 2\sqrt{(-5)})(1 - 2\sqrt{(-5)}).$$

Now $3, 7, 1 + 2\sqrt{(-5)}, 1 - 2\sqrt{(-5)}$ cannot be further factorized in $\mathcal{O}_K$. Suppose, for instance, $3 = \alpha\beta$ with α, β in $\mathcal{O}_K$. Then $N_{K/\mathbb{Q}}(\alpha)N_{K/\mathbb{Q}}(\beta) = 9$ and so if neither α nor β were ± 1 (thus if neither has norm 1) we would have $N_{K/\mathbb{Q}}(\alpha) = 3$, which is impossible since $x^2 + 5y^2 = 3$ has no solution in integers x, y. Similarly $7, 1 + 2\sqrt{(-5)}$ and $1 - 2\sqrt{(-5)}$ cannot be factorized. The situation was restored by Kummer; he proceeded by way of a mapping of the ring $\mathcal{O}_K$ into a multiplicative semigroup of elements which he called 'ideals'.

11.2 Definitions

Let K be an algebraic number field with ring of integers $\mathcal{O}_K$. By an ideal in K we mean a non-empty subset $\mathfrak{a}$ of $\mathcal{O}_K$ such that

 (i) if $\alpha, \alpha' \in \mathfrak{a}$ then $\alpha - \alpha' \in \mathfrak{a}$,
(ii) if $\alpha \in \mathfrak{a}$ and $\beta \in \mathcal{O}_K$ then $\alpha\beta \in \mathfrak{a}$.

For any elements $\alpha_1, \ldots, \alpha_m$ in $\mathcal{O}_K$ the set of all numbers $\alpha_1\beta_1 + \cdots + \alpha_m\beta_m$ with $\beta_1, \ldots, \beta_m$ in $\mathcal{O}_K$ is plainly an ideal, say $\mathfrak{a}$, and we write $\mathfrak{a} = [\alpha_1, \ldots, \alpha_m]$. Then $\alpha_1, \ldots, \alpha_m$ are called generators for $\mathfrak{a}$. Conversely, for any ideal $\mathfrak{a}$, we

111

have $\mathfrak{a} = [\alpha_1, \ldots, \alpha_m]$ for some generators $\alpha_1, \ldots, \alpha_m$. Indeed we shall prove the following.

Theorem 11.1 *There exists a basis for $\mathfrak{a}$; that is, there exist elements $\gamma_1, \ldots, \gamma_n$ in $\mathfrak{a}$ such that, if $\alpha \in \mathfrak{a}$, then $\alpha = u_1\gamma_1 + \cdots + u_n\gamma_n$ with $u_1, \ldots, u_n$ in $\mathbb{Z}$.*

Proof Let $\omega_1, \ldots, \omega_n$ be an integral basis for K. If $\alpha \in \mathfrak{a}$ and $\alpha \neq 0$ then $\alpha\omega_1, \ldots, \alpha\omega_n$ is an $\mathfrak{a}$-basis for K over $\mathbb{Q}$, that is, a basis for K as a vector space over $\mathbb{Q}$ with elements in $\mathfrak{a}$. The theorem follows on arguing as in Section 10.7; indeed we can take for $\gamma_1, \ldots, \gamma_n$ any set of elements of $\mathfrak{a}$ such that $|\Delta(\gamma_1, \ldots, \gamma_n)|$ assumes its smallest value. $\qquad\square$

Note that, if $\gamma_1, \ldots, \gamma_n$ is a basis for $\mathfrak{a}$, then $\mathfrak{a} = [\gamma_1, \ldots, \gamma_n]$ and so we have exhibited a set of generators for the ideal in accordance with the assertion above.

We define the product $\mathfrak{ab}$ of ideals $\mathfrak{a}$ and $\mathfrak{b}$ as the ideal consisting of all elements $a_1 b_1 + \cdots + a_j b_j$ with the a in $\mathfrak{a}$ and the b in $\mathfrak{b}$. Hence if $\mathfrak{a} = [\alpha_1, \ldots, \alpha_l]$ and $\mathfrak{b} = [\beta_1, \ldots, \beta_m]$ then $\mathfrak{ab} = [\alpha_1\beta_1, \ldots, \alpha_l\beta_m]$, that is, the ideal with generators $\alpha_r\beta_s$ ($1 \leq r \leq l$, $1 \leq s \leq m$). Plainly multiplication of ideals is commutative ($\mathfrak{ab} = \mathfrak{ba}$) and associative ($(\mathfrak{ab})\mathfrak{c} = \mathfrak{a}(\mathfrak{bc})$). As for integers, we say that $\mathfrak{a}$ divides $\mathfrak{b}$ (written $\mathfrak{a}|\mathfrak{b}$) if $\mathfrak{b} = \mathfrak{ac}$ for some ideal $\mathfrak{c}$. We write $\mathfrak{a}^m = \mathfrak{a} \ldots \mathfrak{a}$ (m factors) and $\mathfrak{a}^0 = \mathfrak{e} = [1]$, that is, the ideal $\mathcal{O}_K$.

11.3 Principal ideals

An ideal $\mathfrak{a}$ in K is called principal if $\mathfrak{a} = [\alpha]$ for some $\alpha \in \mathcal{O}_K$. The mapping $\alpha \rightarrow [\alpha]$ is the embedding of $\mathcal{O}_K$ into the ideals referred to earlier.

The following result shows that every ideal $\mathfrak{a}$ has an inverse in terms of 'fractional ideals'; though we indicate the meaning of the concept here we shall not need it in a formal sense until Section 11.7 and we give the precise definition there.

Theorem 11.2 *For every ideal $\mathfrak{a}$ there is an ideal $\mathfrak{b}$ such that $\mathfrak{ab}$ is principal. In fact there exists $\mathfrak{b}$ such that $\mathfrak{ab} = [c]$ with $c \in \mathbb{Z}$. Thus if $\mathfrak{b} = [\beta_0, \ldots, \beta_m]$ we have $\mathfrak{aa}^{-1} = \mathfrak{e}$ where $\mathfrak{a}^{-1}$ is the fractional ideal $[\beta_0/c, \ldots, \beta_m/c]$.*

Proof Let $\mathfrak{a} = [\alpha_0, \ldots, \alpha_l]$ and define

$$f(x) = \alpha_0 + \alpha_1 x + \cdots + \alpha_l x^l.$$

Further, define $F(x)$ as the 'norm' of $f(x)$, that is,

$$F(x) = \prod_{j=1}^{n} \left(\sigma_j(\alpha_0) + \cdots + \sigma_j(\alpha_l)x^l \right).$$

Then, by symmetry, we have $F(x) \in \mathbb{Z}[x]$, whence $F(x) = f(x)g(x)$ where $g(x) \in \mathcal{O}_K[x]$, since certainly $g(x) \in \mathcal{O}[x]$. We write

$$g(x) = \beta_0 + \beta_1 x + \cdots + \beta_m x^m$$

and define $\mathfrak{b} = [\beta_0, \ldots, \beta_m]$. Then $\mathfrak{ab} = [c]$ where c is the highest common factor of the coefficients of F. Indeed c is a linear combination of the latter coefficients over $\mathbb{Z}$ and these are themselves linear combinations over $\mathbb{Z}$ of the elements $\alpha_r \beta_s$ $(0 \le r \le l, \ 0 \le s \le m)$, whence $[c] \subset \mathfrak{ab}$. Conversely, we have $\mathfrak{ab} \subset [c]$, that is, $c^{-1}\alpha_r \beta_s \in \mathcal{O}_K$ for all r, s; for we can write $\alpha_r \beta_s = \alpha_l \beta_m \gamma_{rs}$, where γ_{rs} is a product of elementary symmetric functions in the zeros of f, g, and these zeros are precisely those of $c^{-1}F$. Now $c^{-1}\alpha_l \beta_m$ is the leading coefficient in $c^{-1}F$ and hence, from Section 10.6, we obtain $(c^{-1}\alpha_l \beta_m)\gamma_{rs} \in \mathcal{O}$. This gives $c^{-1}\alpha_r \beta_s \in \mathcal{O}_K$ and thus $\mathfrak{ab} = [c]$, as required. $\qquad \square$

Corollary 11.1 (Cancellation) *The equation* $\mathfrak{ac} = \mathfrak{bc}$ *implies that* $\mathfrak{a} = \mathfrak{b}$.

Proof Either multiply both sides by $\mathfrak{c}^{-1}$ (fractional ideal) or observe that there exists an ideal $\mathfrak{d}$ such that $\mathfrak{cd} = [c]$, whence $\mathfrak{a}[c] = \mathfrak{b}[c]$ and the desired conclusion $\mathfrak{a} = \mathfrak{b}$ follows on taking generators. $\qquad \square$

Corollary 11.2 (Division criterion) *We have* $\mathfrak{a}|\mathfrak{b}$ *if and only if* $\mathfrak{b} \subset \mathfrak{a}$.

Proof Plainly, $\mathfrak{a}|\mathfrak{b}$ implies that $\mathfrak{b} = \mathfrak{ac}$ for some $\mathfrak{c}$, whence $\mathfrak{b} \subset \mathfrak{a}$. Conversely, if $\mathfrak{b} \subset \mathfrak{a}$, we can find $\mathfrak{a}'$ such that $\mathfrak{aa}' = [c]$, whence $\mathfrak{ba}' \subset [c]$. By considering generators, we see that this gives $[c]$ divides $\mathfrak{ba}'$, that is, $\mathfrak{aa}'|\mathfrak{ba}'$, and so, by cancellation, we have $\mathfrak{a}|\mathfrak{b}$. (Alternatively, from $\mathfrak{b} \subset \mathfrak{a}$, we obtain $\mathfrak{ba}^{-1} \subset \mathfrak{e}$, that is, $\mathfrak{ba}^{-1} = \mathfrak{c}$ and so $\mathfrak{b} = \mathfrak{ac}$.) $\qquad \square$

11.4 Prime ideals

An ideal $\mathfrak{p} \ne \mathfrak{e}$ is said to be prime if it is divisible only by itself and $\mathfrak{e}$. Our object here is to establish the unique factorization of any ideal as a product of prime ideals, that is, to demonstrate the analogue of the fundamental theorem of arithmetic.

Lemma 11.1 (Generalized division algorithm) *Suppose $\alpha \in \mathcal{O}_K$ and $c \in \mathbb{Z}$, $c \neq 0$. Then $\alpha = c\beta + \gamma$ with $\beta, \gamma \in \mathcal{O}_K$ and with γ in a finite set depending only on c and having c^n elements where n is the degree of K.*

Proof The result follows on writing $\alpha = u_1\omega_1 + \cdots + u_n\omega_n$ in terms of an integral basis for K. By the division algorithm $u_j = cv_j + w_j$ with $v_j, w_j \in \mathbb{Z}$ and $0 \leq w_j < c \, (1 \leq j \leq n)$; the lemma follows on taking $\beta = v_1\omega_1 + \cdots + v_n\omega_n$ and $\gamma = w_1\omega_1 + \cdots + w_n\omega_n$. $\square$

Lemma 11.2 (Finiteness property) *Every ideal $\mathfrak{a}$ has only finitely many divisors.*

Proof We have $\mathfrak{a}\mathfrak{c} = [c]$ for some $c \in \mathbb{Z}$. Hence every divisor $\mathfrak{a}'$ of $\mathfrak{a}$ divides $[c]$. But, by the division criterion, we obtain $c \in \mathfrak{a}'$ and so, by the generalized division algorithm, we see that $\mathfrak{a}' = [\alpha_1, \ldots, \alpha_m] = [\gamma_1, \ldots, \gamma_m, c]$ with only c^{mn} possibilities for $\gamma_1, \ldots, \gamma_m$. The lemma follows. $\square$

We now define the sum of ideals $\mathfrak{a}$ and $\mathfrak{b}$ as the ideal $\mathfrak{a} + \mathfrak{b}$ consisting of all elements $a + b$ with a in $\mathfrak{a}$ and b in $\mathfrak{b}$. Then if $\mathfrak{a} = [\alpha_1, \ldots, \alpha_l]$, $\mathfrak{b} = [\beta_1, \ldots, \beta_m]$ we have $\mathfrak{a} + \mathfrak{b} = [\alpha_1, \ldots, \alpha_l, \beta_1, \ldots, \beta_m]$. Moreover, by the division criterion, we see that $\mathfrak{a} + \mathfrak{b}$ is the highest common factor of $\mathfrak{a}$ and $\mathfrak{b}$ in the usual sense. We can now argue as in the proof of the fundamental theorem of arithmetic (see Section 1.5); if $\mathfrak{p}$ is a prime ideal and $\mathfrak{p}|\mathfrak{a}\mathfrak{b}$ then $\mathfrak{p}|\mathfrak{a}$ or $\mathfrak{p}|\mathfrak{b}$ (for if $\mathfrak{p} \nmid \mathfrak{a}$ then $\mathfrak{a} + \mathfrak{p} = \mathfrak{e}$, whence $\mathfrak{a}\mathfrak{b} + \mathfrak{p}\mathfrak{b} = \mathfrak{b}$ and so $\mathfrak{p}|\mathfrak{b}$) and the unique factorization for ideals follows.

11.5 Norm of an ideal

An element $\alpha \in \mathcal{O}_K$ is said to be divisible by an ideal $\mathfrak{a}$ if $\mathfrak{a}$ divides $[\alpha]$, that is, if $\alpha \in \mathfrak{a}$. If $\alpha, \beta \in \mathcal{O}_K$ and $\mathfrak{a}$ divides $\alpha - \beta$ we write $\alpha \equiv \beta \pmod{\mathfrak{a}}$. This is an equivalence relation and the equivalence classes form a ring $\mathcal{O}_K/\mathfrak{a}$ on defining, in the obvious way, the product of classes with elements α and α', say, as the class containing $\alpha\alpha'$; it is called the residue class ring or quotient ring of $\mathcal{O}_K$ by $\mathfrak{a}$. Further, the number of equivalence classes $(\bmod \, \mathfrak{a})$, that is, elements of the quotient ring, is finite; for we have $\mathfrak{a}\mathfrak{b} = [c]$ for some ideal $\mathfrak{b}$ and with $c \in \mathbb{Z}$ (see Section 11.3) and, by the generalized division algorithm, there are only finitely many classes $(\bmod \, [c])$; note here that if $\alpha \equiv \beta \pmod{[c]}$ then certainly $\alpha \equiv \beta \pmod{\mathfrak{a}}$. The number of classes $(\bmod \, \mathfrak{a})$ is called the norm $N\mathfrak{a}$ of $\mathfrak{a}$. We establish the following.

Theorem 11.3 (Multiplicative property) *We have* $N\mathfrak{a}N\mathfrak{b} = N(\mathfrak{a}\mathfrak{b})$.

Proof On writing $\mathfrak{b}$ as a product of prime ideals, we see that it suffices to prove that $N\mathfrak{a}N\mathfrak{p} = N(\mathfrak{a}\mathfrak{p})$ for a prime ideal $\mathfrak{p}$. Since $\mathfrak{a}\mathfrak{p} \nmid \mathfrak{a}$, there exists an α in $\mathfrak{a}$ and not in $\mathfrak{a}\mathfrak{p}$. We show that $\sigma + \alpha\rho$ runs through all representatives of equivalence classes (mod $\mathfrak{a}\mathfrak{p}$) as σ, ρ run through the representatives (mod $\mathfrak{a}$) and (mod $\mathfrak{p}$) respectively; the result follows.

Now the $\sigma + \alpha\rho$ represent all classes (mod $\mathfrak{a}\mathfrak{p}$). For if $\gamma \in \mathcal{O}_K$ then $\gamma \equiv \sigma$ (mod $\mathfrak{a}$) for some σ; we have $\mathfrak{a} = [\alpha] + \mathfrak{a}\mathfrak{p}$ (that is, $\mathfrak{a}$ is the highest common factor of $[\alpha]$ and $\mathfrak{a}\mathfrak{p}$), whence $\gamma - \sigma \equiv \alpha\beta$ (mod $\mathfrak{a}\mathfrak{p}$) for some $\beta \in \mathcal{O}_K$ and so $\gamma \equiv \sigma + \alpha\rho$ (mod $\mathfrak{a}\mathfrak{p}$) where $\beta \equiv \rho$ (mod $\mathfrak{p}$). Further, the $\sigma + \alpha\rho$ are incongruent (mod $\mathfrak{a}\mathfrak{p}$), for if

$$\sigma + \alpha\rho \equiv \sigma' + \alpha\rho' \; (\text{mod } \mathfrak{a}\mathfrak{p})$$

then, since $\mathfrak{a}$ divides $[\alpha]$, we obtain $\sigma \equiv \sigma'$ (mod $\mathfrak{a}$), whence $\sigma = \sigma'$; but then $\mathfrak{a}\mathfrak{p}$ divides $[\alpha][\rho - \rho']$ and, since $[\alpha] = \mathfrak{a}\mathfrak{c}$ where $\mathfrak{p} \nmid \mathfrak{c}$, this gives $\rho = \rho'$. Hence $N\mathfrak{a}N\mathfrak{p} = N(\mathfrak{a}\mathfrak{p})$ as required. $\qquad\square$

11.6 Formula for the norm

If $\gamma_1, \ldots, \gamma_n$ is a basis for $\mathfrak{a}$ and $\omega_1, \ldots, \omega_n$ is an integral basis for K then we have

$$N\mathfrak{a} = (\Delta(\gamma_1, \ldots, \gamma_n)/\Delta(\omega_1, \ldots, \omega_n))^{1/2}.$$

Thus $N\mathfrak{a} = (\Delta/d)^{1/2}$, where $\Delta = \Delta(\gamma_1, \ldots, \gamma_n)$ and d denotes the discriminant of K.

To establish the formula we need first to show that we can construct a basis of triangular form for $\mathfrak{a}$. We have the following result.

Lemma 11.3 *There exists a basis $\gamma_1, \ldots, \gamma_n$ for $\mathfrak{a}$ of the form*

$$\gamma_1 = a_{11}\omega_1$$
$$\gamma_2 = a_{21}\omega_1 + a_{22}\omega_2$$
$$\vdots$$
$$\gamma_n = a_{n1}\omega_1 + a_{n2}\omega_2 + \cdots + a_{nn}\omega_n,$$

where the a_{ij} are rational integers and $a_{jj} > 0$.

Proof To begin with we note that there is an element $v_1\omega_1 + \cdots + v_n\omega_n$ in $\mathfrak{a}$ with $v_n \neq 0$; for certainly $\mathfrak{a}$ contains n linearly independent elements over

Q. We take for γ_n such an element with $v_n > 0$ and minimal. Let now $\alpha = u_1\omega_1 + \cdots + u_n\omega_n$ be any element of $\mathfrak{a}$. On writing $u_n = qv_n + r, 0 \le r < v_n$, and noting that $\alpha - q\gamma_n \in \mathfrak{a}$, we see that v_n divides u_n (that is, $r = 0$) and we have $\alpha = u'_1\omega_1 + \cdots + u'_{n-1}\omega_{n-1} + q\gamma_n$ for some $u'_1, \ldots, u'_{n-1} \in \mathbb{Z}$. Since $u'_{n-1} \ne 0$ for some α (see earlier), we can now argue successively and replace $\omega_{n-1}, \ldots, \omega_1$ by elements $\gamma_{n-1}, \ldots, \gamma_1$ in $\mathfrak{a}$ of the form indicated above with $a_{jj} > 0$ and minimal. $\qquad\square$

Proof of the norm formula We recall that $\Delta(\gamma_1, \ldots, \gamma_n)$ is the same for all bases of $\mathfrak{a}$; we shall establish the result for a particular basis $\gamma_1, \ldots, \gamma_n$ of the above form with a_{jj} minimal. Now if $u_1\omega_1 + \cdots + u_j\omega_j \in \mathfrak{a}$ we see that a_{jj} divides u_j. Hence the numbers $u_1\omega_1 + \cdots + u_n\omega_n$ with $u_j \in \mathbb{Z}$ and $0 \le u_j < a_{jj}$ $(1 \le j \le n)$ are incongruent $(\bmod\, \mathfrak{a})$. Further, they represent all congruence classes; for, taking $\alpha = v_1\omega_1 + \cdots + v_n\omega_n$ in $\mathcal{O}_K$, we have $v_n = a_{nn}q + r, 0 \le r < a_{nn}$, whence

$$\alpha \equiv \alpha - q\gamma_n = v'_1\omega_1 + \cdots + v'_{n-1}\omega_{n-1} + r\omega_n \pmod{\mathfrak{a}}$$

with $v'_1, \ldots, v'_{n-1}$ in $\mathbb{Z}$ and we can now replace the coefficients of $\omega_{n-1}, \ldots, \omega_1$ similarly. Hence we have $N\mathfrak{a} = a_{11}\ldots a_{nn}$. But, by (*) of Section 10.7, we obtain

$$\Delta(\gamma_1, \ldots, \gamma_n) = (a_{11}\ldots a_{nn})^2 \Delta(\omega_1, \ldots, \omega_n)$$

and the desired formula for $N\mathfrak{a}$ follows. $\qquad\square$

Corollary 11.3 *For any $\alpha \ne 0$ in $\mathcal{O}_K$ we have $N[\alpha] = |N_{K/\mathbb{Q}}(\alpha)|$.*

Proof An integral basis for $[\alpha]$ is given by $\gamma_j = \alpha\omega_j$ $(1 \le j \le n)$. We have

$$\det(\sigma_i(\alpha\omega_j)) = \sigma_1(\alpha)\ldots\sigma_n(\alpha)\det(\sigma_i(\omega_j))$$

and the formula for $N[\alpha]$ gives the result. $\qquad\square$

Definition 11.1 We say that an ideal $\mathfrak{a}$ divides a rational integer b if $\mathfrak{a}$ divides $[b]$, that is, if $b \in \mathfrak{a}$, and we write $\mathfrak{a}|b$ (cf. the division criterion of Section 11.5).

Lemma 11.4 *For every ideal $\mathfrak{a}$ we have $\mathfrak{a}$ divides $N\mathfrak{a}$.*

Proof Let $\theta_1, \ldots, \theta_N$ be a set of representatives of the congruence classes $(\bmod\, \mathfrak{a})$ so that $N = N\mathfrak{a}$. Then $\theta_1 + 1, \ldots, \theta_N + 1$ is another set and thus

$$(\theta_1 + 1) + \cdots + (\theta_N + 1) \equiv \theta_1 + \cdots + \theta_N \pmod{\mathfrak{a}},$$

that is, $N \equiv 0 \pmod{\mathfrak{a}}$, whence $\mathfrak{a}|N\mathfrak{a}$. $\qquad\square$

Corollary 11.4 *For every prime ideal* $\mathfrak{p}$ *there is a unique rational prime* p *such that* $\mathfrak{p}|p$; *the prime ideal* $\mathfrak{p}$ *is said to lie above* p.

Proof We have $\mathfrak{p}|N\mathfrak{p}$. Let now p be the least positive integer such that $\mathfrak{p}|p$. Then p is a prime since obviously $p = mn$ implies $\mathfrak{p}|m$ or $\mathfrak{p}|n$. Further, p is unique, for if $\mathfrak{p}|p'$ then $p' = pq + r$, where $0 \le r < p$, and this gives $\mathfrak{p}|r$ so that $r = 0$; hence, if p' is prime, we have $q = 1$ and $p = p'$. $\qquad\square$

We now give some further definitions. By Corollary 11.3 we have $N[p] = p^n$ and so, by Corollary 11.4, we see that $N\mathfrak{p} = p^f$ for some positive integer f. This is defined as the degree of $\mathfrak{p}$. The exponent e to which $\mathfrak{p}$ divides p, that is, $\mathfrak{p}^e|p$ but $\mathfrak{p}^{e+1} \nmid p$, is called the ramification index of $\mathfrak{p}$. If $e = n$ then the prime ideal $\mathfrak{p}$ is said to be totally ramified and if $e = 1$ then it is said to be unramified.

Theorem 11.4 *If* p *is a prime and* $p = \mathfrak{p}_1^{e_1} \ldots \mathfrak{p}_r^{e_r}$ *as a product of prime ideals* (*that is,* e_j *is the ramification index of* $\mathfrak{p}_j$) *and if* f_j *is the degree of* $\mathfrak{p}_j$ *then*

$$e_1 f_1 + \cdots + e_r f_r = n.$$

Proof Obvious on taking norms. $\qquad\square$

The theorem implies in particular that a rational prime can have at most n prime ideal divisors. We remark further that when K is a normal or Galois field, that is, when K contains all its conjugate fields or, equivalently, any conjugate field is identical to K, we have $e_1 = \cdots = e_r = e$ and $f_1 = \cdots = f_r = f$ so that $n = efr$. In fact in this case the Galois group of K permutes the prime ideals that lie over p.

11.7 The different

Let K be an algebraic number field. By a fractional ideal in K we mean an ideal in K as defined in Section 11.2 but with generators $\alpha_1, \ldots, \alpha_m$ in K rather than in $\mathcal{O}_K$. Since, by Section 10.6, there is an integer $c \neq 0$ such that $c\alpha_1, \ldots, c\alpha_m$ are in $\mathcal{O}_K$, we may alternatively give the definition in terms of generators of the form $\beta_1/c, \ldots, \beta_m/c$ with $\beta_1, \ldots, \beta_m$ in $\mathcal{O}_K$, and this is the approach often adopted in the literature. The product of fractional ideals is defined as for ideals in Section 11.2 and we see from Theorem 11.2 that for any fractional ideals $\mathfrak{a}$ and $\mathfrak{b}$ there is a unique fractional ideal $\mathfrak{c}$ such that $\mathfrak{a} = \mathfrak{b}\mathfrak{c}$; we write $\mathfrak{c} = \mathfrak{a}/\mathfrak{b}$. Furthermore every fractional ideal has an integral basis in the obvious sense.

Now let $\omega_1, \ldots, \omega_n$ be an integral basis for K and let Ω be the non-singular matrix S indicated in Section 10.7 with $\omega_1, \ldots, \omega_n$ in place of $\theta_1, \ldots, \theta_n$, that is, $\Omega = (\sigma_j(\omega_i))$; we shall assume that σ_1 is the identity so that the basis is given by the first column of Ω. We define the dual basis $\omega_1^*, \ldots, \omega_n^*$ as the elements of the first row of the matrix Ω^{-1} and we note that they lie in K. For we have $\Omega^{-1} = \Omega'(\Omega\Omega')^{-1}$ and, since $\Omega\Omega' = (T_{K/\mathbb{Q}}(\omega_i\omega_j))$ is a matrix of rational integers, it follows that $(\Omega\Omega')^{-1}$ is a matrix of rationals. Further, since again $\Omega^{-1} = \Omega'(\Omega\Omega')^{-1}$, we have $\Omega^{-1} = (\sigma_i(\omega_j^*))$, whence, from the fact that $\Omega\Omega^{-1}$ is the unit matrix of order n, we obtain

$$T_{K/\mathbb{Q}}(\omega_i\omega_j^*) = \delta_{ij} \quad (1 \le i, j \le n),$$

where δ_{ij} denotes the Kronecker delta, that is, $\delta_{ij} = 1$ if $i = j$ and $\delta_{ij} = 0$ otherwise.

We now define the co-different of K as the the set of γ in K such that $T_{K/\mathbb{Q}}(\gamma\alpha)$ is a rational integer for all α in $\mathcal{O}_K$. Then the co-different is a fractional ideal and it has an integral basis $\omega_1^*, \ldots, \omega_n^*$. For if

$$\gamma = u_1\omega_1^* + \cdots + u_n\omega_n^*$$

with rational integers $u_1, \ldots, u_n$ then $T_{K/\mathbb{Q}}(\gamma\omega_j) = u_j$ for all j, whence γ lies in the co-different; conversely, if γ is in the latter, then, since the dual basis is certainly a basis for K over $\mathbb{Q}$, there exist rationals $u_1, \ldots, u_n$ as above and from $T_{K/\mathbb{Q}}(\gamma\omega_j) = u_j$ again we see that $u_1, \ldots, u_n$ are integers.

The different $\mathcal{D}$ of K is defined as the reciprocal of the co-different. It is in fact an ideal in K, not just a fractional ideal, for clearly $\gamma = 1$ is in the co-different, that is, it is an element of the fractional ideal $\mathcal{D}^{-1}$, and thus $\mathcal{D}$ is contained in $\mathcal{D}\mathcal{D}^{-1} = \mathcal{O}_K$. Now let d denote the discriminant of K; we establish the following result for the norm of the different.

Theorem 11.5 *We have* $N(\mathcal{D}) = |d|$.

Proof By definition, the discriminant d of K is given by

$$d = \Delta(\omega_1, \ldots, \omega_n) = (\det\Omega)^2$$

and thus $(\det\Omega^{-1})^2 = 1/d$. Further, we have

$$(\det\Omega^{-1})^2 = \Delta(\omega_1^*, \ldots, \omega_n^*),$$

whence it suffices to show that the latter is $d/(N(\mathcal{D}))^2$. Now there is a positive integer c such that $c\mathcal{D}^{-1}$ is an ideal in K and from the formula for the norm in Section 11.5 we obtain

$$c^{2n}\Delta(\omega_1^*, \ldots, \omega_n^*) = \Delta(c\omega_1^*, \ldots, c\omega_n^*) = d(N(c\mathcal{D}^{-1}))^2.$$

But $(c\mathcal{D}^{-1})\mathcal{D} = [c]$ and so $N(c\mathcal{D}^{-1})N(\mathcal{D}) = c^n$. This gives the result. $\square$

We now establish the following fundamental property of the different.

Theorem 11.6 *Let $\mathfrak{p}$ be a prime ideal in K with ramification index e. Then $\mathfrak{p}^{e-1}$ divides $\mathcal{D}$.*

Proof Suppose that $\mathfrak{p}$ lies above p. It will suffice to take β as any element of $p/\mathfrak{p}^{e-1}$ and to show that $T_{K/\mathbb{Q}}(\beta)$ is divisible by p. For then $\beta \in p\mathcal{D}^{-1}$, whence $(p/\mathfrak{p}^{e-1})\mathcal{D}$ is contained in $p\mathcal{D}^{-1}\mathcal{D} = [p]$; thus p divides $(p/\mathfrak{p}^{e-1})\mathcal{D}$ and this gives the desired assertion. Now p divides β^r for all integers $r \geq e$ since $p/\mathfrak{p}^{e-1}$ divides β and so $\mathfrak{p}^r(p/\mathfrak{p}^e)$ divides β^r. Further, from the familiar property that p divides all the relevant binomial coefficients, we obtain, by induction on j,

$$(T_{K/\mathbb{Q}}(\beta))^{p^j} \equiv T_{K/\mathbb{Q}}(\beta^{p^j}) \,(\mathrm{mod}\, p).$$

Thus, for $p^j \geq e$, it follows that p divides $T_{K/\mathbb{Q}}(\beta)$ and the theorem is proved. $\square$

As a corollary we see that if a prime p ramifies in K, that is, if in the prime ideal decomposition of p at least one of the ramification indices exceeds 1, then p divides d; for from $\mathfrak{p}|\mathcal{D}$ for some $\mathfrak{p}$ we get $N(\mathfrak{p})|d$ and the assertion follows since $N\mathfrak{p} = p^f$. The corollary is in fact a celebrated result of Dedekind and it implies in particular that, for any number field K, there are only finitely many ramified primes.

Finally we study the particular case when the field $K = \mathbb{Q}(\alpha)$ has an integral basis $1, \alpha, ..., \alpha^{n-1}$. Let $f(x)$ be the minimum polynomial for α. Then $f(x) = (x - \alpha)g(x)$, where

$$g(x) = \beta_0 + \beta_1 x + \cdots + \beta_{n-1}x^{n-1}$$

is in $\mathcal{O}_K[x]$; for certainly $f(x)/(x - \alpha)$ is in $K[x]$ and it is also in $\mathcal{O}[x]$ since it is a product of factors $(x - \sigma_i(\alpha))(1 \leq i \leq n)$ with $(x - \alpha)$ omitted. Now the dual basis to $1, \alpha, ..., \alpha^{n-1}$ is given by

$$\beta_0/f'(\alpha), \beta_1/f'(\alpha), ..., \beta_{n-1}/f'(\alpha).$$

For on taking $g_i(x) = f(x)/(x - \sigma_i(\alpha))$, we obtain $g_i(\sigma_j(\alpha)) = 0$ if $\sigma_j \neq \sigma_i$ and $g_i(\sigma_j(\alpha)) = \sigma_i(f'(\alpha))$ otherwise; hence if $\Omega = ((\sigma_j(\alpha))^{i-1})$ then $\Omega^{-1} = (\sigma_i(\beta_{j-1}/f'(\alpha)))$ and the dual basis is the row of Ω^{-1} corresponding to the identity monomorphism.

The dual basis is an integral basis of the co-different $\mathcal{D}^{-1}$. Further, $f'(\alpha)\mathcal{D}^{-1}$, that is, the module generated over $\mathbb{Z}$ by $\beta_0, \beta_1, ..., \beta_{n-1}$, is just $\mathbb{Z}(\alpha) = \mathcal{O}_K$. For since $f(x)$ is monic we have $\beta_{n-1} = 1$ and so certainly $\mathbb{Z} \subset f'(\alpha)\mathcal{D}^{-1}$.

Since furthermore $f(x) = (x - \alpha)g(x)$ has integer coefficients which are expressible as $\beta_{n-1}, \beta_{n-2} - \alpha\beta_{n-1}, \ldots, -\alpha\beta_0$, it follows inductively, taking $j = n-1, n-2, \ldots, 0$, that β_j is a monic polynomial in α with integer coefficients and degree $n - j - 1$. Clearly then each α^j is in $f'(\alpha)\mathcal{D}^{-1}$ whence $\mathbb{Z}(\alpha) \subset f'(\alpha)\mathcal{D}^{-1}$ and each β_j is in $\mathbb{Z}(\alpha)$ whence $f'(\alpha)\mathcal{D}^{-1} \subset \mathbb{Z}(\alpha)$. Hence we have $\mathcal{D}^{-1} = \mathcal{O}_K/f'(\alpha)$ and this gives $\mathcal{D} = [f'(\alpha)]$.

11.8 Further reading

The theory here is entirely classical. It is covered in every substantial text on number fields including all the relevant books referred to earlier in Sections 7.7 and 10.9. To these may be added Neukirch's *Algebraic Number Theory* (Springer, 1999), which is an especially authoritative work on the subject.

The result of Dedekind referred to in Section 11.7 was also proved by him in the converse sense, that is, he showed that if p divides d then p ramifies; a demonstration is given in the book by Marcus mentioned in Section 10.9.

11.9 Exercises

(i) Show that, in the field $\mathbb{Q}(\sqrt{(-6)})$, the ideal $[2]$ factorizes as $[2, \sqrt{(-6)}]^2$ as a product of prime ideals. Factorize the ideal $[6]$ similarly.

(ii) Find single generators for the ideals $[2613, 2171]$ in $\mathbb{Z}$ and $[51 - 5i, 43 + 7i]$ in the Gaussian field $\mathbb{Q}(i)$.

(iii) Show that the natural definition of the lowest common multiple of ideals $\mathfrak{a}$ and $\mathfrak{b}$ is the ideal $\mathfrak{a} \cap \mathfrak{b}$.

(iv) Show that, if m and n are the exponents to which a prime ideal $\mathfrak{p}$ divides $\mathfrak{a}$ and $\mathfrak{b}$ respectively in their canonical prime decompositions, then the exponent to which $\mathfrak{p}$ divides $\mathfrak{a} \cap \mathfrak{b}$ is $\max(m, n)$. Show further that the exponent to which $\mathfrak{p}$ divides $\mathfrak{a} + \mathfrak{b}$ is $\min(m, n)$ and establish the relation $(\mathfrak{a} \cap \mathfrak{b})(\mathfrak{a} + \mathfrak{b}) = \mathfrak{a}\mathfrak{b}$.

(v) Verify that the different of the quadratic field $\mathbb{Q}(\sqrt{d})$, with d a square-free integer, is $[2\sqrt{d}]$ when $d \equiv 2$ or $3 \pmod 4$ and $[\sqrt{d}]$ when $d \equiv 1 \pmod 4$.

(vi) What are the differents of the fields $\mathbb{Q}(\sqrt[4]{2})$ and $\mathbb{Q}(\alpha)$ where α is a zero of the polynomial $x^3 - x - 1$?

(vii) Let $K = \mathbb{Q}(\alpha)$ where $\alpha = \sqrt[3]{d}$ with $d \neq 1$ a square-free integer satisfying $d \equiv 1 \pmod 9$. By Section 10.10 Exercise (ix), an integral basis for K is

$1, \alpha, \beta$ with $\beta = \frac{1}{3}(1 + \alpha + \alpha^2)$. Prove that the dual basis is $(\alpha/3d)(\alpha^2 - 1, \alpha - 1, 3)$. Verify that the co-different of K is $(\alpha/3d)[3, \alpha - 1]$ and hence, by the construction of Theorem 11.2, show that the different is $\alpha^2[3, \alpha - 1, \beta]$.

(viii) Prove that if f is the sum of the degrees of the prime ideals that lie above a prime p then p^{n-f}, where n is the degree of the field, divides the discriminant.

12

Units and ideal classes

12.1 Units

An algebraic integer ε is said to be a unit if $1/\varepsilon$ is an algebraic integer. This is equivalent to the condition $N\varepsilon = \pm 1$. Indeed the conjugates of an algebraic integer are again algebraic integers, whence, if ε is a unit, then $N\varepsilon$ and $1/N\varepsilon$ are rational integers and so ± 1. Conversely if $N\varepsilon = \pm 1$ then $1/\varepsilon = \pm N\varepsilon/\varepsilon$ which is clearly an algebraic integer. The set of all units form a group $\mathcal{U}$ under multiplication and the set of units in a number field K form a subgroup $\mathcal{U}_K$. Further, we see that if $[\alpha]$, $[\beta]$ are principal ideals in K then we have $[\alpha] = [\beta]$ if and only if $\alpha/\beta \in \mathcal{U}_K$. In general, we say that non-zero algebraic numbers α, β are associated if $\alpha/\beta \in \mathcal{U}$.

The units in $\mathbb{Q}$ are plainly ± 1 whence they are all the roots of unity in $\mathbb{Q}$, that is, the solutions of an equation $x^l = 1$ for some positive integer l.[†] The units of the quadratic field $K = \mathbb{Q}(\sqrt{d})$, where $d \neq 1$ is a square-free integer, were discussed in Section 7.3. It was shown there that for the imaginary quadratic field K with $d < 0$ the units are again all the roots of unity in the field; they are given by the zeros of $x^2 - 1$ for $D < -4$, of $x^4 - 1$ for $D = -4$ and of $x^6 - 1$ for $D = -3$ where D denotes the discriminant of the field, that is, $D = 4d$ for $d \equiv 2, 3 \pmod 4$ and $D = d$ for $d \equiv 1 \pmod 4$. Further, it was shown that for the real quadratic field K with $d > 0$ there are infinitely many units given by $\pm\varepsilon^m$ with $m = 0, \pm 1, \pm 2, \ldots$ for some unit ε. As mentioned in Section 7.3, these results

---

[†] An lth root of unity in any particular field is defined as a solution of an equation as here and it is said to be primitive if it does not satisfy another equation of the kind with smaller l.

are special cases of a famous theorem of Dirichlet on units in an arbitrary number field and we proceed now to give a detailed account of this theorem.

12.2 Dirichlet's unit theorem

We begin by recalling Minkowski's theorem in the geometry of numbers discussed in Section 6.7; as we shall see it plays an important role in the theory of algebraic number fields. Let S be a convex body symmetrical about the origin and let Λ be the lattice with determinant $d(\Lambda)$ as in Section 6.7; see Fig. 12.1.

Theorem 12.1 (Minkowski) *If S has volume V satisfying $V > 2^n d(\Lambda)$ then S contains a point of Λ other than the origin.*

Proof A proof of the theorem was given in Section 6.7 in the case when Λ is the integer lattice and it carries over easily to the general lattice. The result of Blichfeldt now states that if $\mathcal{R}$ is a bounded set with volume $V > d(\Lambda)$ then there exist x, y in $\mathcal{R}$ such that $x \neq y$ and $x - y \in \Lambda$; Minkowski's theorem follows by taking $\mathcal{R} = \frac{1}{2} S$ as before. For the proof of Blichfeldt's result we define $\mathcal{R}_u$ as the intersection of $\mathcal{R}$ with the cell of Λ with vertex $u = (u_1, \ldots, u_n)$, that is, the points x with $x_i = \sum a_{ij} v_j$ and $u_j \leq v_j < u_j + 1$. Thus if $\mathcal{R}'_u$ is the

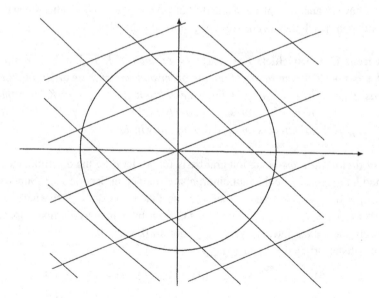

Fig. 12.1 Intersections of diagonal lines give points of Λ; the circular disc is an example of S.

translation of $\mathcal{R}_u$ to the unit cell, that is, the cell given by $u = 0$, and if V_u is the volume of $\mathcal{R}_u$ then we have $\sum V_u = V > d(\Lambda)$. Hence at least two of the $\mathcal{R}'_u$, say $\mathcal{R}'_v$, $\mathcal{R}'_w$, must overlap and there exist points v', w' in $\mathcal{R}_v$, $\mathcal{R}_w$ such that $v' - w' \in \Lambda$. The result follows. $\qquad\square$

As observed in Section 6.7, one of the main applications of Minkowski's theorem is his linear forms theorem and, as shown there, a slightly refined version can be given as follows.

Theorem 12.2 (Minkowski's linear forms theorem) *Let*

$$L_i = L_i(y_1, \ldots, y_n) = \sum_{j=1}^n a_{ij} y_j \quad (1 \le i \le n)$$

be real linear forms with $\Delta = \det(a_{ij}) \neq 0$. *Suppose that* $\lambda_1 > 0, \ldots, \lambda_n > 0$ *and that* $\lambda_1 \cdots \lambda_n = |\Delta|$. *Then there exist integers* $y_1, \ldots, y_n$, *not all* 0, *such that*

$$|L_1| \le \lambda_1 \quad and \quad |L_i| < \lambda_i \quad (2 \le i \le n).$$

Now let $K = \mathbb{Q}(\alpha)$ be any algebraic number field. We shall suppose that K has precisely s real conjugate fields and $2t$ complex conjugate fields so that $n = s + 2t$; thus it is assumed that s of the conjugates of α are real and $2t$ are complex in conjugate pairs. This differs from the notation in the classic texts which have r_1 and r_2 in place of s and t but we prefer to use the latter since the formulae in which they occur are typographically simpler.

Theorem 12.3 (Dirichlet) *The group* $\mathcal{U}_K$ *of units of* K *is finitely generated with* $r = s + t - 1$ *generators of infinite order and one of finite order. In other words, there exist* $r = s + t - 1$ *fundamental units* $\varepsilon_1, \ldots, \varepsilon_r$ *in* K *such that every unit* ε *in* K *can be expressed uniquely in the form* $\varepsilon = \rho \varepsilon_1^{j_1} \cdots \varepsilon_r^{j_r}$ *with* $j_1, \ldots, j_r$ *rational integers and* ρ *a root of unity in* K.

Proof Let $\omega_1, \ldots, \omega_n$ be an integral basis for K, let d be the discriminant of K and let $\sigma_1, \ldots, \sigma_n$ be the monomorphisms defined in Section 10.3. Suppose that $\lambda_1 > 0, \ldots, \lambda_r > 0$ and define λ_{r+1} so that $\lambda_1 \cdots \lambda_n = \sqrt{|d|}$ where, for $t \neq 0$, we put $\lambda_{s+t+j} = \lambda_{s+j} \ (1 \le j \le t)$. Then by Minkowski's linear forms theorem there exist integers $y_1, \ldots, y_n$, not all 0, such that $\theta = y_1 \omega_1 + \cdots + y_n \omega_n$ satisfies $|\sigma_j(\theta)| \le \lambda_j \ (1 \le j \le s)$ and

$$|\text{Re}\,\sigma_j(\theta)| \le \lambda_j/\sqrt{2}, \quad |\text{Im}\,\sigma_j(\theta)| \le \lambda_j/\sqrt{2} \quad (s+1 \le j \le s+t).$$

These give $|\sigma_j(\theta)| \le \lambda_j (1 \le j \le n)$, whence $|N\theta| \le \sqrt{|d|}$ where N denotes the field norm of K, that is, $N = N_{K/\mathbb{Q}}$. Now since $|N\theta| \ge 1$ we have

$$|\sigma_j(\theta)| \ge \lambda_j/(\lambda_1 \ldots \lambda_n) = \lambda_j/\sqrt{|d|}$$

and thus

$$\lambda_j / \sqrt{|d|} \le |\sigma_j(\theta)| \le \lambda_j \quad (1 \le j \le n).$$

We proceed to construct a multiplicatively independent set of units $\eta_1, \dots, \eta_r$ in K, that is, a set of units such that if $\eta_1^{j_1} \cdots \eta_r^{j_r} = 1$ with integers $j_1, \dots, j_r$ then $j_1 = \cdots = j_r = 0$. Accordingly, for $k = 1, 2, \dots, r$ and $l = 1, 2, \dots$, let θ_{kl} denote the θ as above corresponding to $\lambda_j = 1$ $(1 \le j \le r,\ j \ne k)$ and $\lambda_k = \mu^l$ for some $\mu > 1$. Since all the θ_{kl} have bounded norm, there exist $l', l''(l' > l'')$ such that $\theta' = \theta_{kl'}$, $\theta'' = \theta_{kl''}$ satisfy $|N\theta'| = |N\theta''| = N$, say, and the corresponding integers y'_j, y''_j satisfy $y'_j \equiv y''_j \pmod{N}$ for $1 \le j \le n$. We now define $\eta_k = \theta'/\theta''$. Then $N\eta_k = \pm 1$ and since $(\theta' - \theta'')/N$ and $N/\theta'' \in \mathcal{O}_K$ we have $\eta_k - 1 \in \mathcal{O}_K$. Hence $\eta_k \in \mathcal{U}_K$. Further, since $l' > l''$, we see that $|\sigma_k(\eta_k)| > \mu/\sqrt{|d|}$. Thus, on observing that $|\sigma_i(\eta_j)| \le \sqrt{|d|}$ when $1 \le i, j \le r$ and $i \ne j$, we obtain a lower bound for $\det(\log|\sigma_i(\eta_j)|)$ that is asymptotic to $(\log(\mu/\sqrt{|d|}))^r$ as $\mu \to \infty$. In particular the determinant is not 0 for μ sufficiently large and this implies that the units $\eta_1, \dots, \eta_r$ are multiplicatively independent.

Now as in the proof of Corollary 12.1 below we see that every unit ε in K can be written as $\varepsilon = \xi \eta_1^{u_1} \cdots \eta_r^{u_r}$ with ξ belonging to a finite set and with $u_1, \dots, u_r$ integers. By considering $\varepsilon, \varepsilon^2, \dots$ we obtain the same ξ for some distinct powers, whence $\varepsilon^m = \eta_1^{m_1} \cdots \eta_r^{m_r}$ for some integer m independent of ε and some integers $m_1, \dots, m_r$. Then by the argument of Section 11.6 we can find generators $\varepsilon_1^m, \dots, \varepsilon_r^m$ for the set of ε^m of the form

$$\varepsilon_j^m = \eta_1^{m_{j1}} \cdots \eta_j^{m_{jj}},$$

where the m_{jk} are integers with $m_{jj} > 0$ and minimal, and we have $\varepsilon^m = (\varepsilon_1^m)^{j_1} \cdots (\varepsilon_r^m)^{j_r}$ for some integers $j_1, \dots, j_r$. This gives $\varepsilon = \rho \varepsilon_1^{j_1} \cdots \varepsilon_r^{j_r}$ for some root of unity ρ as required. $\qquad\square$

The roots of unity ρ in K form a finite cyclic group. For the minimum polynomial for ρ has degree at most n, the degree of K, and the coefficients are bounded in terms of n since the leading coefficient is 1, the remainder are expressible, apart from sign, as elementary symmetric functions of $\sigma_i(\rho)$ $(1 \le i \le n)$ and these have modulus 1; hence there are only finitely many ρ. Each of the latter satisfies an equation of the form $x^l = 1$ for some integer l, and if we take q as the product of all the l then the roots of unity in K form a subgroup of the cyclic group generated by a primitive qth root of unity and the assertion follows.

Arising from Dirichlet's unit theorem we have the important concept of the regulator of a number field.

Definition 12.1 The quantity $\det(\log|\sigma_i(\varepsilon_j)|)$, where $1 \le i, j \le r$, is the same for all sets of fundamental units $\varepsilon_1, \dots, \varepsilon_r$ and it is called the regulator of K.

For the rationals and also for an imaginary quadratic field the regulator is understood to be 1. For a real quadratic field, it is, by definition, $\log \varepsilon$ where ε is the fundamental unit in the field exceeding 1.

The following application of Theorem 12.3 is needed in Section 17.5 and it is in fact often invoked, implicitly or explicitly, in the study of Diophantine equations.

Corollary 12.1 *There are only finitely many non-associated algebraic integers α in a number field K with $|N_{K/\mathbb{Q}}(\alpha)|$ bounded.*

Proof First we observe that every point $(x_1, \ldots, x_r)$ in Euclidean r-space lies within a distance rc, in the sense of the maximum norm, of a point of the lattice with basis

$$(\log |(\sigma_1 \varepsilon_j)|, \ldots, \log |(\sigma_r \varepsilon_j)|) \quad (1 \le j \le r),$$

where $c = \max |\log |\sigma_i(\varepsilon_j)||$. For, since the regulator of K is non-zero, there exist $v_1, \ldots, v_r$ satisfying

$$x_i = \sum_{j=1}^{r} v_j \log |\sigma_i(\varepsilon_j)| \quad (1 \le i \le r)$$

and the desired lattice point is obtained on taking $u_j = [v_j]\,(1 \le j \le r)$. Now putting $x_i = \log |\sigma_i(\alpha)|$ and defining $\varepsilon = \varepsilon_1^{u_1} \cdots \varepsilon_r^{u_r}$ we see that $\beta = \alpha/\varepsilon$ is an associate of α with $|\log |\sigma_i(\beta)|| \le rc\,(1 \le i \le r)$. Since $r = s + t - 1$ the latter in fact holds for all i with $1 \le i \le n$ except possibly for $i = s + t$ and $i = s + 2t$. But if $|N_{K/\mathbb{Q}}(\alpha)| \le C$ then also $|N_{K/\mathbb{Q}}(\beta)| \le C$, whence

$$\sum_{i=1}^{n} \log |\sigma_i(\beta)| \le \log C.$$

This implies that $|\log |\sigma_i(\beta)|| \le nrc + \log C$ for all i. Thus the field polynomial for β has bounded coefficients and the result follows. $\qquad\square$

12.3 Ideal classes

Let K be an algebraic number field with discriminant d. We say that two ideals $\mathfrak{a}, \mathfrak{b}$ in K are equivalent (written $\mathfrak{a} \sim \mathfrak{b}$) if $[\theta]\mathfrak{a} = [\phi]\mathfrak{b}$ for some principal ideals $[\theta], [\phi]$. This is an equivalence relation. Further, the equivalence classes form a group on taking the product $\mathcal{A}\mathcal{B}$ of classes $\mathcal{A}$ and $\mathcal{B}$ as the class containing $\mathfrak{a}\mathfrak{b}$ where $\mathfrak{a}, \mathfrak{b}$ are any ideals in $\mathcal{A}, \mathcal{B}$ respectively; the inverse $\mathcal{A}^{-1}$ exists by

Theorem 11.2. The group is called the ideal class group of K and we proceed to prove that it is finite. We shall need the following lemma.

Lemma 12.1 *In every ideal $\mathfrak{a}$ there is an element $\theta \neq 0$ such that $|N_{K/\mathbb{Q}}(\theta)| \leq N\mathfrak{a}\sqrt{|d|}$.*

Proof Let $K = \mathbb{Q}(\alpha)$ and suppose first that K is totally real, that is, all the $\sigma_j(\alpha)$ $(1 \leq j \leq n)$ are real or, equivalently, all the conjugate fields of K are real. Let $\gamma_1, \ldots, \gamma_n$ be a basis for $\mathfrak{a}$ and let $\lambda_1, \ldots, \lambda_n$ be positive numbers satisfying $\lambda_1 \ldots \lambda_n = \sqrt{|\Delta|}$, where $\Delta = \Delta(\gamma_1, \ldots, \gamma_n)$. Then by Minkowski's linear forms theorem there exist integers $y_1, \ldots, y_n$, not all 0, such that the element $\theta = y_1\gamma_1 + \cdots + y_n\gamma_n$ in $\mathfrak{a}$ satisfies $|\sigma_j(\theta)| \leq \lambda_j$ $(1 \leq j \leq n)$. This gives $|N_{K/\mathbb{Q}}(\theta)| \leq \lambda_1 \ldots \lambda_n$ and the lemma follows from the formula for $N\mathfrak{a}$ in Section 11.6.

The general result is obtained similarly. We assume, as in Section 12.2, that $\sigma_j(\alpha)$, with $1 \leq j \leq s$, is real and $\sigma_{s+j}(\alpha)$, with $1 \leq j \leq t$, is the complex conjugate of $\sigma_{s+j+t}(\alpha)$ so that $n = s + 2t$. We now take $\lambda_1, \ldots, \lambda_n$ as before satisfying $\lambda_{s+j} = \lambda_{s+j+t}$ $(1 \leq j \leq t)$. Then there are integers $y_1, \ldots, y_n$, not all 0, such that $|\sigma_j(\theta)| \leq \lambda_j$ $(1 \leq j \leq s)$ and

$$|\operatorname{Re}\sigma_j(\theta)| \leq \lambda_j/\sqrt{2}, \quad |\operatorname{Im}\sigma_j(\theta)| \leq \lambda_j/\sqrt{2} \quad (s+1 \leq j \leq s+t);$$

for the system of linear forms indicated here has determinant $2^{-t}\sqrt{|\Delta|}$. This gives $|N_{K/\mathbb{Q}}(\theta)| \leq \lambda_1 \ldots \lambda_n$ and the lemma follows. $\qquad\square$

Theorem 12.4 *The ideal class group is finite. Thus if h denotes the order of the group, that is, the class number of K, then $\mathfrak{a}^h$ is principal for every ideal $\mathfrak{a}$. In particular if $h = 1$ then every ideal is principal and K has unique factorization.*

Proof We shall show that for every ideal $\mathfrak{a}$ in K there is an ideal $\mathfrak{b}$ such that $\mathfrak{a} \sim \mathfrak{b}$ and $N\mathfrak{b} \leq \sqrt{|d|}$. This will establish the theorem since, by Lemma 11.4, there are only finitely many possible $\mathfrak{b}$. Now, by Theorem 11.2, there is an ideal $\mathfrak{c}$ such that $\mathfrak{ac}$ is principal, say $[\phi]$. Further, by Lemma 12.1, there is an element $\theta \neq 0$ in $\mathfrak{c}$ with $|N_{K/\mathbb{Q}}(\theta)| \leq N\mathfrak{c}\sqrt{|d|}$. But then $\mathfrak{c}$ divides $[\theta]$, that is, $[\theta] = \mathfrak{bc}$ for some ideal $\mathfrak{b}$, and we see that $N\mathfrak{b} \leq \sqrt{|d|}$. On noting that $\mathfrak{a}(\mathfrak{bc}) = \mathfrak{b}(\mathfrak{ac})$ and that $\mathfrak{bc} = [\theta]$, $\mathfrak{ac} = [\phi]$ we get $\mathfrak{a} \sim \mathfrak{b}$ as required. $\qquad\square$

12.4 Minkowski's constant

We have seen that in every ideal class of K there is an ideal $\mathfrak{b}$ such that $N\mathfrak{b} \le c\sqrt{|d|}$ with $c = 1$. The value of c can be improved to $c = (4/\pi)^t n!/n^n$ and this is called the Minkowski constant. In particular, we have $c = 2/\pi$ for an imaginary quadratic field.

To obtain the Minkowski constant we apply Minkowski's theorem with the lattice Λ given by the system $\sigma_j(\theta)$, $\operatorname{Re}\sigma_j(\theta)$ and $\operatorname{Im}\sigma_j(\theta)$ as in Sections 12.2 and 12.3. Let S be the convex body defined by

$$\sum_{j=1}^{s} |x_j| + 2\sum_{j=1}^{t} \sqrt{(x_{s+j}^2 + x_{s+j+t}^2)} \le \rho.$$

Then $d(\Lambda) = 2^{-t}\sqrt{|\Delta|}$ and the volume V of S is $2^s(\frac{1}{2}\pi)^t \rho^n/n!$. Further, by the inequality of the arithmetic and geometric means, that is,

$$(\psi_1 \ldots \psi_n)^{1/n} \le (\psi_1 + \cdots + \psi_n)/n,$$

we have, for $\mathbf{x}$ in S,

$$\prod_{j=1}^{s} |x_j| \prod_{j=1}^{t} (x_{s+j}^2 + x_{s+j+t}^2) \le (\rho/n)^n.$$

Now taking ρ so that $\rho^n > (4/\pi)^t n! \, N\mathfrak{a}\sqrt{|d|}$, whence $V > 2^n d(\Lambda)$, we obtain, for any $\varepsilon > 0$, an element θ_ε in $\mathfrak{a}$ with $|N_{K/\mathbb{Q}}(\theta_\varepsilon)| < (c+\varepsilon)N\mathfrak{a}\sqrt{|d|}$, where c is the Minkowski constant. A compactness argument as used in Section 6.7 and alluded to again in Section 12.2 then completes the proof.

The Minkowski constant is important since it facilitates the determination of the ideal class group of K in specific instances. The typical application uses the fact that the group is generated by the classes that contain the prime ideals which lie over the rational primes $p \le c\sqrt{|d|}$ and plainly a smaller value of c means that potentially fewer p need to be examined. Consider as an example the quadratic field $K = \mathbb{Q}(\sqrt{(-5)})$. Here $t = 1$, $d = -20$ and so every ideal in K is equivalent to an ideal with norm at most $(2/\pi)\sqrt{20} < 3$. Now we have

$$2 = [2, 1 + \sqrt{(-5)}]^2$$

as a product of prime ideals; this follows from the general theory of the next section but it may be verified directly by noting that the expression on the right is $[4, 2\sqrt{(-5)} + 2, 2\sqrt{(-5)} - 4]$ and this is both divisible by 2 and contains 2. Further, we see that $[2, 1 + \sqrt{(-5)}]$ has norm 2 and it is not principal, for we cannot have $N(u + v\sqrt{(-5)}) = 2$, that is, $u^2 + 5v^2 = 2$, for integers u, v. It follows that every ideal in K is either principal or equivalent to $[2, 1 + \sqrt{(-5)}]$.

This gives $h = 2$ whence we can say that the ideal class group of K is cyclic of order 2. We shall give more examples of this kind in Section 12.8.

12.5 Dedekind's theorem

Let K be an algebraic number field and let $\mathcal{O}_K$ be the ring of integers of K. We suppose that $\mathcal{O}_K$ has a power integral basis, that is, $\mathcal{O}_K = \mathbb{Z}(\alpha)$ for some $\alpha \in \mathcal{O}_K$. We denote by g the minimum polynomial for α.

Now let p be any rational prime. We signify by $\overline{g}$ the reduction of $g \pmod{p}$, that is, the polynomial obtained by replacing each coefficient of g by its residue $\pmod{p}$; we say that $\overline{g}$ is defined over $\overline{\mathbb{Z}} = \mathbb{F}_p$. Then the following holds.

Theorem 12.5 (Dedekind) *If $\overline{g} = \overline{p}_1^{e_1} \cdots \overline{p}_r^{e_r}$ as a product of irreducible monic polynomials $\overline{p}_1, \ldots, \overline{p}_r$ defined over $\overline{\mathbb{Z}}$ then we have $p = \mathfrak{p}_1^{e_1} \cdots \mathfrak{p}_r^{e_r}$ as a product of prime ideals $\mathfrak{p}_1, \ldots, \mathfrak{p}_r$, and here $\mathfrak{p}_j = [p, \overline{p}_j(\alpha)]$.*

Proof The ideal $\mathfrak{p} = \mathfrak{p}_j = [p, \overline{p}_j(\alpha)]$ is the kernel of the mapping $\mathbb{Z}(\alpha) \to R$ where $R = \overline{\mathbb{Z}}(\overline{\alpha}_j)$ and $\overline{\alpha}_j$ is any zero of $\overline{p}_j$. For clearly $\mathfrak{p}$ is contained in the kernel and if, conversely, we have $q(x) \in \mathbb{Z}[x]$ and $\overline{q}(\overline{\alpha}_j) = 0$, with $\overline{q}$ defined like $\overline{g}$ above, then, since $\overline{p}_j$ is irreducible, it follows that $\overline{p}_j$ divides $\overline{q}$, whence $\overline{q}(\alpha) \in [\overline{p}_j(\alpha)]$ and $q(\alpha) \in \mathfrak{p}$. Hence $\mathfrak{p}$ is a prime ideal. This is classic algebra; in fact R is a ring isomorphic to $\mathcal{O}_K/\mathfrak{p}$ and it has no zero divisors, that is, it is an integral domain. Thus if $\mathfrak{p} = \mathfrak{a}\mathfrak{b}$ and if $a(\alpha) \in \mathfrak{a}$, $b(\alpha) \in \mathfrak{b}$ then $\overline{a}(\overline{\alpha}_j) = 0$ or $\overline{b}(\overline{\alpha}_j) = 0$ and so $a(\alpha)$ or $b(\alpha) \in \mathfrak{p}$; this implies that if $\mathfrak{p}$ does not divide $\mathfrak{a}$, so that there exists $a(\alpha) \in \mathfrak{a}$ not in $\mathfrak{p}$, then $b(\alpha) \in \mathfrak{p}$ for all $b(\alpha) \in \mathfrak{b}$, whence $\mathfrak{p}$ divides $\mathfrak{b}$.

Now we have $\mathfrak{p}_j^{e_j} \subset [p, (\overline{p}_j(\alpha))^{e_j}]$ and hence $\mathfrak{p}_1^{e_1} \cdots \mathfrak{p}_r^{e_r} \subset [p, \overline{g}(\alpha)]$. Further, since $g(\alpha) = 0$ we see that $[p, \overline{g}(\alpha)] = [p]$. Furthermore we have $N\mathfrak{p}_j = p^{f_j}$ where f_j is the degree of $\overline{p}_j$; for the elements $a_0 + a_1\alpha + \cdots + a_{f-1}\alpha^{f-1}$ with $f = f_j$ and $0 \leqslant a_i < p$ $(0 \leq i < f)$ give a complete set of representatives $\pmod{\mathfrak{p}_j}$. Now g is monic and so the degree of $\overline{g}$ is the degree of K, say n. This gives $e_1 f_1 + \cdots + e_r f_r = n$ and $N[p] = p^n$ whence $p = \mathfrak{p}_1^{e_1} \ldots \mathfrak{p}_r^{e_r}$ as asserted. $\qquad\square$

We shall apply Dedekind's theorem to the quadratic field $K = \mathbb{Q}(\sqrt{d})$ where $d \neq 1$ is a square-free integer. We recall that $\mathcal{O}_K = \mathbb{Z}(\sqrt{d})$ when $d \equiv 2, 3 \pmod{4}$ and $\mathcal{O}_K = \mathbb{Z}((\frac{1}{2}(1 + \sqrt{d}))$ when $d \equiv 1 \pmod{4}$ and so the quadratic field has a power integral basis as in the hypothesis of the theorem.

The minimum polynomial of $\sqrt{d}$ is $g(x) = x^2 - d$ and, when $d \equiv 2, 3 \pmod 4$, Dedekind's theorem gives the following three possibilities, where $\mathfrak{p}$ and $\mathfrak{p}'$ denote prime ideals.

(i) $g(x)$ reduces $\pmod p$ into distinct linear factors. Then $p = \mathfrak{p}\mathfrak{p}'$ $(\mathfrak{p} \neq \mathfrak{p}')$ and $N\mathfrak{p} = N\mathfrak{p}' = p$.

(ii) $g(x)$ reduces $\pmod p$ to a square. Then $p = \mathfrak{p}^2$ and $N\mathfrak{p} = p$.

(iii) $g(x)$ is irreducible $\pmod p$. Then $p = \mathfrak{p}$ and $N\mathfrak{p} = p^2$.

Now suppose that $d \equiv 1 \pmod 4$. The minimum polynomial for $\frac{1}{2}(1 + \sqrt{d})$ is $g(x) = x^2 - x + \frac{1}{4}(1 - d)$ and we have $4g(x) = (2x - 1)^2 - d$. Hence, if p is odd, then we get the possibilities (i), (ii) and (iii) as above. When $p = 2$ and $d \equiv 1 \pmod 8$ we have $g(x) \equiv x(x - 1) \pmod 2$ and so $p = \mathfrak{p}\mathfrak{p}'$ as in (i). When $p = 2$ and $d \equiv 5 \pmod 8$ we see that $g(x)$ is irreducible $\pmod 2$ and so $p = \mathfrak{p}$ as in (iii).

Thus, when $d \equiv 2, 3 \pmod 4$, we have, in case (i), d a quadratic residue $\pmod p$ whence the Legendre symbol $\left(\frac{d}{p}\right) = 1$. Similarly we have $\left(\frac{4d}{p}\right) = 0$ in case (ii) and $\left(\frac{d}{p}\right) = -1$ in case (iii). Hence, on taking $D = 4d$, that is, defining D as the discriminant of K, we obtain $\left(\frac{D}{p}\right) = 1$, 0 or -1 in the three cases respectively. Moreover the same holds for the discriminant $D = d$ of K when $d \equiv 1 \pmod 4$ provided that, for $p = 2$, we take the Jacobi symbol $\left(\frac{2}{|D|}\right)$ rather than the Legendre symbol.

The results are intimately related to the theory of characters and L-functions of quadratic fields; see Chapter 15. Indeed, for $d \equiv 2, 3 \pmod 4$ we can define a quadratic character of K by $\chi(p) = \left(\frac{D}{p}\right)$ and then, for any $s > 1$, our results show that

$$\prod_{\mathfrak{p}|p} (1 - (N\mathfrak{p})^{-s}) = (1 - p^{-s})(1 - \chi(p)p^{-s}).$$

Moreover the same holds in the case $d \equiv 1 \pmod 4$ provided we define $\chi(2) = \left(\frac{2}{|D|}\right)$ as above. Hence we get the important relation

$$\zeta_K(s) = \zeta(s)L(s, \chi),$$

where $\zeta_K(s)$ is the Dedekind zeta-function defined by

$$\sum_{\mathfrak{a}} (N\mathfrak{a})^{-s} = \prod_{\mathfrak{p}} (1 - (N\mathfrak{p})^{-s})^{-1},$$

$\zeta(s)$ denotes the Riemann zeta-function given by

$$\sum 1/n^s = \prod_p (1 - p^{-s})^{-1}$$

and $L(s, \chi)$ is the Dirichlet L-function

$$\sum_{n=1}^{\infty} \chi(n)/n^s = \prod_p (1 - \chi(p)p^{-s})^{-1}.$$

All the sums and products here converge for $s > 1$ and indeed for any complex variable $s = \sigma + it$ with $\sigma > 1$. In the case of the L-function we have convergence for $s = 1$ and there are classical results of Dirichlet giving the values of $L(1, \chi)$ in terms of the class number of K; see Section 15.6.

12.6 The cyclotomic field

We shall need Eisenstein's criterion relating to a polynomial

$$f(x) = x^n + a_{n-1}x^{n-1} + \cdots + a_1 x + a_0$$

with integer coefficients. It is stated as follows.

Theorem 12.6 (Eisenstein) *If there exists a prime p such that p divides $a_0, a_1,$ $\ldots, a_{n-1}$ but p^2 does not divide a_0 then $f(x)$ is irreducible.*

Proof Suppose on the contrary that $f(x) = g(x)h(x)$ where $g(x)$ and $h(x)$ are polynomials with integer coefficients and $m < n$ is the degree of $g(x)$. Let $\overline{\mathbb{Z}} = \mathbb{F}_p$ be the field with p elements as in Section 12.5 and let $\overline{f}, \overline{g}, \overline{h}$ be the reductions of $f, g, h \pmod{p}$. Then $\overline{f}(x) = \overline{g}(x)\overline{h}(x)$ and $\overline{f} = x^n$. But, by hypothesis, $p^2 \nmid a_0$, whence the constant terms in $\overline{g}(x), \overline{h}(x)$ cannot both be 0, and this plainly gives a contradiction; indeed $\overline{\mathbb{Z}}[x]$ is a unique factorization domain, whence we have $\overline{g} = x^m, \overline{h} = x^{n-m}$. This proves the theorem. $\square$

The qth cyclotomic field is defined as $K = \mathbb{Q}(\zeta)$ where $\zeta = e^{2\pi i/q}$ and q is an integer > 2. We shall discuss here only the case when q is a prime. Then, since $\zeta^q = 1$, we see that ζ is a zero of the cyclotomic polynomial given by $\Phi_q(x) = x^{q-1} + x^{q-2} + \cdots + 1$ and that this is irreducible; for plainly Eisenstein's criterion applies to

$$\Phi_q(x+1) = \frac{(x+1)^q - 1}{x} = x^{q-1} + \binom{q}{1}x^{q-2} + \cdots + \binom{q}{q-1}.$$

Thus $\Phi_q(x)$ is the minimum polynomial for ζ, whence ζ has degree $q - 1$ and conjugates $\zeta, \zeta^2, \ldots, \zeta^{q-1}$. This gives at once the basic cyclotomic property of ζ, namely that, for any integer k,

$$T_{K/\mathbb{Q}}(\zeta^k) = \begin{cases} -1 & \text{if } q \nmid k, \\ q - 1 & \text{if } q \mid k. \end{cases}$$

Furthermore the following holds.

Theorem 12.7 (Basis) *An integral basis for K is given by* $1, \zeta, \ldots, \zeta^{q-2}$.

Proof First we observe that from the factorization

$$\Phi_q(x) = (x - \zeta)(x - \zeta^2) \ldots (x - \zeta^{q-1})$$

we obtain $N_{K/\mathbb{Q}}(1 - \zeta) = \Phi_q(1) = q$. Further, we observe that $(1 - \zeta^j)/(1 - \zeta)$ is a unit in K for certainly it is in $\mathcal{O}_K$ and we have

$$N_{K/\mathbb{Q}}(1 - \zeta) = N_{K/\mathbb{Q}}(1 - \zeta^j) \quad (1 \le j \le q - 1).$$

Taking the product over all j, it follows that $q/(1 - \zeta)^{q-1}$ is a unit in K.

We now suppose that $\alpha \in \mathcal{O}_K$. Then, for some rationals $a_0, \ldots, a_{q-2}$,

$$\alpha = a_0 + a_1 \zeta + \cdots + a_{q-2} \zeta^{q-2}.$$

By the cyclotomic property of ζ recorded above, this gives

$$T_{K/\mathbb{Q}}(\alpha \zeta^{-j} - \alpha \zeta) = q a_j \quad (0 \le j \le q - 2),$$

whence $q a_j$ is a rational integer. Hence we have

$$q\alpha = b_0 + b_1(1 - \zeta) + \cdots + b_{q-2}(1 - \zeta)^{q-2} \qquad (**)$$

for some integers $b_0, \ldots, b_{q-2}$. We proceed to show that q divides b_j for all j and this will suffice to establish the result.

Now if we assume that $b_0, \ldots, b_{r-1}$ are divisible by q, where $1 \le r \le q - 2$, then $(**)$ shows that $b_r/(1 - \zeta) \in \mathcal{O}_K$; thus q divides $N_{K/\mathbb{Q}}(b_r) = b_r^{q-1}$ and so, since q is a prime, it divides b_r. Plainly, by $(**)$, $b_0/(1 - \zeta) \in \mathcal{O}_K$, whence q divides b_0 and the assertion follows by induction. $\square$

Theorem 12.8 (Discriminant) *The discriminant of K is* $(-1)^{\frac{1}{2}(q-1)} q^{q-2}$.

Proof We shall use the integral basis $\zeta, \zeta^2, \ldots, \zeta^{q-1}$. By the cyclotomic property of ζ referred to above we obtain, for $1 \le i, j \le q - 1$,

$$\det(T_{K/\mathbb{Q}}(\zeta^{i+j})) = \begin{vmatrix} -1 & -1 & \cdots & -1 & q-1 \\ -1 & -1 & \cdots & q-1 & -1 \\ & & \vdots & & \\ -1 & q-1 & \cdots & -1 & -1 \\ q-1 & -1 & \cdots & -1 & -1 \end{vmatrix}.$$

Subtracting the first row from the others and then adding the first $q - 2$ columns to the last we get

$$
\begin{vmatrix}
-1 & -1 & \cdots & -1 & q-1 \\
0 & 0 & \cdots & q & -q \\
& & \vdots & & \\
0 & q & \cdots & 0 & -q \\
q & 0 & \cdots & 0 & -q
\end{vmatrix}
=
\begin{vmatrix}
-1 & -1 & \cdots & -1 & 1 \\
0 & 0 & \cdots & q & 0 \\
& & \vdots & & \\
0 & q & \cdots & 0 & 0 \\
q & 0 & \cdots & 0 & 0
\end{vmatrix}.
$$

Now adding the last column to the others we obtain zeros except on the diagonal and the result follows. $\qquad\square$

Theorem 12.9 (Factorization) *Let p be a prime. If $p \neq q$ then $p = \mathfrak{p}_1 \ldots \mathfrak{p}_l$ where $\mathfrak{p}_1, \ldots, \mathfrak{p}_l$ are distinct prime ideals. If $p = q$ then $p = \mathfrak{p}^{q-1}$ where $\mathfrak{p} = [1 - \zeta]$.*

Proof We have $\mathcal{O}_K = \mathbb{Z}(\zeta)$ and so the hypothesis of Dedekind's theorem is satisfied. Suppose first that $p \neq q$. Then the polynomial $x^q - 1$ in $\overline{\mathbb{Z}}[x]$, where $\overline{\mathbb{Z}} = \mathbb{F}_p$, is relatively prime to its derivative and hence the cyclotomic polynomial $\Phi_q(x)$ has no repeated factor (mod p). This gives the desired result. When $p = q$ we have $x^q - 1 \equiv (x - 1)^q \pmod{q}$ and, since $q/(1 - \zeta)^{q-1}$ is a unit in K, we see that $[q, 1 - \zeta] = [1 - \zeta]$. The assertion follows. $\qquad\square$

We note now that K is a normal field; indeed its Galois group is isomorphic to the multiplicative subgroup of $\mathbb{F}_q$. Hence in the case when $p \neq q$ all prime divisors $\mathfrak{p}$ of p have the same degree, say f, and we have $l = (q - 1)/f$ (cf. Section 11.6). In fact we can calculate f explicitly.

Theorem 12.10 *We have $N\mathfrak{p} = p^f$ for each prime divisor $\mathfrak{p}$ of p where f is the least positive integer such that $p^f \equiv 1 \pmod{q}$.*

Proof It is easily seen that $1, \zeta, \ldots, \zeta^{q-1}$ are in distinct residue classes (mod $\mathfrak{p}$). Indeed otherwise $\mathfrak{p}$ would divide $1 - \zeta^j$ for some j with $1 \leq j < q$, whence it would divide $N_{K/\mathbb{Q}}(1 - \zeta) = q$ contrary to $p \neq q$. The classes form a subgroup of the multiplicative group of the field $\mathcal{O}_K/\mathfrak{p}$ and this has order $N\mathfrak{p} - 1$. Hence $N\mathfrak{p} \equiv 1 \pmod{q}$ and so $N\mathfrak{p} = p^{f'}$ where $f' \geq f$. But since $p^f \equiv 1 \pmod{q}$ we have $\zeta^{p^f} = \zeta$, whence $\alpha^{p^f} \equiv \alpha \pmod{\mathfrak{p}}$ for all α in $\mathcal{O}_K$. Thus each of the $N\mathfrak{p}$ elements of $\mathcal{O}_K/\mathfrak{p}$ is a zero of $x^{p^f} - x$ and this gives $p^{f'} \leq p^f$. We conclude that $f' = f$. $\qquad\square$

It follows as a corollary to Theorem 12.10 that p splits completely in K if and only if $p \equiv 1 \pmod{q}$; the result has connections with class field theory and it is sometimes referred to as the cyclotomic reciprocity law.

Theorem 12.11 (Units) *There are $\frac{1}{2}(q-3)$ elements in any fundamental set of units of K. The roots of unity are given by $(-\zeta)^j$ with $0 \le j < 2q$.*

Proof The first part follows at once from Dirichlet's theorem since here $s = 0$, $t = \frac{1}{2}(q-1)$ and so $r = s + t - 1 = \frac{1}{2}(q-3)$. The roots of unity form a finite cyclic group which includes the elements indicated above as a subgroup. Thus $2q$ divides the order of the group, say m. Then $m = 2q$ follows since the group is generated by a primitive mth root of unity and its degree $\phi(m)$ cannot exceed the degree $\phi(q)$ of K. Alternatively, one can argue that if there were a primitive nth root of unity in the field with $(n, q) = 1$ and $n > 2$, then the product with ζ would be a primitive nqth root of unity in K; but this has degree $\phi(nq) = \phi(n)\phi(q)$ which exceeds the degree $\phi(q)$ of K. Plainly if $(n, q) > 1$, then the degree $\phi(n)$ of a primitive nth root of unity exceeds $\phi(q)$ unless $n = q$ or $n = 2q$, in which case the root is just a power of ζ or $-\zeta$. This completes the proof except for a verification that a primitive mth root of unity with composite m has degree $\phi(m)$, and for this we refer to the recommended books. ☐

It was already observed in the proof of Theorem 12.7 that $(1 - \zeta^j)/(1 - \zeta)$ is a unit in K for $j = 2, \ldots, \frac{1}{2}(q-1)$. Now we have

$$\frac{1 - \zeta^j}{1 - \zeta} = \zeta^{(j-1)/2} \frac{\zeta^{j/2} - \zeta^{-j/2}}{\zeta^{1/2} - \zeta^{-1/2}} = \zeta^{(j-1)/2} \frac{\sin(j\pi/q)}{\sin(\pi/q)}.$$

Further, since $\zeta^{1/2} = \pm \zeta^{(q+1)/2}$ and q is odd, we see that $\zeta^{(j-1)/2}$ is a root of unity in K. Hence the numbers $\sin(j\pi/q)/\sin(\pi/q)$ are real positive units in K; they can be verified as being multiplicatively independent but they do not necessarily comprise a fundamental system.

We note that the units referred to above all lie in the field $K^+ = \mathbb{Q}(\zeta + \zeta^{-1})$. Indeed ζ is a zero of the polynomial $x^2 - (\zeta + \zeta^{-1})x + 1$, whence $[K : K^+] = 2$. Thus $[K^+ : \mathbb{Q}] = \frac{1}{2}(q-1)$ and, since $\zeta + \zeta^{-1} = 2\cos(2\pi/q)$, it follows that K^+ is the maximal real subfield of K. It plays a significant role in studies on the class number of K. One result relating to K^+ can be obtained easily.

Theorem 12.12 *An integral basis for K^+ is $1, \eta, \ldots, \eta^m$ with $\eta = \zeta + \zeta^{-1}$ and $m = \frac{1}{2}(q-1) - 1$.*

Proof Let α be in the ring of integers of K^+. Plainly we have $\alpha = a_0 + a_1\eta + \cdots + a_m\eta^m$ for some rationals $a_0, a_1, \ldots, a_m$. Suppose that these are not all integers. Then there exists a suffix k such that the a_j with $j > k$ are integers and a_k is not. But we have

$$\zeta^k(\alpha - a_{k+1}\eta^{k+1} - \cdots - a_m\eta^m) = \zeta^k(a_0 + a_1\eta + \cdots + a_k\eta^k)$$

and this is in the ring of integers of K. Thus, since $1, \zeta, \ldots, \zeta^{q-2}$ is an integral basis for K and $2k < q - 2$, the coefficient a_k of ζ^{2k} is an integer and this is a contradiction. $\qquad\square$

The Dedekind zeta-function $\zeta_K(s)$ for K, where $s = \sigma + it$ and $\sigma > 1$, is

$$\sum_{\mathfrak{a}}(N\mathfrak{a})^{-s} = \prod_{\mathfrak{p}}(1 - (N\mathfrak{p})^{-s})^{-1}$$

and from Theorems 12.9 and 12.10 we deduce the following.

Theorem 12.13 *We have*

$$\zeta_K(s) = \zeta(s)\prod_{\chi\neq\chi_0}L(s, \chi),$$

where

$$L(s, \chi) = \sum_{n=1}^{\infty}\chi(n)/n^{-s} = \prod_p(1 - \chi(p)p^{-s})^{-1}.$$

Here the characters $\chi = \chi_j$ with $0 \le j \le q - 2$ are defined by

$$\chi_j(n) = \begin{cases} e^{2\pi ij\nu/(q-1)} & \text{if } q \nmid n, \\ 0 & \text{if } q \mid n, \end{cases}$$

where ν denotes the index of n with respect to a primitive root $(\bmod\, q)$; see Section 15.3. The character χ_0 is called the principal character and it satisfies $\chi_0(n) = 0$ if $q|n$ and $\chi_0(n) = 1$ otherwise.

Proof of Theorem 12.13 We have

$$(1 - q^{-s})\zeta_K(s) = \prod_{p\neq q, \mathfrak{p}|p}(1 - (N\mathfrak{p})^{-s})^{-1} = \prod_{p\neq q}(1 - p^{-fs})^{-l}$$

and

$$(1 - q^{-s})\zeta(s)\prod_{\chi\neq\chi_0}L(s, \chi) = \prod_{p\neq q}\prod_{j=0}^{q-2}(1 - \chi_j(p)p^{-s})^{-1}.$$

Now $q - 1 = fl$ and if $p \neq q$ then $p^f \equiv 1 \pmod{q}$ with f minimal. Thus $\nu = kl$ for some integer k when $n = p$ and here $(k, f) = 1$; for $p^{f/(k, f)} \equiv 1 \pmod{q}$

since $vf/(k, f) = (k/(k, f))(q - 1)$. This gives $\chi_j(p) = e^{2\pi ijk/f}$, whence the last product is

$$\prod_{j=0}^{f-1}(1 - e^{2\pi ij/f} p^{-s})^{-l} = (1 - p^{-fs})^{-l}$$

and the theorem follows. □

We remark finally that the L-functions converge for $s = 1$ and we have

$$\prod_{\chi \neq \chi_0} L(1, \chi) = \frac{1}{2}hq^{-q/2}(2\pi)^{(q-1)/2}R,$$

where h is the class number of K and R is the regulator of K.

12.7 Calculation of class numbers

We now give some examples to show how the preceding theories can be used to compute class numbers and to solve certain related Diophantine equations.

Example 12.1 Consider the cyclotomic field $K = \mathbb{Q}(\zeta)$ where $\zeta = e^{2\pi i/5}$. By Theorem 12.7 an integral basis for K is $\zeta, \zeta^2, \zeta^3, \zeta^4$ and by Theorem 12.8 the discriminant of K is 125. Further, the Minkowski constant of K is $3/(2\pi^2)$. Hence every ideal class of K contains an ideal with norm at most $3\sqrt{125}/(2\pi^2) < 2$. This implies that every ideal in K is equivalent to a principal ideal and so is principal. Thus the class number of K is 1 and the field has unique factorization.

Example 12.2 Consider the cubic field $K = \mathbb{Q}(\sqrt[3]{2})$. By Section 10.10 Exercise (ix), an integral basis for K is $1, \sqrt[3]{2}, \sqrt[3]{4}$. Thus by Section 10.7 the discriminant of K is the same as the discriminant of $x^3 - 2$ and this is -108. The Minkowski constant for K is $8/(9\pi)$, whence every ideal class of K contains an ideal with norm at most $8\sqrt{108}/(9\pi) < 3$. Now we have $2 = [\sqrt[3]{2}]^3$ where $[\sqrt[3]{2}]$ has norm 2 and so is a prime ideal as well as being principal. It follows that K has class number 1. Similarly we can treat $K = \mathbb{Q}(\sqrt[3]{3})$. An integral basis for K is $1, \sqrt[3]{3}, \sqrt[3]{9}$ and the discriminant of K is -243. Thus every ideal class contains an ideal with norm at most $8\sqrt{243}/(9\pi) < 5$. We have $3 = [\sqrt[3]{3}]^3$ and $[\sqrt[3]{3}]$ is a prime ideal. Further, we see that $x^3 - 3$ can be expressed as $(x - 1)(1 + x + x^2) - 2$ and the factors here are irreducible mod 2. It follows that $2 = [\sqrt[3]{3} - 1][1 + \sqrt[3]{3} + \sqrt[3]{9}]$ and, by Dedekind's theorem, this gives the factorization of 2 into prime ideals. We conclude that K has class number 1.

Example 12.3 We consider the quadratic field $K = \mathbb{Q}(\sqrt{26})$ and show that this has class number 2. An integral basis for K is $1, \sqrt{26}$ and the minimum polynomial for $\sqrt{26}$, namely $x^2 - 26$, reduces to $x^2 \bmod 2$ and to $(x+1)(x-1) \bmod 5$. Hence by Dedekind's theorem, on taking $\varepsilon = 5 + \sqrt{26}$, we have

$$2 = [2, \varepsilon + 1]^2, \qquad 5 = [5, \varepsilon + 1][5, \varepsilon - 1].$$

Here $[2, \varepsilon + 1]$ and $[5, \varepsilon \pm 1]$ are prime ideals in K with norms 2 and 5. Now $[2, \varepsilon + 1]$ and $[5, \varepsilon + 1]$ divide $[\varepsilon + 1]$ and since $N[\varepsilon + 1] = 10$ we obtain

$$[\varepsilon + 1] = [2, \varepsilon + 1][5, \varepsilon + 1].$$

Further, $[2, \varepsilon + 1]$ is not principal for otherwise we would have $[2, \varepsilon + 1] = [x + \sqrt{26}y]$ for some integers x, y, whence $x^2 - 26y^2 = 2$ and, since a square is congruent to 0 or 1 (mod 4), this is insoluble. We now use the fact that the Minkowski constant for K is $\frac{1}{2}$. This implies that every ideal class in K has an element with norm at most $\frac{1}{2}\sqrt{d}$ where d is the discriminant of K. Since $26 \equiv 2 \pmod 4$ we have $d = 4 \times 26$, whence $\frac{1}{2}\sqrt{d} < 6$ and we need consider only the primes 2, 3 and 5. But $x^2 - 26$ is irreducible mod 3 and so, by Dedekind's theorem, 3 remains prime in K. Thus the ideal class group of K is generated by the class containing $[2, \varepsilon + 1]$ and K has class number 2.

Example 12.4 We note now that ε is the fundamental unit in $K = \mathbb{Q}(\sqrt{26})$. For $x = 5$, $y = 1$ is a solution of $x^2 - 26y^2 = -1$ and so certainly ε is a unit in K. To show that it is fundamental we observe that any solution of $x^2 - 26y^2 = \pm 1$ with $x + \sqrt{26}y > 1$ gives $|x - \sqrt{26}y| < 1$ whence $x > 0$ and $y > 0$; thus if also $x + \sqrt{26}y \leq \varepsilon$ then $y = 1$ and the assertion follows. We apply this to solve the equation

$$x^2 - 26y^2 = \pm 10$$

in integers x, y. The latter gives $[x + \sqrt{26}y][x - \sqrt{26}y] = [10]$ and $[10] = [\varepsilon + 1][\varepsilon - 1]$ is the unique factorization of 10 into principal ideals with the same norm. Thus $[x + \sqrt{26}y] = [\varepsilon \pm 1]$ and Dirichlet's unit theorem now shows that the complete solution is given by

$$x + \sqrt{26}y = \pm \varepsilon^n(\varepsilon \pm 1) \quad (n = 0, \pm 1, \pm 2, \ldots).$$

Example 12.5 Consider the field $K = \mathbb{Q}(\sqrt{(-34)})$. We proceed to show that the ideal class group of K is cyclic of order 4. An integral basis for K is $1, \sqrt{(-34)}$ and the minimum polynomial for $\sqrt{(-34)}$ is $x^2 + 34$. This reduces to $x^2 \bmod 2$, to $x^2 + 1 \bmod 3$ and to $(x + 1)(x - 1) \bmod 5$ and $\bmod 7$. Hence by Dedekind's theorem we have

$$2 = [2,\ \omega - 1]^2, \quad 5 = [5,\ \omega][5,\ \overline{\omega}], \quad 7 = [7,\ \omega][7,\ \overline{\omega}],$$

where $\omega = 1 + \sqrt{(-34)}$, $\overline{\omega} = 1 - \sqrt{(-34)}$ and the factors here are prime ideals. Further, it is obvious on taking $x = 0$, ± 1 that $x^2 + 1$ is irreducible $\bmod 3$ whence 3 is a prime ideal in K. Now $[5,\ \omega]$ and $[7,\ \omega]$ divide $[\omega]$ and we have $N[5,\ \omega] = 5$, $N[7,\ \omega] = 7$ and $N[\omega] = 35$. Thus we obtain $[\omega] = [5,\ \omega][7,\ \omega]$. Similarly $[2,\ \omega + 3]$ and $[5,\ \omega + 3]$ divide $[\omega + 3]$ and we have $N[2,\ \omega + 3] = N[2,\ \omega - 1] = 2$, $N[5,\ \omega + 3] = N[5,\ \overline{\omega}] = 5$ and $N[\omega + 3] = 4^2 + 34 = 50$. Furthermore $[5,\ \omega]$ cannot divide $[\omega + 3]$ for otherwise it would divide $[5,\ \omega + 3]$ contrary to $[5,\ \omega]$ and $[5,\ \overline{\omega}]$ being distinct prime ideals. Thus we obtain

$$[\omega + 3] = [2,\ \omega + 3][5,\ \omega + 3]^2.$$

Now we have $2 = [2,\ \omega + 3]^2$ whence $[5,\ \overline{\omega}]^4 \sim 1$ and $[2,\ \omega + 3] \sim [5,\ \overline{\omega}]^2$. Further, from $5 = [5,\ \omega][5,\ \overline{\omega}]$ we get $[5,\ \omega] \sim [5,\ \overline{\omega}]^3$ and from $[\omega] = [5,\ \omega][7,\ \omega]$ we get $[7,\ \omega] \sim [5,\ \omega]^3 \sim [5,\ \overline{\omega}]$. Furthermore from $7 = [7,\ \omega][7,\ \overline{\omega}]$ we get $[7,\ \overline{\omega}] \sim [5,\ \omega]$. Note that none of the prime factors of 2, 5 and 7 are principal since, for instance, we cannot have $N(a + b\sqrt{(-34)}) = 7$, that is, $a^2 + 34b^2 = 7$ for integers a, b. Now the Minkowski constant for K is $2/\pi$ and so every ideal class of K contains an ideal with norm at most $(2/\pi)\sqrt{|d|}$ where $d = -4 \times 34$ is the discriminant of K. We have $(4/\pi)\sqrt{34} < 8$ and the only non-principal prime ideals in K with norm < 8 are $[2,\ \omega + 3]$, $[5,\ \omega]$, $[5,\ \overline{\omega}]$, $[7,\ \omega]$ and $[7,\ \overline{\omega}]$. We conclude that the class containing $[5,\ \omega]$ generates the ideal class group of K and the latter is cyclic of order 4.

Example 12.6 We shall apply the latter result to solve the equation

$$y^2 = x^3 - 34$$

in integers x, y. First we note that, since 34 is square-free, we have x and y relatively prime and odd. Now $(y + \sqrt{(-34)})(y - \sqrt{(-34)}) = x^3$ and the ideals $[y + \sqrt{(-34)}]$ and $[y - \sqrt{(-34)}]$ are relatively prime; for if they were divisible by a prime ideal $\mathfrak{p}$ then $\mathfrak{p}$ would divide x and $2y$ and this is impossible since, with x odd, $\mathfrak{p}$ cannot divide 2. Hence we obtain $[y + \sqrt{(-34)}] = \mathfrak{a}^3$ for some ideal $\mathfrak{a}$. Now the class number of K is 4 whence $\mathfrak{a}$ must be principal. Further, by Section 7.3, the units in K are just ± 1; indeed it is clear that the equation $a^2 + 34b^2 = \pm 1$ has no solution in integers a, b except with $b = 0$. Thus $y + \sqrt{(-34)} = \pm(u + \sqrt{(-34)}v)^3$ for some integers u, v. But on equating the coefficients of $\sqrt{(-34)}$ we get $v(3u^2 - 34v^2) = 1$ whence $v = \pm 1$

and $3u^2 - 34 = \pm 1$. The latter equation is plainly impossible and we conclude that $y^2 = x^3 - 34$ has no solutions in integers x, y. The argument here is a simple instance of studies on the famous Mordell equation $y^2 = x^3 + k$ which has played a pivotal role in the overall development of number theory; see Section 8.3.

12.8 Local fields

An account of the theory of algebraic numbers would not be complete without a brief indication of the properties of local fields of which p-adic fields are a special case. We shall give here a short description of the rudiments of the subject and refer to the books recommended in Section 12.9 for detailed treatments.

We begin with some valuation theory. Let K be any field. A valuation on K is defined as a real function w such that, for all a, b in K, we have

   (i) $w(a) > 0 \, (a \neq 0)$, $w(0) = 0$,
   (ii) $w(ab) = w(a)w(b)$,
   (iii) $w(a \pm b) \leq w(a) + w(b)$.

If (iii) can be replaced by $w(a \pm b) \leq \max(w(a), w(b))$ then we say that w is non-Archimedean, otherwise Archimedean. Further, we say that valuations w and w' are equivalent if, for some $\lambda > 0$, we have $w(a) = (w'(a))^\lambda$ for all a in K. Now the function $d(x, y) = w(x - y)$ is a metric on K. Let K_w be the completion of K with respect to the metric, that is, the set of residue classes of Cauchy sequences modulo the null sequences. Then w can be extended to the completion by defining $w(\alpha) = \lim w(a_n)$, where $a_1, a_2, \ldots$ is an arbitrary Cauchy sequence in α.

We take now $K = \mathbb{Q}$ the rational field. By the trivial valuation on $\mathbb{Q}$ we mean $w_0(a) = 1 \, (a \neq 0)$ and $w_0(0) = 0$. Another valuation on $\mathbb{Q}$ is given by $w(a) = |a|$ and then the completion of $\mathbb{Q}$ with respect to the valuation, that is, $\mathbb{Q}_w$, is the field of real numbers. Further, if p is any prime, then for $a \in \mathbb{Q}$ we have $a = p^k u/v$ for some integer k and some integers u, v not divisible by p and the function $w(a) = p^{-k}$ defines a further valuation on $\mathbb{Q}$. It is non-Archimedean; indeed in place of (iii) we have

   (iv) $w(a \pm b) \leq \max (w(a), w(b))$ with equality when $w(a) \neq w(b)$.

We write $w(a) = |a|_p$ and we call $|a|_p$ the p-adic valuation on $\mathbb{Q}$. The corresponding completion is denoted by $\mathbb{Q}_p$ and it is referred to as the p-adic number field. A result of Ostrowski of 1918 asserts that the set consisting of

w_0, $|a|$ and the $|a|_p$ as p runs through the primes gives the totality of valuations on $\mathbb{Q}$ up to equivalence.

The elements α in $\mathbb{Q}_p$ with $|\alpha|_p \leq 1$ are called p-adic integers. We prove the following.

Theorem 12.14 *If α is a p-adic integer then there exist rational integers a_0, $a_1, \ldots$ such that $0 \leq a_j < p$ and*

$$\alpha = \sum_{j=0}^{\infty} a_j p^j, \quad \text{that is,} \quad |\alpha - \sum_{j=0}^{N} a_j p^j|_p \to 0 \quad \text{as } N \to \infty.$$

Proof Suppose that $\alpha \neq 0$. Let $\alpha_1, \alpha_2, \ldots$ be any Cauchy sequence in α so that $\alpha_m = p^{k_m} u_m / v_m$ $(m = 1, 2, \ldots)$ for some integer k_m and integers u_m, v_m not divisible by p. Then by (iv) we have $k_m = k$ for some fixed integer $k \geq 0$ and all sufficiently large m. If $k > 0$ we put $a_0 = 0$. If $k = 0$ we write $u_m = u'_m v_m + p u''_m$ with $0 \leq u'_m < p$, as we may since $p \nmid v_m$. Then $u'_m = a_0$ for some fixed a_0 and all sufficiently large m. In either case we have $\alpha = a_0 + p \alpha^{(0)}$, where $\alpha^{(0)}$ is a p-adic integer, and we now continue by expanding $\alpha^{(0)}$ similarly.   □

The p-adic integers form a ring and since, in any given sequence, there are only finitely many choices for the coefficients of the powers of p, we deduce that it is compact.

Now let K be an algebraic number field with degree $n = s + 2t$ as earlier. There exist $s + t$ Archimedean valuations on K given by $w_j(\alpha) = |\sigma_j(\alpha)|$ with $1 \leq j \leq s + t$. The completion of K with respect to w_j is the real or complex field according as $j \leq s$ or $j > s$. Further, if $\mathfrak{p}$ is any prime ideal in K that lies over a rational prime p then, by the theory of fractional ideals, for each $\alpha \neq 0$ in K we have $[\alpha] = \mathfrak{p}^k \mathfrak{u} / \mathfrak{v}$ for some integer k and some ideals $\mathfrak{u}, \mathfrak{v}$ not divisible by $\mathfrak{p}$. The function $w(\alpha) = |\alpha|_{\mathfrak{p}} = p^{-k/e}$, where e is the ramification index of $\mathfrak{p}$, defines a non-Archimedean valuation on K and the completion K_w of K with respect to $|\alpha|_{\mathfrak{p}}$ is called the $\mathfrak{p}$-adic number field. As for $\mathbb{Q}$, it can be shown that, up to equivalence, $w_0(\alpha)$, the $|\sigma_j(\alpha)|$ and the $|\alpha|_{\mathfrak{p}}$, as $\mathfrak{p}$ runs through the prime ideals, are the only valuations on K. The valuations are sometimes denoted simply by $|\alpha|_v$ and one describes v as infinite, written $v \mid \infty$, if the completion is the real or complex field and finite, written $v \mid p$, if the completion is a $\mathfrak{p}$-adic field.

We now take π to be an element in $\mathfrak{p}$ but not in $\mathfrak{p}^2$. Then the elements in the ring of integers of $K_{\mathfrak{p}}$ can be written in the form $\sum_{j=0}^{\infty} \alpha_j \pi^j$ for some $\alpha_0, \alpha_1, \ldots$ in a set of representatives of $\mathcal{O}_K / \mathfrak{p}$. We have $p = \varepsilon \pi^e$ where ε is a $\mathfrak{p}$-adic unit in $\mathcal{O}_K$, that is, an element of $\mathcal{O}_K$ not divisible by π, and

$|\pi^e|_{\mathfrak{p}} = p^{-1}$; hence we can identify $\mathbb{Q}_p$ with a subfield of $K_{\mathfrak{p}}$ by mapping $\sum a_j p^j$ to $\sum a_j \varepsilon^j \pi^{ej}$. Then we have $[K_{\mathfrak{p}} : \mathbb{Q}_p] = ef$ where f is the degree of $\mathfrak{p}$; indeed a basis for $K_{\mathfrak{p}}$ as a vector space over $\mathbb{Q}_p$ is given by $\omega_i \pi^l$ with $1 \le i \le f$, $0 \le l < e$, where $\omega_1, \ldots, \omega_f$ are representative elements of a basis for $\mathcal{O}_K / \mathfrak{p}$ as an algebraic extension of $\mathbb{F}_p$. We have the following result.

Theorem 12.15 (Product formula) *For $\alpha \ne 0$ in K we have*

$$\prod_v |\alpha|_v^{d_v} = 1,$$

where the product is over all inequivalent valuations v of K and d_v is the degree of the corresponding completion (so that $d_v = 1$ or 2 if the completion is real or complex).

Proof Let p be a rational prime. It is easily seen that

$$|N(\alpha)| \prod_p |N(\alpha)|_p = 1 \quad \text{and} \quad |N(\alpha)| = \prod_{v \mid \infty} |\alpha|_v^{d_v}.$$

Now let $\mathfrak{p}$ be a prime ideal that lies over p and let $f_{\mathfrak{p}}$ be the degree of $\mathfrak{p}$ so that $N\mathfrak{p} = p^{f_{\mathfrak{p}}}$. Further, let $k_{\mathfrak{p}}$ be the exponent of $\mathfrak{p}$ when the fractional ideal $[\alpha]$ is expressed as a canonical product of prime ideals; here $k_{\mathfrak{p}}$ may be positive, negative or 0. Then we have

$$|N(\alpha)|_p = \prod_{\mathfrak{p} \mid p} |N\mathfrak{p}|_p^{k_{\mathfrak{p}}} = \prod_{\mathfrak{p} \mid p} p^{-f_{\mathfrak{p}} k_{\mathfrak{p}}}.$$

Thus if we write $d_{\mathfrak{p}} = e_{\mathfrak{p}} f_{\mathfrak{p}}$ with $e_{\mathfrak{p}}$ the ramification index of $\mathfrak{p}$, so that $d_{\mathfrak{p}} = [K_{\mathfrak{p}} : \mathbb{Q}_p]$, then we obtain

$$|N(\alpha)|_p = \prod_{\mathfrak{p} \mid p} |\alpha|_{\mathfrak{p}}^{d_{\mathfrak{p}}} = \prod_{v \mid p} |\alpha|_v^{d_v}$$

and the result follows. $\qquad\square$

The subject of local fields was originated by Hensel in the early part of the twentieth century and it has now become an important instrument in many branches of number theory. Not least among these are studies in Diophantine analysis and transcendence theory, in cyclotomic fields and p-adic L-functions and in the theory of quadratic forms. There is a particular result that pays homage to Hensel and it is known as Hensel's lemma. As Hensel was aware, the theory of local fields is intimately related to the theory of higher congruences; indeed if $f(x)$ is a polynomial with integer coefficients then the congruence $f(x) \equiv 0 \pmod{p^m}$, where p is any prime, has a solution for all positive integers m if and only if the equation $f(x)=0$ has a solution in p-adic integers; this follows easily from the definition of the p-adic valuation and the compactness property of the p-adic integers. Hensel's lemma concerns approximations to roots of polynomials and shows that in certain circumstances it suffices if the congruence has a solution in the case $m = 1$; thus, as it is sometimes said, one can Hensel one's way upwards. We have already met an example of this kind in Section 3.6.

Theorem 12.16 (Hensel's lemma) *Let $K_{\mathfrak{p}}$ be a $\mathfrak{p}$-adic field and let $\mathcal{O}_{\mathfrak{p}}$ be the ring of $\mathfrak{p}$-adic integers in $K_{\mathfrak{p}}$. Suppose that $f(x)$ is a polynomial in $\mathcal{O}_{\mathfrak{p}}[x]$ and that there exists an element a in $\mathcal{O}_{\mathfrak{p}}$ such that $|f(a)|_{\mathfrak{p}} < |f'(a)|_{\mathfrak{p}}^2$ where $f'(x)$ denotes the derivative of $f(x)$. Then the sequence $a_0, a_1, a_2, \ldots$ where $a_0 = a$ and*

$$a_{j+1} = a_j - f(a_j)/f'(a_j) \qquad (j=0,1,2,\ldots)$$

converges $\mathfrak{p}$-adically to an element of $\mathcal{O}_{\mathfrak{p}}$ and this is a zero of $f(x)$.

Proof The argument is an analogue of the Newton approximation method for real numbers. For simplicity we shall omit the suffix $\mathfrak{p}$ in writing the valuations. We assume that $|f(a_i)| < |f'(a_i)|^2$ for all $i \le j$ and we proceed to prove this with j replaced by $j+1$. Denoting by $f^{(r)}(x)$ the rth derivative of $f(x)$, we have

$$f(a_{j+1}) = \sum_{r=2}^{\infty} \frac{f^{(r)}(a_j)}{r!}(a_{j+1} - a_j)^r;$$

for obviously the terms with $r=0$ and $r=1$ cancel. Now $f^{(r)}(x)/r!$ has coefficients in $\mathcal{O}_{\mathfrak{p}}$ and, by the inductive hypothesis, we have

$$|a_{j+1} - a_j| < |f'(a_j)| \le 1.$$

Hence we obtain

$$|f(a_{j+1})| \le |a_{j+1} - a_j|^2 < |f'(a_j)|^2.$$

Further, from properties of polynomials, we see that

$$|f'(a_{j+1}) - f'(a_j)| \le |a_{j+1} - a_j| < |f'(a_j)|,$$

whence $|f'(a_{j+1})| = |f'(a_j)|$ and so $|f(a_{j+1})| < |f'(a_{j+1})|^2$. But, from the above series with j replaced by $j - 1$, we obtain

$$|a_{j+1} - a_j| = |f(a_j)/f'(a_j)| \le |a_j - a_{j-1}|^2/|f'(a)|.$$

Thus we conclude by induction that $|a_{j+1} - a_j| \to 0$ as $j \to \infty$, whence $a_0, a_1, a_2, \ldots$ is a Cauchy sequence and it converges to a zero of $f(x)$ as required. $\qquad\square$

To give an example, suppose that $p \ne 5$ is a prime and that $u \equiv v^5 \pmod{p}$ for some integers u, v not divisible by p. Then on applying Hensel's lemma to the polynomial $x^5 - u$ we deduce that $u = a^5$ for some p-adic integer a in $\mathbb{Q}_p$.

Local fields play an important role in connection with studies on the Hasse principle or, as it is alternatively called, the local–global principle. It concerns Diophantine equations and it is said to hold if a sufficient condition for the existence of a non-trivial solution in an algebraic number field K is that the equation has such a solution in all the completions of K both Archimedean and non-Archimedean; in other words if solutions everywhere locally imply a solution globally. The condition is certainly necessary and, although there are known counter-examples and indeed in most situations the principle is thought not to hold, it has nevertheless been verified in a few notable instances. The main one is due to Hasse himself, who showed it to be true for quadratic forms defined over K in any number of variables. Here we shall prove the following special case of Hasse's result which actually goes back to work of Legendre.

Theorem 12.17 *Let a, b, c be integers with abc odd and square-free and suppose that for every prime p the congruence $ax^2 + by^2 + cz^2 \equiv 0 \pmod{p^2}$ is soluble in integers x, y, z not all divisible by p. Then there exist integers x, y, z not all 0 such that $ax^2 + by^2 + cz^2 = 0$.*

Proof We first show that the hypotheses imply that there exist integers r, s, t such that

$$ar^2 + b \equiv 0 \pmod{c}, \quad bs^2 + c \equiv 0 \pmod{a}, \quad ct^2 + a \equiv 0 \pmod{b}.$$

Suppose, for instance, that p divides c. Then there exist integers x, y not both divisible by p such that $ax^2 + by^2 \equiv 0 \pmod{p}$. Now since abc is square-free it follows that neither x nor y is divisible by p, whence there exists an integer r_p such that $r_p y \equiv x \pmod{p}$. Thus we have $ar_p^2 + b \equiv 0 \pmod{p}$. By the Chinese remainder theorem there exists an integer r such that $r \equiv r_p \pmod{p}$ for all p that divide c; this gives the first of the asserted congruences and the other two follow similarly. Note that, since again abc is square-free, we have $(r, c) = (s, a) = (t, b) = 1$. Note also that, by hypothesis, we have $ax^2 + by^2 + cz^2$ divisible by 4 for some integers x, y, z not all even and, since a square is congruent to 0 or 1(mod 4), it follows that just one of them is even; we shall assume that it is z whence $a + b \equiv 0 \pmod 4$.

Now let l be a solution of the congruences $bl \equiv ra \pmod c$ and $bl \equiv a \pmod 2$; this exists by another application of the Chinese remainder theorem. Further let m be a solution of $tm \equiv 2 \pmod b$ and let n be a solution of the congruences $n \equiv 2s \pmod a$, $rn \equiv m \pmod c$ and $n \equiv m \pmod 2$. Then the points (x, y, x) of the lattice Λ in $\mathbb{R}^3$ with basis

$$(2bc, 0, 0), \quad (bl, a, 0), \quad (m, n, 2)$$

satisfy

$$ry \equiv x \pmod c, \quad sz \equiv y \pmod a, \quad tx \equiv z \pmod b.$$

and $x \equiv y, z \equiv 0 \pmod 2$. Further, the determinant $d(\Lambda)$ of the lattice is $4|abc|$. We take S as the convex body, symmetric about the origin, given by

$$|a|x^2 + |b|y^2 + |c|z^2 < 4|abc|.$$

It has volume $(8\pi/3)(4|abc|) > 8d(\Lambda)$. Thus, by Minkowski's theorem, S contains a point (x, y, z) of Λ other than the origin. But $ax^2 + by^2 + cz^2$ is congruent to $(ar^2 + b)y^2 \equiv 0 \pmod c$ and similar congruences hold $\pmod a$ and $\pmod b$ whence, since x and y have the same parity, z is even and, by assumption, $a + b \equiv 0 \pmod 4$, it is divisible by $4abc$ and the theorem follows. $\square$

12.9 Further reading

There are several substantial books by Henri Cohen that emphasize the computational aspects of algebraic number theory. The most recent and perhaps closest in spirit to the text here are *Number Theory*, Vols I and II (Springer, 2007). As regards Section 12.8, especially recommended is Cassels, *Local Fields* (Cambridge University Press, 1986). We mention also Cassels, *Rational Quadratic Forms* (Academic Press, 1978), already cited in Section 5.6, which covers more advanced topics in the area. The book by Borevich and Shafarevich referred to in Section 8.7 contains a full and accessible account of the subject. For a classic and very sophisticated treatment see Serre, *Local Fields* (Springer, 1979).

The proof given above of Theorem 12.17 is due to Cassels; an exposition is included in both of his books referred to above. We note, however, that only the case when abc is odd is treated here and consequently our argument is

slightly simpler. To cover the even case one needs to impose the additional
hypothesis that the quadratic form has a non-trivial solution (mod 8). Further,
as Cassels points out, the fact that a, b, c cannot all be of the same sign must
necessarily be implied by the hypotheses since there is no assumption about
the existence of a non-trivial real solution. The explicit basis for the lattice
derived in the proof of Theorem 12.17 is not from Cassels; it comes from a
paper of Davenport and Hall in *Quart. J. Math. Oxford* **19** (1948), 189–192.

12.10 Exercises

(i) Show that in a number field K there are only finitely many ideals with a
given norm. Hence verify the assertion in Section 8.1 about the solutions
of the equation $x^2 - dy^2 = k$.

(ii) By Section 10.10 Exercise (ix), an integral basis for $K = \mathbb{Q}(\sqrt[3]{2})$ is
$1, \sqrt[3]{2}, \sqrt[3]{4}$. Verify that $\varepsilon = 1 + \sqrt[3]{2} + \sqrt[3]{4}$ is a unit in K. Assuming
that it is the fundamental unit, show that all solutions in integers x, y, z
of the equation

$$x^3 + 2y^3 + 4z^3 - 6xyz = 1$$

are given by $x + \sqrt[3]{2}y + \sqrt[3]{4}z = \varepsilon^n$ for $n = 0, \pm1, \pm2, \ldots$. Calculate the
particular solution for $n = 2$.

(iii) Let $K = \mathbb{Q}(\sqrt{(-23)})$. Factorize the primes 2 and 3 in K and verify that

$$[2, \omega][3, \omega] = [\omega],$$

where $\omega = \frac{1}{2}(1 + \sqrt{(-23)})$. Hence prove that K has class number 3.

(iv) Show that, in the field $K = \mathbb{Q}(\sqrt{11})$, the prime ideal factors of 2 and 3
are principal. Deduce that K has class number 1.

(v) Verify that $\varepsilon = 10 + 3\sqrt{11}$ is the fundamental unit in $K = \mathbb{Q}(\sqrt{11})$.
Hence show that all solutions in integers x, y of the equation $x^2 - 11y^2 = -2$ are given by

$$x - \sqrt{11}y = \pm\varepsilon^n(\sqrt{11} \pm 3) \quad (n = 0, \pm1, \pm2, \ldots).$$

(vi) Factorize the primes 2 and 3 in the field $K = \mathbb{Q}(\sqrt{(-17)})$. Verify that
5 remains prime in K. Show that $[\omega] = [2, \omega][3, \omega]^2$, where $\omega = 1 + \sqrt{(-17)}$. Hence prove that the ideal class group of K is cyclic of order 4.

(vii) Show that, in the quadratic field $K = \mathbb{Q}(\sqrt{d})$, a prime p ramifies if and
only if p divides the discriminant of K. For a given prime q determine
all quadratic fields K in which q is the only ramified prime.

(viii) Show that $\mathbb{Q}(\sqrt{(-5)})$ has class number 2. Hence prove that the equation $y^2 + 5 = x^3$ has no solutions in integers x, y.

(ix) Using the Minkowski constant, show that the discriminant of any algebraic number field other than $\mathbb{Q}$ exceeds 1.

(x) Let $\alpha = e^{2\pi i/7} + e^{-2\pi i/7}$. Show that the minimum polynomial of α is $f(x) = x^3 + x^2 - 2x - 1$. Verify that $27 f(x) = y^3 - 21y - 7$ with $y = 3x + 1$ and hence show that the discriminant of $K = \mathbb{Q}(\alpha)$ is 49. Deduce that K has class number 1.

(xi) Let α be a zero of $f(x) = x^3 - 4x + 1$. Show that the discriminant of f is positive and square-free. Hence, from Section 10.10 Exercise (viii) and Section 10.7, verify that $K = \mathbb{Q}(\alpha)$ is totally real and that it has an integral basis $1, \alpha, \alpha^2$. Using the Minkowski constant, show that K has class number 1.

(xii) Show that if p is a prime and $p \neq 2$ or 17 then one at least of 2, 17 and 34 is a quadratic residue (mod p). Hence, using Hensel's lemma, verify that the equation $(x^2 - 2)(x^2 - 17)(x^2 - 34) = 0$ has a solution in the reals and in every p-adic field but not in the rationals.

13

Analytic number theory

13.1 Introduction

Analytic number theory in its classical form is concerned with studies on the distribution of the primes. However, any technique that involves the application of mathematical analysis to the solution of number-theoretical problems can come under this heading. We have already introduced the subject in Sections 1.6 and 2.8. In particular we referred there to the prime-number theorem, to the Riemann zeta-function, to primes in arithmetical progressions and to sieve methods and their applications. This and subsequent chapters will be devoted to expanded accounts of these topics; we begin here with a brief history to help set them in context.

Euclid (c. 300 BC): existence of infinitely many primes.

Legendre (1788): asserted that every arithmetical progression $a, a + q, a + 2q, \ldots$ with $(a, q) = 1$ includes infinitely many primes. He gave no proof. Also conjectured (1808) that $\pi(x)$, the number of primes $\leq x$, is 'approximately' $x/(\log x - 1.08\ldots)$ so that $\pi(x) \log x/x \to 1$ as $x \to \infty$.

Dirichlet (1839): established Legendre's assertion on arithmetical progressions; the work introduced L-functions, characters, class number formulae etc.

Gauss (1849): modified Legendre's conjecture of 1808 to $\pi(x)/\mathrm{li}\, x \to 1$ as $x \to \infty$, where $\mathrm{li}\, x = \int_2^x dt/\log t$.

Tchebychev (1852): proved that $\pi(x) \log x/x$ lies between two bounds close to 1 for sufficiently large x. Thus verified Bertrand's postulate that, for any positive integer n, there exists a prime p such that $n < p \leq 2n$.

Riemann (1860): introduced the zeta-function $\zeta(s)$ for a complex variable s
and demonstrated the fundamental connection with the distribution of
the primes.

Mertens (1874): applied Tchebychev's results to sums and products involving
primes.

Hadamard and de la Vallée Poussin (1896): proved, independently, Gauss'
conjecture and so, in particular, the prime-number theorem.

von Mangoldt (1905): established a conjecture of Riemann on the number of
zeros of $\zeta(s)$.

Landau (1909): laid the modern foundations of analytic number theory with
the publication of his *Handbuch*.

Hardy and Littlewood (early 1900s): numerous studies on number-theoretical
functions. Particularly noted for the introduction of the circle method
based at least in part on ideas of Ramanujan.

Vinogradov (1937): refined the circle method and proved that every sufficiently large odd integer is the sum of three primes.

Selberg and Erdős (1948): gave an 'elementary' proof of the prime-number
theorem.

Brun (1920): derived a new method inspired by the sieve of Eratosthenes.
Following further works of Selberg, Linnik, Rényi and others, sieve
methods now feature routinely throughout analytic number theory.
They provide, through work of Chen Jing-Run, the best approach to
date to the famous Goldbach and twin-prime conjectures.

Davis, Robinson, Putnam and Matiyasevich (1971): showed by techniques
from logic that the set of prime numbers is Diophantine, that is, there
exists a polynomial of several variables, with integer coefficients, such
that the primes are precisely the set of positive values assumed by the
polynomial as the variables run through all integers ≥ 0.

Green and Tao (2004): applied combinatorial methods to show that there exist
arbitrarily long arithmetic progressions consisting only of primes.

13.2 Dirichlet series

By a Dirichlet series we mean an expression of the form $\sum_{n=1}^{\infty} a_n/n^s$, where
$a_1, a_2, \ldots$ are real or complex numbers and $s = \sigma + it$ is a complex variable.
There is a unique σ_0, possibly $\pm\infty$, such that the series converges for every
s with $\sigma > \sigma_0$ and no s with $\sigma < \sigma_0$; the quantity σ_0 is called the abscissa of
convergence and $\sigma = \sigma_0$ is called the line of convergence. Similarly there exists
a unique abscissa σ_0^* of absolute convergence. Any pair of Dirichlet series

$\sum a_n/n^s$ and $\sum b_n/n^s$ can be multiplied together to give $\sum c_n/n^s$, where $c_n = \sum_{d|n} a_d b_{n/d}$, provided that $\sigma > \sigma^*$ where σ^* is the maximum of the abscissae of absolute convergence.

The Riemann zeta-function, introduced in Section 2.8, is the basic example of a Dirichlet series and plainly it has $\sigma_0 = \sigma_0^* = 1$. Further, as noted in Section 2.8, there are some simple relations between $\zeta(s)$ and other Dirichlet series. For instance we have, for $\sigma > 1$,

$$\frac{\zeta(2s)}{\zeta(s)} = \sum_{n=1}^{\infty} \frac{\lambda(n)}{n^s},$$

where λ is the Liouville function defined as $(-1)^{\Omega(n)}$ with $\Omega(n)$ given by the total number of prime divisors of n for $n > 1$ and $\Omega(1) = 0$.

The particular Dirichlet series where $a_n = 1$ if n is a prime p and 0 otherwise has much historical interest. Indeed, as already remarked in Section 1.6, the following theorem relating to it was first proved by Euler in 1737 and provided another verification of the infinity of the sequence of primes. Our demonstration follows an argument due to Clarkson given in 1966.

Theorem 13.1 *The series $\sum(1/p)$ summed over all primes p diverges.*

Proof Suppose that the series converges. Then there is an integer k such that $\sum_{n>k}(1/p_n) < \frac{1}{2}$, where, as usual, p_n denotes the nth prime. Let $P = p_1 \cdots p_k$. Since none of the numbers $1 + nP$ for $n = 1, 2, 3, \ldots$ is divisible by any of $p_1, \ldots, p_k$ it follows that all their prime factors occur in the sequence p_{k+1}, $p_{k+2}, \ldots$. Thus, for each integer $m \geq 1$, we have

$$\sum_{n=1}^{m} \frac{1}{1+nP} \leq \sum_{j=1}^{\infty} \left(\sum_{n=k+1}^{\infty} \frac{1}{p_n} \right)^j.$$

The right-hand side is less than $\sum_{j=1}^{\infty} (\frac{1}{2})^j = 1$, whence the series $\sum_{n=1}^{\infty} 1/(1+nP)$ has bounded partial sums and so converges. But by the comparison or integral test, for example, the series in fact diverges and the contradiction establishes the theorem. $\qquad\square$

The first general results on the distribution of the primes were obtained by Tchebychev[†] in 1852. They were based on a study of the binomial coefficient

$$\binom{2n}{n} = \frac{(2n)!}{(n!)^2}$$

[†] The name has been transliterated in various ways; Chebyshev is another common form.

and depended on the formula $\sum_{j=1}^{\infty}[n/p^j]$ given in Section 2.1 for the exponent to which a prime p divides $n!$. Thus Tchebychev proved that $\pi(x)$, the
number of primes $p \leq x$, satisfies

$$ax/\log x < \pi(x) < bx/\log x,$$

where a and b are positive constants. A notable application was obtained by
Mertens in 1874 who showed that $\sum_{p\leq x}(1/p) \sim \log\log x$ and indeed gave
the more precise expression $\log\log x + c + O(1/\log x)$ as recorded in Section
1.6. The Tchebychev estimates, in a refined form, also yielded a proof of the
famous Bertrand postulate to the effect that if n is a positive integer then there
is always a prime p satisfying $n < p \leq 2n$.[‡] We shall discuss Tchebychev's
work and its ramifications in Sections 13.3–13.6.

Another example of a Dirichlet series is the L-function

$$L(s, \chi) = \sum_{n=1}^{\infty} \frac{\chi(n)}{n^s},$$

where χ is a Dirichlet character $(\bmod\, q)$, that is, a completely multiplicative function defined on the positive integers, periodic with period q and with
$\chi(n) = 0$ for $(n, q) > 1$. An account of the theory of L-functions will be given
in Chapter 15. It is closely linked to studies on primes in arithmetical progressions; in particular, it furnishes Dirichlet's famous theorem that there exist infinitely many primes in any arithmetical progression $a, a + q, a + 2q, \ldots$ with
$(a, q) = 1$. There is an analogue of the prime-number theorem to the effect that
the number of primes p in the arithmetical progression with $p \leq x$ is asymptotic to $(1/\phi(q))x/\log x$. Moreover, the sum of the reciprocals of the primes
in the arithmetical progression diverges, for we have, for some c depending on
a and q,

$$\sum_{\substack{p \leq x \\ p \equiv a(\bmod q)}} \frac{1}{p} = \frac{1}{\phi(q)} \log\log x + c + O\left(\frac{1}{\log x}\right).$$

The theory of zeta-functions and L-functions has been widely generalized;
in particular, the functions can be defined for any algebraic number field.
We have already given some indication of their properties in the case of the
quadratic and cyclotomic fields in Sections 12.5 and 12.6.

[‡] A well-known rhyme due to N. J. Fine remarks 'Tchebychev said it and I'll say it again:
there's always a prime between n and $2n$'.

13.3 Tchebychev's estimates

The basic result of Tchebychev, which he established in 1852, is as follows.

Theorem 13.2 (Tchebychev) *There exist $a > 0$ and $b > 0$ such that for all $x \geq 2$ we have*

$$ax/\log x < \pi(x) < bx/\log x.$$

Proof For any positive integer n the binomial coefficient

$$C = \binom{2n}{n} = \frac{(2n)!}{(n!)^2}$$

is the largest term in the expansion of $(1+1)^{2n}$ and so

$$2^{2n}/(2n+1) \leq C \leq 2^{2n}.$$

Now by Section 2.1 the exponent to which any prime p divides $n!$ is $\sum_{j=1}^{\infty}$ $[n/p^j]$. Thus the exponent to which p divides C is

$$\sum_{j=1}^{\infty} \{[2n/p^j] - 2[n/p^j]\}.$$

Each term in the above sum is either 0 or 1 according as the fractional part of n/p^j is or is not less than $\frac{1}{2}$. Further, the number of non-zero terms is at most $(\log(2n))/\log p$ and the value of the sum is 1 if p satisfies $n < p \leq 2n$. It follows that C cannot exceed

$$\prod_{p \leq 2n} p^{(\log 2n)/\log p} = (2n)^{\pi(2n)}$$

but is certainly divisible by each prime p with $n < p \leq 2n$. Hence we obtain

$$n^{\pi(2n)-\pi(n)} \leq C \leq (2n)^{\pi(2n)}.$$

Comparison with the estimates for C recorded at the beginning gives

$$\pi(2n) \geq (2n\log 2 - \log(2n+1))/\log(2n)$$

and

$$\pi(2n) - \pi(n) \leq 2n\log 2/\log n.$$

The first inequality implies that $\pi(x) > ax/\log x$ for some $a > 0$ and from the second inequality we get $\pi(2x) - \pi(x) < cx/\log x$ for some $c > 0$.

To complete the proof we note that

$$\pi(x) = \sum_{j=1}^{\infty} \{\pi(x/2^{j-1}) - \pi(x/2^{j})\}.$$

Now if $2^j \le \sqrt{x}$ then the jth term in the sum is at most $cx/(2^{j-1}\log x)$ and the number of non-zero terms is at most $\log x/\log 2$; if $2^j > \sqrt{x}$ we have

$$\pi(x/2^{j-1}) < x/2^{j-1} < 2\sqrt{x}.$$

Thus we see that

$$\pi(x) < (cx/\log x)\sum_{j=1}^{\infty}(1/2^{j-1}) + 2\sqrt{x}(\log x/\log 2) < bx/\log x$$

for some $b > 0$ as required. ☐

Let now p_n denote the nth prime in ascending order of magnitude so that $p_1 = 2$, $p_2 = 3, \ldots$. An immediate corollary to Theorem 13.2 is the following.

Theorem 13.3 *There exist $a' > 0$ and $b' > 0$ such that for all n we have*

$$a'n\log n < p_n < b'n\log n.$$

Proof We take $x = p_n$ in Theorem 13.2. Since $\pi(p_n) = n$ we have

$$ap_n/\log p_n < n < bp_n/\log p_n,$$

whence

$$(1/b)n\log p_n < p_n < (1/a)n\log p_n.$$

But, for sufficiently large n, we have $\log p_n < \sqrt{p_n}$ and thus $\sqrt{p_n} < (1/a)n$, that is, $p_n < (n/a)^2$. This gives $\log p_n < c'\log n$ for some $c' > 0$ and, since certainly $p_n > n$ whence $\log p_n > \log n$ for all n, the theorem follows. ☐

Bertrand conjectured in 1845 that, for every positive integer n, there is always a prime p with $n < p \le 2n$; this is the famous Bertrand's postulate to which we alluded in Section 1.6 and again in Section 13.2. It gives in particular $p_{n+1} < 2p_n$. The conjecture was first proved by Tchebychev. In fact, by Theorem 13.2, we have

$$\pi(2x) - \pi(x) > (2a - b - 2a\log 2/\log(2x))x/\log x$$

and Tchebychev showed that one can take $a = 0 \cdot 92129\ldots$ and $b = 1 \cdot 1055\ldots$ if $x \ge 30$. Hence, in the latter range, we have $\pi(2x) > \pi(x)$ and there exists a prime p with $x < p \le 2x$; it is readily checked that this holds also for $2 \le x < 30$.

A relatively short proof of the postulate along these lines, due to Erdős, is given in the book by Hardy and Wright referred to in Section 1.7, and another proof, due to Pillai, can be found in the book of Chandrasekharan of 1968 referred to in Section 2.9. As mentioned in Section 1.6, Bertrand's postulate has initiated a large body of research on the difference $p_{n+1} - p_n$ between consecutive primes.

13.4 Partial summation formula

In order to discuss further applications of Theorem 13.2 we shall need the following result from real analysis.

Theorem 13.4 *Let $a_1, a_2, \ldots$ be any real sequence and let $s(x) = \sum_{n \le x} a_n$. Further, let $f(x)$ be a real function with continuous derivative $f'(x)$ for all real $x > 0$. Then*

$$\sum_{n \le x} a_n f(n) = s(x) f(x) - \int_1^x s(u) f'(u) \, du.$$

Proof We have $s(0) = 0$ and $s(x) = s([x])$. Hence

$$\sum_{n \le x} a_n f(n) = \sum_{n \le x} (s(n) - s(n-1)) f(n)$$

$$= \sum_{n=1}^{[x]-1} s(n)(f(n) - f(n+1)) + s(x) f([x]).$$

The latter can be expressed in the form

$$-\sum_{n=1}^{[x]-1} \int_n^{n+1} s(u) f'(u) \, du - s(x) \int_{[x]}^x f'(u) \, du + s(x) f(x)$$

and the result follows. ☐

Alternatively, we can argue that

$$s(x) f(x) - \sum_{n \le x} a_n f(n) = \sum_{n \le x} a_n (f(x) - f(n)) = \sum_{n \le x} \int_n^x a_n f'(u) \, du.$$

On defining $f_n(u)$ as $a_n f'(u)$ if $u \ge n$ and as 0 otherwise, the last sum becomes

$$\sum_{n \le x} \int_1^x f_n(u) \, du = \int_1^x \left(\sum_{n \le x} f_n(u) \right) du$$

and the expression on the right is

$$\int_1^x \left(\sum_{n \le u} a_n f'(u) \right) du = \int_1^x s(u) f'(u) \, du.$$

As a simple application of Theorem 13.4 let us take $f(x) = \log x$ and $a_1 = a_2 = \cdots = 1$. Then $s(x) = [x]$ and we obtain

$$\sum_{n \le x} \log n = [x] \log x - \int_1^x ([u]/u) \, du = x \log x - x + O(\log x).$$

13.5 Mertens' results

In 1874 Mertens applied Tchebychev's estimates to give asymptotic expressions for certain sums and products involving primes.

Theorem 13.5 *We have*

$$\sum_{p \le x} \frac{\log p}{p} = \log x + O(1).$$

Proof By Section 2.1, since $[[x]/p^j] = [x/p^j]$, the exponent to which a prime p divides $[x]!$ is $\sum_{j=1}^{\infty}[x/p^j]$. It follows that

$$\sum_{n \le x} \log n = \log[x]! = \sum_{p \le x} \sum_{j=1}^{\infty} \log p \, [x/p^j].$$

The contribution from the term $j = 1$ is

$$\sum_{p \le x} \log p \, [x/p] = x \sum_{p \le x} (\log p)/p + O(\pi(x) \log x)$$

and, by Theorem 13.2, the error term here is $O(x)$. The remaining terms contribute

$$\sum_{p \le x} \sum_{j=2}^{\infty} \log p \, [x/p^j] \le x \sum_{2 \le n \le x} (\log n)/(n(n-1))$$

which is $O(x)$. Thus

$$\sum_{n \le x} \log n = x \sum_{p \le x} (\log p)/p + O(x).$$

But from the example given at the end of the preceding section the sum on the left is $x \log x + O(x)$ and the result follows. $\qquad\square$

Theorem 13.6 *For some constant c we have, as $x \to \infty$,*

$$\sum_{p \leq x} 1/p = \log \log x + c + O(1/\log x).$$

Proof We shall apply Theorem 13.4 with $f(x) = 1/\log x$ and with $a_n = (\log n)/n$ if n is a prime p and 0 otherwise; then $\sum_{n \leq x} a_n f(n) = \sum_{p \leq x} 1/p$. It will be noted that $f(x)$ is defined and has continuous derivative only for $x > 1$ (and not for $x > 0$ as in the theorem) but since $a_1 = 0$ this is immaterial. Now by Theorem 13.5 we have $s(x) = \log x + \tau(x)$ where $\tau(x) = O(1)$ and so $s(x)f(x) = 1 + O(1/\log x)$. Further, we have

$$-\int_1^x s(u)f'(u)\,du = \int_2^x \frac{s(u)du}{u(\log u)^2} = \int_2^x \frac{du}{u \log u} + \int_2^x \frac{\tau(u)du}{u(\log u)^2}.$$

The first integral on the right evaluates to $\log \log x - \log \log 2$ and the second is $c' + O(1/\log x)$ where

$$c' = \int_2^\infty \frac{\tau(u)du}{u(\log u)^2}.$$

This proves the theorem with $c = 1 - \log \log 2 + c'$. □

The same method of proof shows that, for any $\delta > 0$, the series $\sum 1/(p(\log p)^\delta)$ summed over all primes p converges. Indeed, on defining a_n as above and $f(x) = (\log x)^{-(1+\delta)}$, the partial summation formula gives

$$\sum_{p \leq x} \frac{1}{p(\log p)^\delta} = \frac{s(x)}{(\log x)^{1+\delta}} + (1+\delta) \int_2^x \frac{s(u)du}{u(\log u)^{2+\delta}}$$

and, since $s(u) = O(\log u)$ and

$$\delta \int_2^x \frac{du}{u(\log u)^{1+\delta}} = \frac{1}{(\log 2)^\delta} - \frac{1}{(\log x)^\delta} = O(1),$$

the series in question has positive terms and is bounded above.

Theorem 13.7 *For some $b > 0$, we have*

$$\prod_{p \leq x} (1 - 1/p)^{-1} = b \log x + O(1).$$

Proof Clearly we have

$$\prod_{p \leq x} (1 - 1/p)^{-1} = \prod_{p \leq x} \exp(-\log(1 - 1/p)) = \exp\left(\sum_{p \leq x} (1/p + \rho_p)\right),$$

where $\rho_p = 1/(2p^2) + 1/(3p^3) + \cdots$. Now the series $\sum_p \rho_p$ converges and indeed $\sum_{p>x} \rho_p$ does not exceed $\sum_{n>x} 1/(n(n-1)) = O(1/x)$. On applying Theorem 13.6, we obtain

$$\exp\left(\sum_{p \le x} 1/p\right) = e^c \log x(1 + O(1/\log x))$$

and the theorem follows.   □

Precise values for the c and b appearing in Theorems 13.6 and 13.7 can be determined. In fact we have $b = e^\gamma$ and

$$c = \gamma + \sum_p \{\log(1 - 1/p) + 1/p\},$$

where γ is Euler's constant.

Theorem 13.8 *If $\pi(x) \log x/x$ tends to a limit as $x \to \infty$ then the limit must be* 1.

Proof By partial summation, we have

$$\sum_{p \le x} \frac{1}{p} = \frac{\pi(x)}{x} + \int_2^x \frac{\pi(u)}{u^2}\, du.$$

If $\pi(x) \log x/x \to l$ as $x \to \infty$ then the right-hand side is asymptotic to

$$l \int_2^x \frac{du}{u \log u} \sim l \log \log x$$

as $x \to \infty$. But, by Theorem 13.6, the left-hand side is asymptotic to $\log \log x$. Hence $l = 1$ as required.   □

13.6 The Tchebychev functions

In his original memoirs Tchebychev introduced functions $\theta(x)$ and $\psi(x)$ which are now classical and commonly feature in studies involving primes. They are defined by

$$\theta(x) = \sum_{p \le x} \log p, \quad \psi(x) = \sum_{p^m \le x} \log p,$$

where the first sum is over all primes $p \le x$ and the second is over all primes p and positive integers m with $p^m \le x$. Clearly, since $p \ge 2$, we have $m \le \log x/\log 2$ and thus

$$\psi(x) = \sum_{m \le \log x/\log 2} \theta(x^{1/m}).$$

Further, we observe that $e^{\psi(x)}$ is the lowest common multiple of the integers $1, 2, \ldots, [x]$.

Theorem 13.2 gives at once analogous results for the Tchebychev functions. In this context it is convenient to use Vinogradov's notation, introduced in Section 8.2: by $f \ll g$ for functions f and g we shall mean $f < cg$ for some constant $c > 0$. Then for all $x \geq 2$ we have $x \ll \theta(x) \ll x$ and $x \ll \psi(x) \ll x$. For certainly $\theta(x) \leq \pi(x) \log x$ and, for any δ with $0 < \delta < 1$, we see that

$$\theta(x) \geq \sum_{x^\delta < p \leq x} \log p \geq \delta \log x \, \{\pi(x) - \pi(x^\delta)\}.$$

Now, by Theorem 13.2, we have $\pi(x^\delta) < x^\delta = o(\pi(x))$. Further, for $m \geq 2$, we have $\theta(x^{1/m}) \leq \sqrt{x} \log x$ and so

$$\psi(x) - \theta(x) \ll \sqrt{x}(\log x)^2.$$

Plainly the inequalities here show that, as $x \to \infty$,

$$\pi(x) \sim \theta(x)/\log x \sim \psi(x)/\log x.$$

Finally we introduce the von Mangoldt function $\Lambda(n)$, which is defined as $\log p$ if n is a power of a prime p and as 0 otherwise. Then we have

$$\psi(x) = \sum_{n \leq x} \Lambda(n).$$

We note also that

$$\sum_{d \mid n} \Lambda(d) = \sum_{p^m \mid n} \log p = \log n.$$

Hence from the properties of Dirichlet series referred to in Section 13.2 we obtain, for $\sigma > 1$,

$$\frac{\zeta'(s)}{\zeta(s)} = -\sum_{n=1}^{\infty} \frac{\Lambda(n)}{n^s}.$$

The expression on the left is called the logarithmic derivative of the Riemann zeta-function.

13.7 The irrationality of $\zeta(3)$

As remarked in Section 6.6, Apéry proved in 1978 that $\zeta(3)$ is irrational; it was a surprising and notable achievement and, so far at least, it has turned out to be a rather singular result; no one has yet established the irrationality of $\zeta(5)$

or $\zeta(2n+1)$ for integers $n > 1$. In contrast, as is well known, the values of the Riemann zeta-function at positive even integers are given by

$$\zeta(2n) = (-1)^{n-1} 2^{2n-1} \pi^{2n} B_{2n}/(2n)!,$$

where B_{2n} is the Bernoulli number defined by $t/(e^t - 1) = \sum_{n=0}^{\infty} B_n t^n/n!$. Thus $\zeta(2) = \pi^2/6$, $\zeta(4) = \pi^4/90$ and in general $\zeta(2n)$ is a rational multiple of π^{2n} and so, in particular, transcendental. We shall give here a short proof of Apéry's result following work of Beukers of around the same time.

The proof rests on a study of the integral

$$I = \int_0^1 \int_0^1 \int_0^1 \frac{f^m dx dy dz}{1 - (1-xy)z},$$

where m is a positive integer and

$$f = \frac{xyz(1-x)(1-y)(1-z)}{1 - (1-xy)z}.$$

We shall show that the value of I for $m = 0$ is $2\zeta(3)$ and that in general we have

$$I = (a_m + b_m \zeta(3))/c_m^3$$

for some integers a_m and b_m, where c_m is the lowest common multiple of $1, 2, \ldots, m$. By Tchebychev's estimates with refined explicit constants we obtain $c_m = e^{\psi(m)} \leq 3^m$. Further, by solving the equations

$$\frac{\partial f}{\partial x} = \frac{\partial f}{\partial y} = \frac{\partial f}{\partial z} = 0,$$

we find that the maximum of the function f over the range of integration is $(\sqrt{2}-1)^4$; the bound is attained when $x = y = \sqrt{2}-1$ and $z = 1/\sqrt{2}$. Thus we get $I \leq 2\zeta(3)(\sqrt{2}-1)^{4m}$. Since $27(\sqrt{2}-1)^4 < 1$ and clearly $I > 0$ it follows that $a_m + b_m \zeta(3) \to 0$ as $m \to \infty$. But if $\zeta(3)$ were rational, say a/b with a, b positive integers, then the expression would be at least $1/b$. The contradiction establishes the result.

To verify the assertions about I, we observe that integrating partially m times with respect to y gives

$$I = \int_0^1 \int_0^1 \int_0^1 \frac{(1-x)^m (1-z)^m P_m(y)}{1 - (1-xy)z} dx dy dz,$$

where

$$P_m(y) = \frac{1}{m!} \frac{d^m}{dy^m} (y^m (1-y)^m)$$

so that $(-1)^m P_m(\frac{1}{2}(y+1))$ is the Legendre polynomial. The substitution

$$w = \frac{1-z}{1-(1-xy)z}$$

then gives

$$I = \int_0^1 \int_0^1 \int_0^1 \frac{(xyw(1-x))^m P_m(y)}{(1-(1-xy)w)^{m+1}} dxdydw$$

and on integrating partially m times with respect to x we obtain

$$I = \int_0^1 \int_0^1 \int_0^1 \frac{P_m(x)P_m(y)}{1-(1-xy)w} dxdydw.$$

Now on integrating with respect to w we get

$$I = -\int_0^1 \int_0^1 \frac{P_m(x)P_m(y)\log(xy)}{1-xy} dxdy.$$

Since the polynomials $P_m(x)$ and $P_m(y)$ have integer coefficients and degree m it follows that I is a linear combination with integer coefficients of the integrals

$$I_{rs} = \int_0^1 \int_0^1 \frac{x^r y^s \log(xy)}{1-xy} dxdy \quad (0 \leq r, s \leq m).$$

But I_{rs} is the derivative with respect to t of the integral

$$J_{rs} = \int_0^1 \int_0^1 \frac{x^{r+t} y^{s+t}}{1-xy} dxdy$$

evaluated at $t=0$ and, on writing $(1-xy)^{-1} = 1 + xy + (xy)^2 + \cdots$, we see that

$$J_{rs} = \sum_{j=1}^\infty (r+j+t)^{-1}(s+j+t)^{-1}.$$

Hence if $r=s$ then $I_{rs} = -2(\zeta(3) - \sum_{j=1}^r j^{-3})$. If $r > s$ then J_{rs} can be expressed as a finite sum, namely $(r-s)^{-1} \sum_{j=1}^{r-s}(s+j+t)^{-1}$; this gives $I_{rs} = -(r-s)^{-1} \sum_{j=1}^{r-s}(s+j)^{-2}$ and a similar expression holds if $s > r$. We have $I = -I_{00}$ when $m=0$ and the initial assertions concerning I are now clear.

13.8 Further reading

Several books on the subject matter here have already been mentioned in Sections 1.7 and 2.9. We now add the works of W. and F. Ellison, *Prime*

Numbers (Wiley, Hermann, 1985); M. N. Huxley, *The Distribution of the Prime Numbers* (Clarendon Press, 1972); A. E. Ingham, *The Distribution of Prime Numbers* (Cambridge University Press, 1990); H. Iwaniec and E. Kowalski, *Analytic Number Theory* (AMS Colloquium Publications, 2004); and M. R. Murty, *Problems in Analytic Number Theory* (Springer, 2008). The classic *Handbuch* to which we referred in Section 13.1 is Landau's *Handbuch der Lehre von der Verteilung der Primzahlen* (Chelsea Publishing, 1953). This contains references to all fundamental material including the papers of Tchebychev.

The proof of Theorem 13.1 is due to Clarkson in *Proc. Amer. Math. Soc.* **17** (1966), 541; it is reproduced in the book by Apostol referred to in Section 2.9. As regards Theorem 13.2, improved values for the constants a and b were given by Sylvester in 1892 and by Rosser and Schoenfeld in *Illinois J. Math.* **6** (1962), 64–94; *Math. Comp.* **29** (1975), 243–269. For an excellent compilation of results of this kind see the book by P. Ribenboim, *The New Book of Prime Number Records* (Springer, 1996). He mentions, for instance, that $\pi(x)$ has been computed for certain very large x and that in particular, in 1994, M. Deléglise and J. Rivat gave the value $\pi(10^{18}) = 24\,739\,954\,287\,740\,860$.

The paper of Beukers referred to in Section 13.8 appeared in *Bull. London Math. Soc.* **11** (1979), 268–272. Rivoal, in *C. R. Acad. Sci. Paris Sér I Math.* **331** (2000), 267–270, and Ball and Rivoal, in *Invent Math.* **146** (2001), 193–207, have used a new construction of transcendence type to show that there are infinitely many irrational values among the $\zeta(2n + 1)$ with $n > 1$. Zudilin, in *Uspekhi Mat. Nauk* **56** (2001), 149–150 (translated as *Russian Math. Surveys* **56** (2001), 774–776), has subsequently developed this sphere of ideas to prove that at least one of $\zeta(5)$, $\zeta(7)$, $\zeta(9)$ and $\zeta(11)$ is irrational.

13.9 Exercises

(i) Let $\sigma_k(n) = \sum_{d|n} d^k$ for an integer k. Show that if $\sigma > \max(1, k + 1)$ then

$$\sum_{n=1}^{\infty} \frac{\sigma_k(n)}{n^s} = \zeta(s)\zeta(s - k).$$

(ii) Show that the Liouville function $\lambda(n)$ is multiplicative. Hence verify the assertion about $\sum \lambda(n)/n^s$ in Section 13.2.

(iii) Prove that, for $\kappa > 1$,

$$\sum_{p \leq x} \frac{(\log p)^{\kappa}}{p} = \frac{1}{\kappa}(\log x)^{\kappa} + O((\log x)^{\kappa - 1}).$$

Verify that the latter holds also for $0 < \kappa < 1$ with an error term $O(1)$.

(iv) Show that $\sum (1/p(\log \log p)^\delta)$ converges for $\delta > 1$.

(v) Prove that, as $x \to \infty$,

$$\sum_{p \le x} 1/(p \log \log p) \sim \log \log \log x.$$

(vi) Let $\omega(n)$ denote the number of distinct prime divisors of n. Prove that, for some constant c,

$$\sum_{n \le x} \omega(n) = x \log \log x + cx + O(x/\log x).$$

(vii) A function $\rho(n)$ is defined on the positive integers by $\rho(1) = 0$ and, for $n > 1$, by $\rho(n) = \log(p_1 \cdots p_k)$, where $p_1, \ldots, p_k$ denote the distinct prime factors of n. Prove that

$$\sum_{n \le x} \rho(n) = x \log x + O(x).$$

(viii) Show that the number of primes q that divide an integer $n > 3$ and exceed $\log n$ is at most $\log n / \log \log n$. Hence verify that $\prod_q (1 - 1/q) > c > 0$ for some constant c. Deduce that Euler's totient function $\phi(n)$ satisfies $\phi(n) > c'n/\log \log n$ for some constant $c' > 0$.

(ix) Using the Euler product proved in Section 2.8 and the value for b referred to after Theorem 13.7, show that, as $x \to \infty$,

$$\prod_{p \le x} (1 + 1/p) \sim (6e^\gamma/\pi^2) \log x.$$

(x) Prove that

$$\sum_{n \le x} \log n = \sum_{n \le x} [x/n] \Lambda(n).$$

Hence show that

$$\sum_{n \le x} \frac{\Lambda(n)}{n} = \log x + O(1).$$

14

On the zeros of the zeta-function

14.1 Introduction

We recall from Section 2.8 that the Riemann zeta-function is defined for any complex number $s = \sigma + it$ with $\sigma > 1$ by the equation

$$\zeta(s) = \sum_{n=1}^{\infty} \frac{1}{n^s}.$$

The series converges in the region of definition and indeed uniformly for $\sigma > 1 + \delta$ with $\delta > 0$. The function can be continued analytically to the region $\sigma > -1$ by the equation

$$\zeta(s) = s \int_1^{\infty} \frac{f(x)}{x^{s+1}} dx + \frac{1}{s-1} + \frac{1}{2},$$

where $f(x)$ is the 'saw-tooth' function given by

$$f(x) = [x] - x + \tfrac{1}{2}$$

and $x^{s+1} = \exp((s+1)\log x)$ with $\log x$ real; the integral converges uniformly for $\sigma > \delta > 0$, since clearly $f(x) = \tfrac{1}{2} - \{x\}$, where $\{x\}$ denotes the fractional part of x, whence $|f(x)| \le \tfrac{1}{2}$ for all x. Indeed the integral converges for all s with $\sigma > -1$ and uniformly for $\sigma > -1 + \delta > -1$; it therefore defines an analytic function of s and the equation serves to define $\zeta(s)$ in this range. For we have

$$\int_X^Y \frac{f(x)}{x^{s+1}} dx = \left[\frac{F(x)}{x^{s+1}} \right]_X^Y + (s+1) \int_X^Y \frac{F(x)}{x^{s+2}} dx,$$

162

where $F(x) = \int_1^x f(u)\,du$, and, by virtue of the boundedness of $F(x)$, the terms on the right tend to 0 as $X, Y \to \infty$.

To verify the analytic continuation, that is, to show that the expression above for $\zeta(s)$ reduces to the original series when $\sigma > 1$, we observe that

$$\zeta(s) = s \sum_{n=1}^{\infty} \int_n^{n+1} \frac{f(x)}{x^{s+1}} dx + \frac{1}{s-1} + \frac{1}{2},$$

and the integral here is

$$\int_n^{n+1} \frac{n - x + \frac{1}{2}}{x^{s+1}} dx = \left[\frac{1}{(s-1)x^{s-1}} - \frac{n + \frac{1}{2}}{sx^s} \right]_n^{n+1}.$$

It is easily seen that the number on the right can be written in the form

$$\left(\frac{1}{s} - \frac{1}{s-1} \right) \left(\frac{1}{n^{s-1}} - \frac{1}{(n+1)^{s-1}} \right) + \frac{1}{2s} \left(\frac{1}{n^s} + \frac{1}{(n+1)^s} \right)$$

and on rearranging the absolutely convergent series which thus arise we obtain the required identity

$$\zeta(s) = \sum_{n=1}^{\infty} \frac{1}{n^s} \qquad (\sigma > 1).$$

14.2 The functional equation

We remarked in Section 2.8 that $\zeta(s)$ satisfies a functional equation $\Xi(s) = \Xi(1-s)$, where $\Xi(s) = \pi^{-\frac{1}{2}s} \Gamma(\frac{1}{2}s)\zeta(s)$. This we now establish.

First we observe that if $-1 < \sigma < 0$ we have

$$\zeta(s) = s \int_0^{\infty} \frac{f(x)}{x^{s+1}} dx;$$

for if $\sigma < 0$ then

$$s \int_0^1 \frac{f(x)}{x^{s+1}} dx = \frac{1}{s-1} + \frac{1}{2}.$$

Now $f(x)$ can be expressed as a Fourier series

$$\sum_{n=1}^{\infty} \frac{\sin(2n\pi x)}{n\pi},$$

and, on substituting in the above expression for $\zeta(s)$ and integrating term by term, a process justified since the series is boundedly convergent, we obtain

$$\zeta(s) = \frac{s}{\pi}\sum_{n=1}^{\infty}\frac{1}{n}\int_{0}^{\infty}\frac{\sin(2n\pi x)}{x^{s+1}}\,dx$$

$$= \frac{s}{\pi}\sum_{n=1}^{\infty}\frac{(2n\pi)^{s}}{n}\int_{0}^{\infty}\frac{\sin y}{y^{s+1}}\,dy$$

$$= \frac{s}{\pi}(2\pi)^{s}(-\Gamma(-s))\sin(\tfrac{1}{2}s\pi)\zeta(1-s).$$

On using $z\Gamma(z)=\Gamma(z+1)$, this simplifies to

$$\zeta(s) = 2^{s}\pi^{s-1}\sin(\tfrac{1}{2}s\pi)\Gamma(1-s)\zeta(1-s).$$

Now recalling that $\Xi(s)=\pi^{-\frac{1}{2}s}\Gamma(\tfrac{1}{2}s)\zeta(s)$ and using the fact that

$$\Gamma(\tfrac{1}{2}-\tfrac{1}{2}s)/\Gamma(\tfrac{1}{2}s)=\pi^{-1/2}2^{s}\sin(\tfrac{1}{2}s\pi)\Gamma(1-s)$$

we obtain the famous functional equation

$$\Xi(s) = \Xi(1-s)$$

satisfied by $\zeta(s)$ as required. We have proved the validity only for $-1<\sigma<0$ but, since the right-hand side is analytic throughout the half-plane $\sigma<0$, the equation defines the analytic continuation of $\zeta(s)$ for $\sigma\le-1$ and it is then valid for all s. The functional equation shows that $\zeta(s)$ is analytic throughout the complex plane except for a simple pole at $s=1$ with residue 1. It is customary to define $\xi(s)=\tfrac{1}{2}s(s-1)\Xi(s)$; then $\xi(s)$ is an entire function and it satisfies $\xi(s)=\xi(1-s)$.

In the proof of the functional equation above, we used the fact that

$$\int_{0}^{\infty}\frac{\sin y}{y^{s+1}}\,dy = -\sin(\tfrac{1}{2}s\pi)\Gamma(-s).$$

This is readily verified by classical complex integration methods. We have

$$\int_{0}^{\infty}\frac{\sin y}{y^{s+1}}\,dy = \frac{1}{2i}\left\{\int_{0}^{\infty}\frac{e^{iy}}{y^{s+1}}\,dy - \int_{0}^{\infty}\frac{e^{-iy}}{y^{s+1}}\,dy\right\}.$$

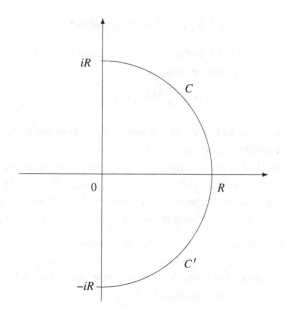

Fig. 14.1 The contour of integration in the evaluation of the integral.

Let C denote the contour consisting of the arc from R to iR of the circle centre the origin and radius R together with the axes from iR to 0 and 0 to R; see Fig. 14.1. Then we have

$$\int_C \frac{e^{iz}}{z^{s+1}}\,dz = \int_0^R \frac{e^{iy}}{y^{s+1}}\,dy - i\int_0^R \frac{e^{-t}}{(it)^{s+1}}\,dt + \int_0^{\frac{\pi}{2}} \frac{e^{iRe^{i\theta}}}{(Re^{i\theta})^{s+1}}iRe^{i\theta}\,d\theta.$$

By Jordan's lemma, the last integral on the right is $o(1)$ as $R \to \infty$. Similarly, defining C' as the reflection of C in the real axis as shown, we obtain

$$\int_{C'} \frac{e^{-iz}}{z^{s+1}}\,dz = \int_0^R \frac{e^{-iy}}{y^{s+1}}\,dy + i\int_0^R \frac{e^{-t}}{(-it)^{s+1}}\,dt + o(1).$$

The integrands on the left are analytic except at the origin, whence

$$\int_0^\infty \frac{\sin y}{y^{s+1}}\,dy = \frac{1}{2}\left\{\int_0^\infty \frac{e^{-t}}{(it)^{s+1}}\,dt + \int_0^\infty \frac{e^{-t}}{(-it)^{s+1}}\,dt\right\}$$

$$= \frac{1}{2}\left\{\frac{1}{i^{s+1}}\Gamma(-s) + \frac{1}{(-i)^{s+1}}\Gamma(-s)\right\}$$

and the result follows since

$$\frac{1}{2i}(i^{-s} - (-i)^{-s}) = \frac{1}{2i}(e^{-\frac{1}{2}is\pi} - e^{\frac{1}{2}is\pi}) = -\sin(\tfrac{1}{2}s\pi).$$

14.3 The Euler product

We observed in Section 2.8 that when $\sigma > 1$ the function $\zeta(s)$ can be expressed as a product over the primes p, namely

$$\zeta(s) = \prod_p (1 - 1/p^s)^{-1}.$$

The expression is due to Euler and it is essentially equivalent to the fundamental theorem of arithmetic; see Section 1.5.

The Euler product converges for $\sigma > 1$ and none of the factors vanish; hence we have $\zeta(s) \neq 0$ for $\sigma > 1$. The zero-free region was extended by Hadamard and de la Vallée Poussin; in particular they obtained the following result which formed the key to their celebrated proofs of the prime-number theorem.

Theorem 14.1   *We have $\zeta(s) \neq 0$ on the line $\sigma = 1$.*

Proof Let t be any real number, let σ be any number such that $1 < \sigma \leq 2$, and put $s = \sigma + it$. By the Euler product we have

$$\log \zeta(s) = -\sum_p \log(1 - 1/p^s) = \sum_p \sum_{j=1}^{\infty} (1/j) p^{-js}$$

and thus

$$\log |\zeta(s)| = \sum_p \sum_{j=1}^{\infty} (1/j) p^{-j\sigma} \cos(jt \log p).$$

Now since

$$3 + 4\cos\theta + \cos 2\theta = 2(1 + \cos\theta)^2 \geq 0$$

we obtain

$$3\log |\zeta(\sigma)| + 4\log |\zeta(\sigma + it)| + \log |\zeta(\sigma + 2it)| \geq 0,$$

that is,

$$|(\zeta(\sigma))^3 (\zeta(\sigma + it))^4 \zeta(\sigma + 2it)| \geq 1.$$

But since $\zeta(s)$ has a simple pole at $s = 1$ we have $|\zeta(\sigma)| < c(\sigma - 1)^{-1}$ for some absolute constant c and since $\zeta(s)$ is regular at $s = \sigma + 2it$ we have $|\zeta(\sigma + 2it)| < c'$ for some $c' = c'(t)$ independent of σ. Further, if $\zeta(s)$ possessed a zero at $s = 1 + it$ we would have $|\zeta(\sigma + it)| < c''(\sigma - 1)$ for some $c'' = c''(t)$ and so the expression on the left of the above inequality would be less than $c^3 c' c''^4 (\sigma - 1)$. This clearly gives a contradiction if σ is sufficiently near to 1, the degree of proximity depending on t, and the contradiction proves the theorem. $\square$

14.4 On the logarithmic derivative of $\zeta(s)$

We recall from Section 13.6 that the logarithmic derivative of $\zeta(s)$ is given by

$$\frac{\zeta'(s)}{\zeta(s)} = -\sum_{n=1}^{\infty} \frac{\Lambda(n)}{n^s} \quad (\sigma > 1),$$

where Λ is the von Mangoldt function introduced in Section 13.6. We now establish some estimates for $\zeta(s)$, $\zeta'(s)$ and $\zeta'(s)/\zeta(s)$ and we proceed to derive a basic relation between the latter and the Tchebychev function $\psi(x)$. We shall use the Vinogradov notation referred to in Sections 8.2 and 14.4 so that by $f(t) \ll g(t)$ for real non-negative functions $f(t)$ and $g(t)$ we shall mean $f(t) < cg(t)$ for some absolute constant $c > 0$ and similarly by $f(t) \gg g(t)$ we shall mean $f(t) > cg(t)$ for some $c > 0$.

Theorem 14.2 *For any* $s = \sigma + it$, *with* $\sigma \geq 1 - 2/\log|t| > \frac{1}{2}$, *we have*

$$|\zeta(s)| \ll \log|t|.$$

Proof It is easily verified by dividing $[1, N]$ into unit intervals and integrating by parts that, for any positive integer N,

$$s \int_1^N \frac{f(x)}{x^{s+1}} dx + \frac{1}{s-1} + \frac{1}{2} = \sum_{n=1}^N \frac{1}{n^s} + \frac{N^{1-s}}{s-1} - \frac{1}{2} N^{-s}$$

and so

$$\zeta(s) = \sum_{n=1}^N \frac{1}{n^s} + \frac{N^{1-s}}{s-1} - \frac{1}{2} N^{-s} + s \int_N^{\infty} \frac{f(x)}{x^{s+1}} dx.$$

Thus we have

$$|\zeta(s)| \leq \sum_{n=1}^N \frac{1}{n^\sigma} + \frac{N^{1-\sigma}}{|t|} + \frac{1}{2} N^{-\sigma} + \frac{1}{2}|s| \int_N^{\infty} \frac{dx}{x^{\sigma+1}}.$$

Now, on assuming that $\sigma > \rho = 1 - 2/\log|t|$ and that $\log|t| > 4$ and then taking N so that $|t| < N < 2|t|$, we see that the sum on the right cannot exceed

$$\sum_{n=1}^N n^{-\rho} \ll N^{1-\rho}/(1-\rho) \ll \log|t|.$$

Further, the last term on the right is given by

$$\tfrac{1}{2}|s|\sigma^{-1} N^{-\sigma} \ll |t| N^{-\sigma} \ll 1$$

and the two remaining terms are also $\ll 1$. The required estimate for $\zeta(s)$ follows. $\qquad\square$

Theorem 14.3 *For any $s = \sigma + it$, with $\sigma \geq 1 - 1/\log|t| > \frac{3}{4}$, we have*

$$|\zeta'(s)| \ll (\log|t|)^2.$$

Proof We use the equation

$$\zeta'(s) = \frac{1}{2\pi i}\int_C \frac{\zeta(s+z)}{z^2}\,dz$$

where C denotes the positively oriented circle, centre the origin, with radius $1/\log|t|$. The equation follows from Cauchy's theorem, on observing that $\zeta(s+z)$ is regular within and on C. For any z on C, it is clear that the real part of $s+z$ exceeds $1 - 2/\log|t|$ and so, by Theorem 14.2, we have $|\zeta(s+z)| \ll \log|t|$. It follows that the integrand has absolute value $\ll (\log|t|)^3$ and this gives the required result. □

Theorem 14.4 *For any $s = \sigma + it$ with $\sigma \geq 1$ and $|t| \gg 1$ we have*

$$|\zeta'(s)/\zeta(s)| \ll (\log|t|)^{10}.$$

Proof By taking the logarithmic derivative of the Euler product we see that

$$\frac{\zeta'(s)}{\zeta(s)} = -\sum_p \sum_{j=1}^\infty p^{-js}\log p$$

for $\sigma > 1$, and so $|\zeta'(s)/\zeta(s)| \leq |\zeta'(2)/\zeta(2)|$ if $\sigma \geq 2$. Hence we can assume that $1 \leq \sigma \leq 2$. We now put $\sigma' = \sigma + (\log|t|)^{-10}$. Then certainly $\sigma' > 1$ and so, as in the proof of Theorem 14.1, we obtain

$$|(\zeta(\sigma'))^3(\zeta(\sigma'+it))^4\zeta(\sigma'+2it)| \geq 1.$$

By Theorem 14.2 together with the inequality $|\zeta(\sigma')| \ll (\sigma'-1)^{-1}$, this gives

$$|\zeta(\sigma'+it)|^4 \gg (\sigma'-1)^3/\log|t|.$$

Hence, since $\sigma'-1 \geq \sigma'-\sigma = (\log|t|)^{-10}$, we get $|\zeta(\sigma'+it)| \gg (\log|t|)^{-\frac{31}{4}}$. Further, from the equation

$$\zeta(s) = \zeta(\sigma'+it) - \int_\sigma^{\sigma'} \zeta'(u+it)\,du,$$

we see from Theorem 14.3 that

$$|\zeta(\sigma'+it)| - |\zeta(s)| \ll (\sigma'-\sigma)(\log|t|)^2 = (\log|t|)^{-8}.$$

Thus we have $|\zeta(s)| \gg (\log|t|)^{-\frac{31}{4}}$ for $|t|$ sufficiently large and this gives $|\zeta'(s)/\zeta(s)| \ll (\log|t|)^{10}$ as required. □

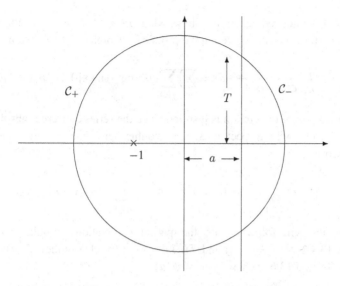

Fig. 14.2   The contours C_+ and C_- in the proof of Theorem 14.5.

Theorem 14.5 *For any $x > 0$ we have*

$$\int_0^x \psi(u)\,du = -\frac{1}{2\pi i}\int_C \frac{\zeta'(s)}{\zeta(s)}\frac{x^{s+1}}{s(s+1)}\,ds,$$

where C denotes the straight line $a + it$ with $a > 1$ and $-\infty < t < \infty$.

Proof We note first that

$$\frac{1}{2\pi i}\int_C \frac{x^{s+1}}{s(s+1)}\,ds = \max(0, x - 1).$$

For by Cauchy's theorem we have, for any $T > 1$,

$$\int_{a-iT}^{a+iT} \frac{x^{s+1}}{s(s+1)}\,ds = 2\pi i\,\max(0, x - 1) + \int_{\mathcal{C}} \frac{x^{s+1}}{s(s+1)}\,ds,$$

where $\mathcal{C}$ ($=C_+$ or C_-) denotes the arc of the circle with centre the origin and radius $\sqrt{(T^2 + a^2)}$, to the left or right of the line C, and described clockwise or anti-clockwise respectively according as $x > 1$ or $x \le 1$; see Fig. 14.2. The absolute value of the integral on the right is at most $2\pi x^{a+1}(T - 1)^{-1}$ in both cases, and the desired equation follows when $T \to \infty$. Now on recalling the expression for $\zeta'(s)/\zeta(s)$ given in the proof of Theorem 14.4 we obtain

$$-\frac{1}{2\pi i} \int_C \frac{\zeta'(s)}{\zeta(s)} \frac{x^{s+1}}{s(s+1)} \, ds = \sum_p \sum_{j=1}^{\infty} (p^j \log p) \max\{0, (x/p^j) - 1\};$$

here term-by-term integration is justified since the series converges absolutely and uniformly with respect to s. The double series on the right can be expressed as

$$\sum_p \sum_{j=1}^{\infty} (\log p) \max(0, x - p^j) = \sum_{n \leq x} \Lambda(n)(x - n)$$

and the theorem follows from the partial summation formula given in Section 13.4 with $a_n = \Lambda(n)$ and $f(u) = x - u$; for clearly then $f'(u) = -1$ and, by Section 13.6, we have $s(u) = \psi(u)$. □

14.5 The Riemann hypothesis

We have already referred to the subject here in Section 2.8; we now give a brief resumé and make some supplementary remarks.

We recall from Section 14.3 that, by virtue of the Euler product, $\zeta(s)$ has no zeros in the half-plane $\sigma > 1$ and, by Theorem 14.1, the function also has no zeros on the line $\sigma = 1$. It follows from the functional equation that the only zeros of $\zeta(s)$ in the region $\sigma \leq 0$ are given by the poles $s = -2, -4, -6, \ldots$ of $\Gamma(\frac{1}{2}s)$ and these are termed the 'trivial' zeros. All other zeros must lie in the 'critical strip' given by $0 \leq \sigma \leq 1$, indeed they must satisfy $0 < \sigma < 1$, and Riemann conjectured in his path-breaking memoir of 1861 that they in fact lie on the line $\sigma = \frac{1}{2}$. This is the now famous Riemann hypothesis. It remains as yet unproved but there is considerable computational and other evidence in favour of it. In particular it is known to be valid for the first trillion zeros above the real axis. Furthermore, in 1915, Hardy showed that there are infinitely many zeros on the line $\sigma = \frac{1}{2}$ and Selberg proved that indeed a positive proportion of the zeros, in a certain sense, have this property; results of Levinson, and later Conrey, have established that this proportion is at least $\frac{2}{5}$.

Nevertheless, only a very limited zero-free region within the critical strip has been demonstrated to date. In 1899 de la Vallée Poussin showed that $\zeta(s) \neq 0$ for $\sigma > 1 - c/\log|t|$ with some $c > 0$ and, apart from an improvement reducing $\log|t|$ to $(\log|t|)^{\frac{2}{3}}(\log\log|t|)^{\frac{1}{3}}$ by Vinogradov and Korobov in 1958, there has been no further progress. Indeed, even the conjecture that $\zeta(s) \neq 0$ for $\sigma > 1 - \delta$

with some $\delta > 0$, the so-called 'quasi-Riemann' hypothesis, is beyond our reach at present.

It is known that the Riemann hypothesis is equivalent to the assertion that

$$\pi(x) = \text{li}\, x + O(\sqrt{x}\log x),$$

where, for any positive integer j,

$$\text{li}\, x = \int_2^x \frac{dt}{\log t} = \frac{x}{\log x} + \frac{x}{(\log x)^2} + \cdots + \frac{j!\, x}{(\log x)^{j+1}} (1 + o(x)).$$

The work of de la Vallée Poussin gave the estimate

$$\pi(x) = \text{li}\, x + O(x e^{-c\sqrt{\log x}})$$

for some constant $c > 0$; Vinogradov and Korobov improved the exponent $\sqrt{\log x}$ to $(\log x)^{\frac{3}{5}}(\log\log x)^{-\frac{1}{5}}$ and this is the best result to date. As remarked in Section 2.8, another assertion known to be equivalent to the Riemann hypothesis is

$$\sum_{n \le x} \mu(n) = O(x^{\frac{1}{2}+\varepsilon})$$

for any $\varepsilon > 0$.

We conclude the section with a cautionary tale. It had been conjectured on the basis of numerical evidence that $\pi(x) < \text{li}\, x$ for all large x. This was disproved by Littlewood in 1914. He showed that

$$\pi(x) - \text{li}\, x = \Omega_{\pm}(\sqrt{x}\log\log\log x / \log x).$$

Here we are using the Ω-notation: $f(x) = \Omega_{\pm}(g(x))$ means that $f(x) \gg g(x)$ for a sequence of values of x tending to infinity and that $f(x) \ll -g(x)$ similarly. Littlewood's proof was divided into two cases according as the Riemann hypothesis is true or false; owing to the indirect character, it did not make it possible to specify a particular x_0 such that $\pi(x) > \text{li}\, x$ for some $x < x_0$. However, in 1955, Skewes gave the value $x_0 = \exp\exp\exp\exp(7 \cdot 7)$ and, for a while, this 'Skewes number' was quite famous. Later works of Lehman and te Riele reduced the bound to $6 \cdot 69 \times 10^{370}$; even so, the story suggests that a zero of $\zeta(s)$ off the line $\sigma = \frac{1}{2}$ with a very large ordinate cannot be discounted.

14.6 Explicit formula for $\zeta'(s)/\zeta(s)$

In Section 14.2 we introduced the function $\xi(s) = \frac{1}{2}s(s-1)\Xi(s)$, where $\Xi(s) = \pi^{-\frac{1}{2}s}\Gamma(\frac{1}{2}s)\zeta(s)$, and we observed that it is entire and satisfies $\xi(s) = \xi(1-s)$.

Now clearly the zeros of $\xi(s)$ are just the non-trivial zeros ρ of $\zeta(s)$. Further, by Stirling's formula, we have $\log|\Gamma(\frac{1}{2}s)| \ll |s|\log|s|$. Furthermore, if $\sigma \geq \frac{1}{2}$, then the integral defining $\zeta(s)$ is uniformly bounded, whence we obtain $|\zeta(s)| \ll |s|$ as $|s| \to \infty$. It follows that $\log|\xi(s)| \ll |s|\log|s|$. This must in fact hold for all s in view of the functional equation and it shows that $\xi(s)$ is an entire function of order 1.

We can now apply Hadamard's factorization theorem and deduce that

$$\xi(s) = e^{A+Bs}\prod_{\rho}(1-s/\rho)e^{s/\rho}$$

for some absolute constants A and B. On taking the logarithmic derivative we get

$$\frac{\xi'(s)}{\xi(s)} = B + \sum_{\rho}\left(\frac{1}{s-\rho} + \frac{1}{\rho}\right).$$

Further, by the definition of $\xi(s)$ and the fact that $z\Gamma(z) = \Gamma(z+1)$, we obtain

$$\frac{\zeta'(s)}{\zeta(s)} = \frac{1}{2}\log\pi - \frac{1}{s-1} - \frac{1}{2}\frac{\Gamma'(\frac{1}{2}s+1)}{\Gamma(\frac{1}{2}s+1)} + \frac{\xi'(s)}{\xi(s)}.$$

Thus we have an explicit formula for $\zeta'(s)/\zeta(s)$ in terms of the set of zeros ρ of $\zeta(s)$ in the critical strip. The Γ-term can be expressed as

$$\frac{1}{2}\gamma + \sum_{n=1}^{\infty}\left(\frac{1}{s+2n} - \frac{1}{2n}\right)$$

and so represents the contribution of the trivial zeros. Moreover the value for B can be calculated as $-\frac{1}{2}\gamma - 1 + \frac{1}{2}\log 4\pi$.

The explicit formula has found wide application. As a first instance, we shall use it to give the classical zero-free region of the zeta-function mentioned in Section 14.5. We shall adopt the customary notation $\rho = \beta + i\gamma$ to denote the real and imaginary parts of ρ. Then the real part of $1/\rho$ is $\beta/(\beta^2+\gamma^2)$ and, in particular, it has the same sign as β.

Theorem 14.6 *We have $\zeta(s) \neq 0$ for $\sigma > 1 - c/\log|t|$, where $c > 0$ is an absolute constant and $|t| > 2$.*

Proof For brevity we shall write $(\zeta'/\zeta)(s)$ for the function $\zeta'(s)/\zeta(s)$. By definition, for $\sigma > 1$, we have

$$\text{Re}\{(\zeta'/\zeta)(s)\} = -\sum_{n=1}^{\infty}\Lambda(n)n^{-\sigma}\cos(t\log n).$$

It follows as in the proof of Theorem 14.1 that

$$-\text{Re}\{3(\zeta'/\zeta)(\sigma) + 4(\zeta'/\zeta)(\sigma + it) + (\zeta'/\zeta)(\sigma + 2it)\} \geq 0.$$

Now $(\zeta'/\zeta)(\sigma)$ is real and, since $(\zeta'/\zeta)(s)$ has a simple pole at $s = 1$ with residue -1, we have $-(\zeta'/\zeta)(\sigma) - 1/(\sigma - 1) \ll 1$ assuming that $1 < \sigma \ll 1$, where, as later, the implied constants are absolute. Further, since the real part of any zero ρ lies between 0 and 1, the real parts of $1/(s - \rho)$ and $1/\rho$ are both positive, whence the real part of the sum over ρ in the formula for $\xi'(s)/\xi(s)$ above is positive. By Stirling's formula, the Γ-term in the associated expression for ζ'/ζ is $O(\log|t|)$. Thus we obtain $-\text{Re}\{(\zeta'/\zeta)(\sigma + 2it)\} \ll \log|t|$. We now restrict t to be the ordinate γ of a zero ρ of $\zeta(s)$ in the critical strip with $|\gamma| > 2$; then $\sigma + it - \rho = \sigma - \beta$ and from the explicit formula again we have

$$-\text{Re}\{(\zeta'/\zeta)(\sigma + it)\} + 1/(\sigma - \beta) \ll \log|t|.$$

On combining the estimates, we get

$$4/(\sigma - \beta) - 3/(\sigma - 1) \ll \log|t|.$$

Then, on taking $\sigma = 5 - 4\beta$, so that $1 < \sigma < 5$, the left-hand side becomes $1/(20(1 - \beta))$; hence we see that $1 \ll (1 - \beta)\log|t|$ and thus $\beta < 1 - c/\log|t|$ for some constant $c > 0$. This proves the theorem. □

14.7 On certain sums

We now give two further applications of the explicit formula for $\zeta'(s)/\zeta(s)$. The results are needed preliminary to the work of the next section.

Lemma 14.1 *We have*

$$\sum_\rho \frac{1}{4 + (t - \gamma)^2} \ll \log|t|.$$

Proof The formulae in Section 14.6 give

$$\frac{\zeta'(2 + it)}{\zeta(2 + it)} = \sum_\rho \left(\frac{1}{2 + it - \rho} + \frac{1}{\rho} \right) + O(\log|t|).$$

The left-hand side is expressible as the Dirichlet series $-\sum \Lambda(n)n^{-2-it}$ and so is $O(1)$. Further, the real parts of the $1/\rho$ are positive. Thus

$$\text{Re} \sum_\rho \frac{1}{2 + it - \rho} \ll \log|t|$$

and the lemma follows since

$$\text{Re}\left(\frac{1}{2+it-\rho}\right) = \frac{2-\beta}{(2-\beta)^2+(t-\gamma)^2} \geq \frac{1}{4+(t-\gamma)^2}.$$

$\square$

Lemma 14.2 *For any $s = \sigma + it$ with $-1 \leq \sigma \leq 2$ and t not coinciding with the imaginary part γ of a zero ρ, we have*

$$\frac{\zeta'(s)}{\zeta(s)} = \sum_{\rho}' \frac{1}{s-\rho} + O(\log|t|),$$

where the summation is over all zeros ρ of $\zeta(s)$ in the critical strip such that $|t-\gamma| < 1$.

Proof On comparing $\zeta'(s)/\zeta(s)$ with $\zeta'(2+it)/\zeta(2+it)$ we obtain

$$\frac{\zeta'(s)}{\zeta(s)} = \sum_{\rho}\left(\frac{1}{s-\rho} - \frac{1}{2+it-\rho}\right) + O(\log|t|).$$

Now the summand here is

$$\frac{2-\sigma}{(s-\rho)(2+it-\rho)}$$

and the imaginary part of both factors in the denominator is $t - \gamma$; hence the absolute value is at most $3/(t-\gamma)^2$. The latter does not exceed $15/(4 + (t-\gamma)^2)$ if $|t-\gamma| \geq 1$. Thus, by Lemma 14.1, it suffices to sum over the ρ satisfying $|t-\gamma| < 1$. But $|2 + it - \rho|^2 = (2-\beta)^2 + (t-\gamma)^2$ and, since $(2-\beta)^2 > 1$, we have $|2+it-\rho| > 1 + \frac{1}{4}(t-\gamma)^2$ if $|t-\gamma| < 1$. Again from Lemma 14.1, this gives $\sum_{\rho}' 1/|2+it-\rho| \ll \log|t|$, and the desired result follows.

$\square$

14.8 The Riemann–von Mangoldt formula

It is customary to denote by $N(T)$ the number of zeros $\rho = \beta + i\gamma$ of $\zeta(s)$ in the critical strip with $0 < \gamma \leq T$. In his original memoir, Riemann gave conjecturally an asymptotic formula for $N(T)$ and, in 1905, this was established precisely by von Mangoldt.

Theorem 14.7 *We have*

$$N(T) = \frac{T}{2\pi}\log\frac{T}{2\pi} - \frac{T}{2\pi} + O(\log T).$$

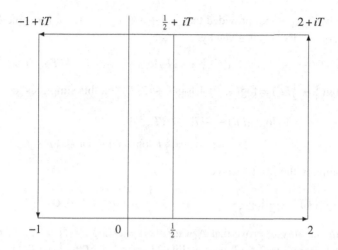

$-1 + iT$ $\frac{1}{2} + iT$ $2 + iT$

-1 0 $\frac{1}{2}$ 2

Fig. 14.3 The rectangle R in the proof of Theorem 14.7.

Proof Let R denote the rectangular contour with vertices $2, 2 + iT, -1 + iT, -1$ described anti-clockwise; see Fig. 14.3. Further, let $\xi(s)$ be the entire function introduced earlier satisfying the functional equation $\xi(s) = \xi(1 - s)$. We assume for simplicity that T does not coincide with an ordinate γ of a zero of $\xi(s)$. Then we have $2\pi N(T) = \Delta_R(\arg \xi(s))$ where the latter signifies the variation of the argument of $\xi(s)$ as s describes R. The variation of the argument of $\xi(s)$ as s describes the line from -1 to 2 is 0 since here $\xi(s)$ is real and does not vanish. Indeed $\xi(0) = \xi(1) = \frac{1}{2}$ and $\xi(\sigma) = \frac{1}{2}\sigma(\sigma - 1)\zeta(\sigma) > 0$ for $0 < \sigma < 1$; see Section 14.10 Exercise (v). Now, since $\xi(\sigma + it) = \xi(1 - \sigma - it)$ and this is just the complex conjugate of $\xi(1 - \sigma + it)$, it follows that the variation of the argument of $\xi(s)$ as s describes the part of R from $\frac{1}{2} + iT$ to -1 is the same as the variation as s describes the part from 2 to $\frac{1}{2} + iT$. With C used to denote the latter, it therefore suffices to give an asymptotic formula for $\pi N(T) = \Delta_C(\arg \xi(s))$.

We recall again that $\xi(s) = \frac{1}{2}s(s - 1)\Xi(s)$, where $\Xi(s) = \pi^{-\frac{1}{2}s}\Gamma(\frac{1}{2}s)\zeta(s)$, and we proceed to calculate the variation of the argument of each factor over C. We have $\Delta_C(\arg s(s - 1)) = \arg(-\frac{1}{4} - T^2) = \pi$ and

$$\Delta_C(\arg \pi^{-\frac{1}{2}s}) = \Delta_C(\arg e^{-\frac{1}{2}s \log \pi}) = -\frac{1}{2}T \log \pi.$$

By Stirling's formula

$$\log \Gamma(z) = (z - \tfrac{1}{2}) \log z - z + \tfrac{1}{2} \log 2\pi + O(|z|^{-1})$$

uniformly as $|z| \to \infty$, provided that $-\pi + \delta < \arg z < \pi - \delta$ for some $\delta > 0$. Hence $\Delta_C(\arg \Gamma(\frac{1}{2}s))$ is given by

$$\text{Im}\{\log \Gamma(\tfrac{1}{4} + \tfrac{1}{2}iT)\} = \text{Im}\{(-\tfrac{1}{4} + \tfrac{1}{2}iT) \log(\tfrac{1}{4} + \tfrac{1}{2}iT) - \tfrac{1}{2}iT + O(T^{-1})\}.$$

Since $\log(\tfrac{1}{4} + \tfrac{1}{2}iT) = \log(\tfrac{1}{2}iT) + \log(1 + (2iT)^{-1})$, this simplifies to

$$\text{Im}\{(-\tfrac{1}{4} + \tfrac{1}{2}iT) \log(\tfrac{1}{2}iT) - \tfrac{1}{2}iT + O(T^{-1})\}$$
$$= \tfrac{1}{2}T \log(\tfrac{1}{2}T) - \tfrac{1}{8}\pi - \tfrac{1}{2}T + O(T^{-1}).$$

On combining the results we get

$$N(T) = \frac{T}{2\pi} \log \frac{T}{2\pi} - \frac{T}{2\pi} + \frac{7}{8} + \frac{1}{\pi} \Delta_C(\arg \zeta(s)) + O(T^{-1}).$$

It remains only to prove that $\Delta_C(\arg \zeta(s)) = O(\log T)$. Now C consists of the line L from 2 to $2 + iT$ and the line M from $2 + iT$ to $\frac{1}{2} + iT$. On L, the real part of $\zeta(s)$ is at least $1 - \sum_{n \geq 2}(1/n^2) > 0$ and hence $\Delta_L(\arg \zeta(s)) < \pi$. By Lemma 14.2 we have

$$\Delta_M(\arg \zeta(s)) = \text{Im} \int_M (\zeta'(s)/\zeta(s)) ds = \sum_\rho{}' \Delta_M(\arg(s - \rho)) + O(\log T),$$

where the sum is over all zeros ρ of $\xi(s)$ with $|t - \gamma| < 1$. From Lemma 14.1 it is easily deduced that the number of these zeros is $O(\log T)$. We have also $\Delta_M(\arg(s - \rho)) < \pi$ and the theorem follows. $\square$

As a corollary to Theorem 14.7 we deduce that if the ordinates of the zeros ρ of $\xi(s)$ above the real axis are written as an increasing sequence $\gamma_1, \gamma_2, \ldots$ then $\gamma_n \sim 2\pi n / \log n$ as $n \to \infty$. For we have $2\pi N(T) \sim T \log T$ as $T \to \infty$ and, since $N(\gamma_n) = n$, this gives $2\pi n \sim \gamma_n \log \gamma_n$ as $n \to \infty$ and thus $\log \gamma_n \sim \log n$.

Apart from $N(T)$, much study has also been made of the function $N(\sigma, T)$ defined as the number of zeros $\rho = \beta + i\gamma$ of $\zeta(s)$ with $\beta \geq \sigma$ and $0 < \gamma \leq T$. We have $N(T) = N(0, T)$ and so Theorem 14.7 gives $N(\sigma, T) \ll T \log T$ for all $\sigma > 0$. Better results in this direction have been obtained by Ingham, Huxley and others; in particular they have shown that $N(\sigma, T) \ll T^{(12/5)(1-\sigma)+\varepsilon}$ for $\frac{1}{2} \leq \sigma \leq 1$ and any $\varepsilon > 0$, where the implied constant depends only on ε. Generally speaking the goal here is to prove the so-called density hypothesis that $N(\sigma, T) \ll T^{2(1-\sigma)+\varepsilon}$. This is clearly implied by the Riemann hypothesis for then $N(\sigma, T) = 0$ for $\sigma > \frac{1}{2}$. But the density hypothesis is also known to hold under a conjecture of Lindelöf of 1908, the famous Lindelöf hypothesis, to the effect that $|\zeta(\frac{1}{2} + it)| \ll t^\varepsilon$ for all $t > 0$. Backlund proved in 1918 that the latter is equivalent to the assertion that the number of zeros of $\zeta(s)$ with

$\sigma > \frac{1}{2}$ and $T < t \leq T + 1$, which we know to be $O(\log T)$ by Theorem 14.7, is in fact $o(\log T)$. In particular the Lindelöf hypothesis is a consequence of the Riemann hypothesis.

14.9 Further reading

The material of this chapter is entirely classical and some version of it can be found in all of the principal books on analytic number theory referred to in Sections 1.7, 2.9 and 13.8. We have mainly followed the expositions of Ingham and Davenport; for further reading we recommend the texts of Titchmarsh and Ivić and also the work of Montgomery and Vaughan, *Multiplicative Number Theory I: Classical Theory* (Cambridge University Press, 2006).

14.10 Exercises

(i) Defining the Gamma function by

$$\Gamma(z) = \lim_{n \to \infty} \frac{n^z n!}{z(z+1)\dots(z+n)} \qquad (z \neq 0, -1, -2, \dots),$$

establish the following properties used in Section 14.2.

(a) $z\Gamma(z) = \Gamma(z+1)$;

(b) $\Gamma(z)\Gamma(1-z) = \pi \operatorname{cosec}(\pi z)$;

(c) $\Gamma(z)\Gamma(z+\frac{1}{2})2^{2z-1} = \Gamma(\frac{1}{2})\Gamma(2z)$;

(d) $\Gamma(\frac{1}{2}) = \sqrt{\pi}$.

(ii) Prove that, if the series $\sum_{n=1}^{\infty} z_n$ of complex numbers converges absolutely and if $z_n \neq 1$ for all n, then $\prod_{n=1}^{\infty}(1 - z_n) \neq 0$. Hence verify the assertion of Section 14.3 that $\zeta(s) \neq 0$ for $\sigma > 1$. Show that the convergence of the series is necessary by proving that $\prod_{n=2}^{\infty}(1 - 1/(n \log n)) = 0$.

(iii) Verify the following for $\sigma > 1$ where the sum is over all primes p.

$$\zeta(s) = s \int_1^{\infty} \frac{[x]}{x^{s+1}} dx, \qquad \sum_p \frac{1}{p^s} = s \int_1^{\infty} \frac{\pi(x)}{x^{s+1}} dx.$$

(iv) Suppose that the Dirichlet series $\eta(s) = \sum_{n=1}^{\infty} a_n/n^s$ converges absolutely for $\sigma > b > 0$. Show that, if $S(x) = \sum_{n \leq x} a_n$, then, for any $c > b$,

$$\frac{1}{2\pi i} \int_{c-i\infty}^{c+i\infty} \frac{\eta(s)x^{s+1}}{s(s+1)} ds = \int_0^x S(u)\, du.$$

(v) Show that, for $\sigma > 1$,

$$(1 - 2^{1-s})\zeta(s) = 1 - 2^{-s} + 3^{-s} - 4^{-s} + \cdots .$$

Verify that, for $\sigma > 0$, both sides are analytic and so the equation remains valid there. Deduce that $\zeta(\sigma) < 0$ for $0 < \sigma < 1$.

(vi) Show that the Riemann–von Mangoldt formula can be written in the form

$$N(T) = \int_1^{T/(2\pi)} (\log t)\, dt + O(\log T).$$

Deduce that $N(T + A) > N(T)$ if A is sufficiently large. Hence prove that, if $\gamma_1, \gamma_2, \ldots$ are the ordinates of the zeros of $\zeta(s)$ above the real axis in ascending order of magnitude, then $\gamma_{n+1} - \gamma_n$ is bounded independently of n.

(vii) With the preceding notation, prove that $N(T + 1) - N(T) = O(\log T)$. Hence verify that there exists a constant $c > 0$ such that $\gamma_{n+1} - \gamma_n > c/\log n$ for an infinite sequence of values of n.

(viii) Prove that

$$\sum_{0 < \gamma_n \leq T} \frac{1}{\gamma_n} = O(\log T)^2.$$

15

On the distribution of the primes

15.1 The prime-number theorem

In this chapter we shall give an account of classical results on the distribution of the primes both in the sequence of the ordinary integers and, more generally, in arithmetical progressions. To begin with we establish the celebrated prime-number theorem originally conjectured by Legendre and first proved by Hadamard and de la Vallée Poussin independently in 1896. The proofs were based on the theory of functions of a complex variable and could not therefore be considered as elementary. An 'elementary' proof was given by Selberg and Erdős in 1948; we shall mention some details relating to it in Section 15.2. Hardy famously predicted in 1921 that if an elementary proof were ever found then it would be time 'for the books to be cast aside and for the theory to be rewritten'. But in reality the original analytic method has remained the basis of all the more precise refinements and corollaries established to date and it is the one that we follow here.

We show first that the prime-number theorem is valid in a certain average sense and we then obtain the theorem itself as a direct corollary.

Theorem 15.1 *We have*

$$\int_0^x \psi(u)\,du \sim \tfrac{1}{2}x^2 \quad as \quad x \to \infty.$$

Proof Suppose that $N > 1$ and let C denote the path consisting of the lines $1 + it$ with $-N \le t \le -T$ and $T \le t \le N$, the line $b + it$ with $-T \le t \le T$, where $0 < b < 1$, and the connecting lines $\sigma \pm iT$ with $b \le \sigma \le 1$. Here the numbers T and b are chosen in that order (see later) so that $\zeta(s)$ has no zeros on or to the right of C; such a choice is possible in view of Theorem 14.1 together with the fact that $\zeta(s)$ has no zeros in $\sigma > 1$ and only a finite number

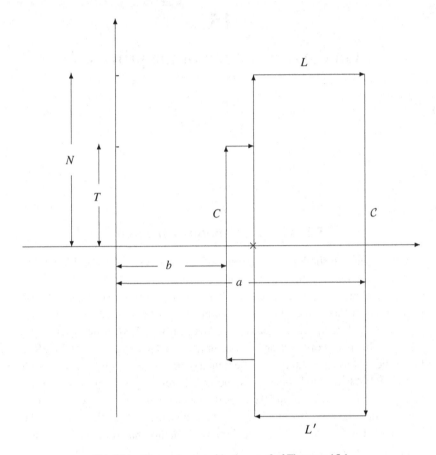

Fig. 15.1   The contour used in the proof of Theorem 15.1.

of zeros in any bounded region. Further, let C denote the line segment $a + it$ with $-N \leq t \leq N$ for some $a > 1$, let L, L' denote the connecting lines $\sigma \pm iN$ with $1 \leq \sigma \leq a$, and suppose that the complete contour C, L, C, L' is described clockwise; see Fig. 15.1.

Now $\zeta'(s)/\zeta(s)$ is analytic within and on the contour, apart from a simple pole at $s = 1$ with residue -1, and thus the function

$$\Phi(s) = \frac{\zeta'(s)}{\zeta(s)} \frac{x^{s+1}}{s(s+1)}$$

is also analytic within and on the contour apart from a simple pole at $s = 1$ with residue $-\frac{1}{2}x^2$. Further, the integrals of $\Phi(s)$ over L and L' have absolute values

$$\ll (\log N)^{10} \max(1, x^{a+1})(a-1)/\{N(N-1)\},$$

which clearly tends to 0 as $N \to \infty$. This, together with Cauchy's theorem, implies that

$$\int_{C'} \Phi(s)\, ds + \int_{\mathcal{C}'} \Phi(s)\, ds = 2\pi i (\tfrac{1}{2} x^2),$$

where $C', \mathcal{C}'$ denote the contours derived from $C, \mathcal{C}$ as $N \to \infty$. But by Theorem 14.5 we have

$$\frac{1}{2\pi i} \int_{C'} \Phi(s)\, ds = \int_0^x \psi(u)\, du$$

and thus to prove Theorem 15.1 it suffices now to show that

$$x^{-2} \int_{C'} \Phi(s)\, ds \to 0 \quad \text{as} \quad x \to \infty.$$

By Theorem 14.4 we see that

$$|\Phi(1+it)| \ll (\log |t|)^{10} x^2 |t|^{-2}$$

if $|t| \gg 1$ and so certainly the integral of $x^{-2}\Phi(s)$ over that part of C' which lies on the line $\sigma = 1$ tends to 0 as $T \to \infty$. In particular, T can be chosen so that the absolute value of the integral is at most $\frac{1}{2}\varepsilon$ for a given $\varepsilon > 0$. Further, we see that $x^{-s-1}\Phi(s)$ is analytic on C' and so, on that part of C' which does not lie on $\sigma = 1$, it is bounded in absolute value by $M = M(T, b)$ independently of x. The integral of $x^{-2}\Phi(s)$ over the latter part has therefore absolute value at most

$$M \int_{-T}^{T} x^{b-1}\, dt + 2M \int_b^1 x^{\sigma-1}\, d\sigma = M \left\{ 2T x^{b-1} + 2(1 - x^{b-1})/\log x \right\}$$

and, with T and b fixed, this clearly tends to 0 as $x \to \infty$. In particular the absolute value of the expression is at most $\frac{1}{2}\varepsilon$ for x sufficiently large and this proves the theorem. $\qquad\square$

Theorem 15.2 (Prime-number theorem) *We have*

$$\pi(x) \sim x/\log x \quad as \quad x \to \infty.$$

Proof It will suffice to show that $\psi(x) \sim x$ as $x \to \infty$; for it was proved in Section 13.6 that $\pi(x) \sim \psi(x)/\log x$. Now $\psi(u)$ increases with u and so, for any $h > 0$, we see that $\psi(x)$ lies between

$$\frac{1}{h} \int_{x-h}^{x} \psi(u)\, du \quad \text{and} \quad \frac{1}{h} \int_{x}^{x+h} \psi(u)\, du.$$

By Theorem 15.1, these expressions have the form $x \pm \frac{1}{2}h + o(x^2/h)$ for x sufficiently large and so, on taking $h = \delta x$ with $\delta > 0$, we deduce that $\psi(x)/x$ lies between $1 - \delta$ and $1 + \delta$ if x is sufficiently large. This gives $\psi(x) \sim x$ as required. $\square$

15.2 Refinements and developments

Following the preceding proofs but modified so that the contour reflects the expanded zero-free region given by Theorem 14.6, we deduce that

$$\psi(x) = x + O(xe^{-c\sqrt{\log x}})$$

for some $c > 0$. Now by the partial summation formula in Theorem 13.4 we have

$$\sum_{n \leq x} \frac{\Lambda(n)}{\log n} = \int_2^x \frac{\psi(u)\,du}{u(\log u)^2} + \frac{\psi(x)}{\log x}.$$

The sum on the left is just

$$\sum_{p^m \leq x} \frac{1}{m} = \sum_{m \leq \log x / \log 2} \frac{\pi(x^{1/m})}{m} = \pi(x) + O(\sqrt{x}).$$

Hence from the basic result $\psi(x) \sim x$ established in Theorem 15.2 we obtain

$$\pi(x) \sim \int_2^x \frac{du}{(\log u)^2} + \frac{x}{\log x} \sim \operatorname{li} x \quad \text{as} \quad x \to \infty.$$

Further, from the more precise expression above we get

$$\pi(x) = \operatorname{li} x + O(xe^{-c\sqrt{\log x}})$$

with $c > 0$. As remarked in Section 14.5, the best result to date is due to Vinogradov and Korobov who gave the exponent $(\log x)^{\frac{3}{5}}(\log \log x)^{-\frac{1}{5}}$.

A precise connection between $\psi(n)$ and $\zeta(s)$ is given by

$$\psi(n) = -\frac{1}{2\pi i} \int_{a-i\infty}^{a+i\infty} \frac{(n+\frac{1}{2})^s}{s} \frac{\zeta'(s)}{\zeta(s)}\,ds,$$

valid for any $a > 1$, and the proof of the prime-number theorem can be based on this result instead of the formula given in Theorem 14.5. But the method involves slight additional complications due to the less rapid convergence of the integral on the right; see, for example, the book by Prachar referred to in Section 15.7. In this way, however, one avoids the need for the Tauberian-type deduction in the proof of Theorem 15.2. Various attempts were made in the early years to reduce the amount of complex-variable theory occurring in the exposition and it was shown, in particular, that one could dispense with Cauchy's theorem. But the proof could still not be regarded as elementary.

The 'elementary' proof of the prime-number theorem of Selberg and Erdős is based on the Selberg formula

$$\psi(x)\log x + \sum_{n\leq x} \Lambda(n)\psi(x/n) = 2x\log x + O(x).$$

This is equivalent to

$$\log x \sum_{p\leq x} \log p + \sum_{p,p'\leq x} \log p \log p' = 2x\log x + O(x),$$

where the sums are over primes p and p'. The proof of the formula depends on a double application of the equation

$$F(x)\log x + \sum_{n\leq x} F(x/n)\Lambda(n) = \sum_{d\leq x} \mu(d)G(x/d),$$

where $G(x) = \sum_{m\leq x} F(x/m)\log x$, first with $F(x) = \psi(x)$ and then with $F(x) = x - \gamma - 1$ where γ is Euler's constant; for details see the book of Hardy and Wright referred to in Section 1.7 or that of Trost referred to in Section 15.7.

The function $N(\sigma, T)$ discussed in Section 14.8 is critical to studies on the difference between consecutive primes. In particular Ingham showed that an estimate of the form $N(\sigma, T) \ll T^{\kappa(1-\sigma)}(\log T)^c$, where c is an absolute constant, implies that $p_{n+1} - p_n \ll p_n^{(1-1/\kappa)+\varepsilon}$ for any $\varepsilon > 0$, the implied constant depending only on ε. After an initial breakthrough by Hoheisel in 1930 yielding the first exponent of p_n less than 1 there have been many subsequent works in this context; especially, in 1972, Huxley gave the value $\kappa = \frac{12}{5}$ and thus the exponent $\frac{7}{12} + \varepsilon$. The density hypothesis implies that $p_{n+1} - p_n \ll p_n^{1/2+\varepsilon}$ and on the Riemann hypothesis it has been shown that p_n^ε can be replaced by $\log p_n$. But these estimates are probably far from best possible; indeed in 1937 Cramér conjectured that $p_{n+1} - p_n \ll (\log p_n)^2$.

We remark finally that there is a so-called approximate functional equation for the Riemann zeta-function which goes further than the observations given in the proof of Theorem 14.2. A simple form, deducible from the latter, is

$$\zeta(s) = \sum_{n=1}^{N} \frac{1}{n^s} + \frac{N^{1-s}}{s-1} + O(|t|N^{-\sigma}) \quad (\sigma > 0).$$

This can be refined to give

$$\zeta(s) = \sum_{n \leq x} \frac{1}{n^s} + \frac{x^{1-s}}{s-1} + O(x^{-\sigma}), \quad 0 < \sigma_0 \leq \sigma \leq 2, \, x \geq |t|/\pi,$$

where the constant implied in the O-term depends on σ_0; see the book by Ivić referred to in Section 2.9. The most famous approximate functional equation, due to Hardy and Littlewood and obtained in 1921, is

$$\zeta(s) = \sum_{n \leq x} n^{-s} + \chi(s) \sum_{n \leq y} n^{s-1} + O(x^{-\sigma}) + O(t^{\frac{1}{2}-\sigma} y^{\sigma-1})$$

where $0 \leq \sigma \leq 1$, $2\pi xy = t$ with $x, y, t > c > 0$, and

$$\chi(s) = 2^s \pi^{s-1} \sin(\tfrac{1}{2} s\pi) \Gamma(1-s).$$

15.3 Dirichlet characters

The remainder of this chapter is devoted to the subject of primes in arithmetical progressions. Our principal objective will be to prove the famous theorem of Dirichlet of 1839 in this context. The proof rests on the concept of Dirichlet characters and we begin with an account of their definition and properties.

A complex-valued function χ defined on an arbitrary group G is called a character of G if χ is multiplicative, that is, $\chi(a)\chi(b) = \chi(ab)$ for all a, b in G, and χ is not identically 0. It is a consequence of group theory that, if G is a finite abelian group of order d, then there are d distinct characters and they themselves form a group with the product $\chi_1 \chi_2$ of χ_1, χ_2 defined by $\chi_1 \chi_2(a) = \chi_1(a)\chi_2(a)$ for all a in G. If one further specializes G to be the group of reduced residue classes modulo an integer $q > 1$ then, on defining $\chi(n)$ for any integer n with $(n, q) = 1$ by $\chi(\tilde{n})$ where $\tilde{n}$ is the residue class that contains n and $\chi(n) = 0$ for $(n, q) > 1$, one obtains a Dirichlet character $(\bmod q)$.

There are $\phi(q)$ Dirichlet characters $\chi \pmod q$, where ϕ is Euler's function, and each of these is periodic with period q, that is, $\chi(n+q) = \chi(n)$ for every n. The particular Dirichlet character χ_0 that satisfies $\chi_0(n) = 1$ when $(n, q) = 1$ is called the principal character and it is the identity element in the character group. We have the basic relations

$$\sum_{n=1}^{q} \chi(n) = \phi(q) \text{ if } \chi = \chi_0, \qquad \sum_{n=1}^{q} \chi(n) = 0 \text{ if } \chi \neq \chi_0.$$

The first of these is clear and the second follows from the observation that, if $\chi \neq \chi_0$, then $\chi(m) \neq 1$ for some m with $(m, q) = 1$, whence as n runs through a reduced set of residues $(\bmod q)$ so does mn, and we obtain

$$\sum_{n=1}^{q} \chi(n) = \sum_{n=1}^{q} \chi(mn) = \chi(m) \sum_{n=1}^{q} \chi(n).$$

Further, we have the relations

$$\sum_{\chi} \chi(n) = \begin{cases} \phi(q) & \text{if } n \equiv 1 \ (\bmod q), \\ 0 & \text{otherwise,} \end{cases}$$

where the sum is over all characters $\chi \ (\bmod q)$. Again the first of these is clear since $\chi(1) = 1$ for all χ; the second follows from the fact that for every n with $(n, q) = 1$ other than $n \equiv 1 \ (\bmod q)$ there exists, by the initial group theory, a character χ' such that $\chi'(n) \neq 1$ and we have

$$\sum_{\chi} \chi(n) = \sum_{\chi} \chi'\chi(n) = \chi'(n) \sum_{\chi} \chi(n).$$

For each Dirichlet character χ we define the conjugate $\overline{\chi}$ in the obvious way by taking $\overline{\chi}(n)$ as the complex conjugate of $\chi(n)$. Then $\overline{\chi}$ is the reciprocal of χ in the character group and the following holds.

Theorem 15.3 *For any integer a with $(a, q) = 1$ we have*

$$\sum_{\chi} \chi(n)\overline{\chi}(a) = \phi(q) \text{ if } n \equiv a \ (\bmod q), \quad \sum_{\chi} \chi(n)\overline{\chi}(a) = 0 \text{ otherwise.}$$

Proof We observe that $\chi(n) = \chi(a)\chi(n')$, where n' satisfies $n \equiv an' \ (\bmod q)$, and also $\chi(a)\overline{\chi}(a) = 1$, whence the result follows at once from the relations above. $\qquad\Box$

The subject of Dirichlet characters can be introduced in another way utilizing properties of primitive roots rather than group theory. For simplicity we restrict our discussion to the case $q = p_1 p_2 \cdots p_k$ where $p_1, p_2, \ldots, p_k$ are distinct odd primes. If n is any integer with $(n, q) = 1$, we denote by $m_j (1 \leq j \leq k)$ the index of n with respect to a primitive root $g_j \ (\bmod p_j)$ so that $n \equiv g_j^{m_j} \ (\bmod p_j)$ and $0 \leq m_j < \phi(p_j)$. For each set of integers $r_1, r_2, \ldots, r_k$ with $0 \leq r_j < \phi(p_j)$ we define a homomorphism χ from the integers n into the complex numbers by $\chi(n) = e^{2\pi i f(n)}$, where

$$f(n) = \sum_{j=1}^{k} \frac{m_j r_j}{\phi(p_j)}.$$

We take $\chi(n) = 0$ if $(n, q) > 1$ and then χ becomes a Dirichlet character to the modulus q. As before, there are $\phi(q)$ distinct Dirichlet characters $\chi \pmod q$ and the principal character χ_0, given by $r_1 = r_2 = \cdots = r_k = 0$, is the identity in the character group. The basic relations stated above are obtained by summing $e^{2\pi i f(n)}$ over all $m_1, \ldots, m_k$ or $r_1, \ldots, r_k$ and using the cyclotomic property of the exponential function. The theory generalizes readily to deal with an arbitrary odd integer $q > 1$; for even q there is a slight complication in that there is no primitive root $\pmod{2^l}$ for $l \geq 3$ but we have $n \equiv (-1)^m 5^{m'} \pmod{2^l}$ where $m = 0$ or 1 and $0 \leq m' < 2^{l-2}$ and the pair m, m' serves as the index of n in this instance.

15.4 Dirichlet *L*-functions

For the complex variable $s = \sigma + it$ with $\sigma > 0$ and the Dirichlet character $\chi \pmod q$ with $\chi \neq \chi_0$ we define the Dirichlet L-function by

$$L(s, \chi) = s \int_1^\infty \left(\sum_{n \leq x} \chi(n) \right) x^{-s-1} \, dx.$$

In view of the basic relation $\sum_{n=1}^q \chi(n) = 0$ for $\chi \neq \chi_0$ and the periodicity of χ, the sum is bounded in terms of q and so the integral converges for $\sigma > 0$ and uniformly for $\sigma > \delta > 0$. When $\sigma > 1$, we have

$$L(s, \chi) = s \sum_{n=1}^\infty \chi(n) \int_n^\infty x^{-s-1} \, dx = \sum_{n=1}^\infty \chi(n) n^{-s}.$$

The Dirichlet series on the right can be expressed as an Euler product, namely

$$L(s, \chi) = \prod_p (1 - \chi(p) p^{-s})^{-1}.$$

For $\chi = \chi_0$ we define

$$L(s, \chi_0) = \zeta(s) \prod_{p|q} (1 - p^{-s}).$$

Then, for $\sigma > 1$, we have

$$L(s, \chi_0) = \sum_{n=1}^\infty \chi_0(n) n^{-s}.$$

The existence of the Euler product shows that $L(s, \chi) \neq 0$ when $s > 1$. We shall prove in the next section that this holds also for $s = 1$. Here we shall establish the following preliminary result.

Lemma 15.1 *When $\chi \neq \overline{\chi}$, we have $L(1, \chi) \neq 0$.*

Proof Again in view of the Euler product, we obtain, for any $s > 1$,

$$\log L(s, \chi) = -\sum_p \log(1 - \chi(p)p^{-s}) = \sum_p \sum_{m=1}^{\infty} \chi(p^m)m^{-1}p^{-ms}.$$

On summing over all characters $\chi \pmod q$, we obtain

$$\sum_\chi \log L(s, \chi) = \sum_p \sum_{m=1}^{\infty} m^{-1}p^{-ms} \sum_\chi \chi(p^m).$$

Now by Theorem 15.3, or simply by the relations which preceded the theorem, we see that the last sum is $\phi(q)$ if $p^m \equiv 1 \pmod q$ and 0 otherwise; in particular, the expression on the right of the equation is positive. Hence we have, for $s > 1$,

$$\prod_\chi |L(s, \chi)| \geq 1.$$

But $L(s, \chi_0)$ has a simple pole at $s = 1$, the remaining factors are all analytic at $s = 1$ and, if $L(1, \chi)$ were 0, then also $L(1, \overline{\chi})$ would be 0. Thus, for $\chi \neq \overline{\chi}$, the left-hand side would tend to 0 as $s \to 1$, contrary to it being bounded below. This proves the lemma. □

15.5 Primes in arithmetical progressions

Dirichlet proved in 1839 that Euclid's result on the existence of infinitely many primes holds more generally in any arithmetical progression. As remarked in Section 1.6, though some special cases were known previously, Dirichlet's proof involved entirely new concepts and has been of far-reaching significance.

Theorem 15.4 *There exist infinitely many primes in the arithmetical progression $a, a + q, a + 2q, \ldots$, where a and q are integers with $q > 0$ and $(a, q) = 1$.*

Proof We shall assume that $q > 1$ and that $s > 1$. Then, as in the proof of Lemma 15.1, we see that

$$\sum_\chi \overline{\chi}(a) \log L(s, \chi) = \sum_p \sum_{m=1}^{\infty} m^{-1}p^{-ms} \sum_\chi \chi(p^m)\overline{\chi}(a).$$

We shall show that the expression on the left tends to infinity as $s \to 1$ and this will suffice to establish the theorem. For, by Theorem 15.3, the last sum on the right is $\phi(q)$ if $p^m \equiv a \pmod{q}$ and 0 otherwise. Hence the part of the expression on the right for $m \geq 2$ is bounded above by $\phi(q) \sum_p 1/(p(p-1))$ and so converges; there remains only $\phi(q) \sum p^{-s}$, where the sum is over all primes $p \equiv a \pmod{q}$. It follows that the latter tends to infinity as $s \to 1$ and so certainly the sequence of primes $p \equiv a \pmod{q}$ cannot terminate.

Now, by Lemma 15.1, the terms in the expression on the left given by those characters with $\chi \neq \overline{\chi}$ remain bounded as $s \to 1$. A fundamental result of Dirichlet, which we establish in Lemma 15.2 below, gives $L(1, \chi) \neq 0$ for real $\chi \neq \chi_0$; it follows that all the terms corresponding to non-principal characters remain bounded as $s \to 1$. On the other hand, the term given by $\chi = \chi_0$ tends to infinity as $s \to 1$. Thus the whole expression tends to infinity as $s \to 1$ and Theorem 15.4 is proved subject to a verification of the following result. $\quad\square$

Lemma 15.2 *We have $L(1, \chi) \neq 0$ for any real non-principal character χ.*

Proof Let $\chi \pmod{q}$ be a real non-principal character and suppose that $L(1, \chi) = 0$. Then the function

$$F(s) = \frac{L(s, \chi)L(s, \chi_0)}{L(2s, \chi_0)}$$

is analytic for $s = \sigma + it$ with $\sigma > \frac{1}{2}$; indeed the pole of $L(s, \chi_0)$ at $s = 1$ is simple and so cancels with the zero of $L(s, \chi)$, and certainly $L(2s, \chi_0)$ is analytic when $\sigma > \frac{1}{2}$. Further, since $L(2s, \chi_0) \to \infty$ as $s \to \frac{1}{2}$, we have $F(s) \to 0$ as $s \to \frac{1}{2}$ from the right.

Now, on expressing the factors of $F(s)$ as Euler products and using the property that a real character takes only the values 0 and ± 1, we obtain

$$F(s) = \prod_{p,(p,q)=1} \frac{1 - p^{-2s}}{(1 - \chi(p)p^{-s})(1 - p^{-s})} = \prod_{p, \chi(p)=1} \frac{1 + p^{-s}}{1 - p^{-s}}.$$

This gives $F(s) = \sum_{n=1}^{\infty} c_n n^{-s}$, where $c_n \geq 0$ for all n and $c_1 = 1$. On denoting the jth derivative of $F(s)$ with respect to s by $F^{(j)}(s)$, we see that

$$F^{(j)}(2) = (-1)^j \sum_{n=1}^{\infty} c_n (\log n)^j n^{-2}.$$

Further, there is the Taylor expansion

$$F(s) = \sum_{j=0}^{\infty} \frac{F^{(j)}(2)}{j!} (s - 2)^j$$

valid for s with $|s - 2| < \frac{3}{2}$. If $\frac{1}{2} < s < 2$ then all the terms on the right are non-negative; for we have $(-1)^j F^{(j)}(2) \geq 0$ and $(-1)^j (s - 2)^j \geq 0$. Thus we obtain $F(s) \geq F(2)$. But the observations above on the c_n give $F(2) \geq 1$ and this contradicts the property that $F(s) \to 0$ as $s \to \frac{1}{2}$. The contradiction implies that $L(1, \chi) \neq 0$ as required. □

15.6 The class number formulae

The proof of Lemma 15.2 given above dates from a paper of de la Vallée Poussin of 1896 and, though relatively short, it may seem unmotivated. Dirichlet himself derived the result from some famous class number formulae which have been very important historically and better explain the lemma; we proceed to discuss them briefly.

A character $\chi \pmod{q}$ may have a period $q_1 < q$; then $\chi(n + q_1) = \chi(n)$ for all n with $(n, q) = 1$, and the least such q_1 is a proper divisor of q. If χ has no period $q_1 < q$ then it is said to be primitive. Now it is easily shown that for any character $\chi \pmod{q}$ there is a divisor q_1 of q and a primitive character $\chi_1 \pmod{q_1}$ such that

$$\chi(n) = \chi_1(n) \text{ if } (n, q) = 1, \qquad \chi(n) = 0 \text{ if } (n, q) > 1;$$

conversely, given $\chi_1 \pmod{q_1}$, these define a character $\chi \pmod{q}$ for each multiple q of q_1 and we say that χ_1 induces χ. It is clear that the Dirichlet L-functions for χ and χ_1 satisfy the relation

$$L(s, \chi) = L(s, \chi_1) \prod_{p \mid q} (1 - \chi_1(p) p^{-s}).$$

Thus, to verify Lemma 15.2, it suffices to consider only real non-principal primitive characters.

Now from the analytic definition in Section 15.3, it is relatively easy to deduce that if $\chi \pmod{q}$ is a real primitive character and if $\chi \neq \chi_0$ then $q = |d|$ where d is the discriminant of a quadratic field $K = \mathbb{Q}(\sqrt{d})$. From this we get $\chi(n) = \left(\frac{d}{n}\right)$ where, for odd $n > 0$, the symbol is Jacobi's; for even $n > 0$ it is the natural extension, termed the Kronecker symbol, given by defining $\chi(2)$ as in Section 12.5. Thus we obtain

$$L(s, \chi) = \sum_{n=1}^{\infty} \left(\frac{d}{n}\right) n^{-s}.$$

Dirichlet proved that, for these particular L-functions, we have

$$L(1, \chi) = \frac{2\pi h(d)}{w\sqrt{|d|}} \ (d < 0), \qquad L(1, \chi) = \frac{h(d) \log \varepsilon}{\sqrt{d}} \ (d > 0).$$

Here $h(d)$ is the class number, w is the number of roots of unity and ε is the fundamental unit in K. The formulae make obvious the validity of Lemma 15.2. Siegel obtained in 1935 an important strengthening, namely $L(1, \chi) \gg q^{-\varepsilon}$ for any $\varepsilon > 0$, where the implied constant depends only on ε. This gives $h(d) \gg |d|^{\frac{1}{2}-\varepsilon}$ when $d < 0$ which is essentially best possible.

Much of the theory discussed earlier relating to the Riemann zeta-function can be generalized in a natural way to Dirichlet L-functions. Thus, there is an analogue of the prime-number theorem, due to Siegel and Walfisz and referred to as the Siegel–Walfisz theorem, to the effect that the number $\pi(x; q, a)$ of primes $\leq x$ in the arithmetical progression $a, a+q, a+2q, \ldots$ with $(a, q) = 1$ satisfies

$$\pi(x; q, a) = \frac{1}{\phi(q)} \operatorname{li} x + O(xe^{-c\sqrt{\log x}})$$

uniformly for $q \leq (\log x)^A$ with any given $A > 0$, where $c > 0$ and the constant implied in the O-term depend only on A. Further, there is an analogue of the Riemann–von Mangoldt formula to the effect that, for primitive χ,

$$\frac{1}{2}N(T, \chi) = \frac{T}{2\pi} \log \frac{qT}{2\pi} - \frac{T}{2\pi} + O(\log(qT)),$$

where $N(T, \chi)$ is the number of zeros of $L(s, \chi)$ satisfying $0 < \sigma < 1$ and $|t| \leq T$. Moreover, again for primitive χ, there is a functional equation for $L(s, \chi)$, namely

$$\Xi(1 - s, \overline{\chi}) = \frac{i^\delta \sqrt{q}}{\tau(\chi)} \Xi(s, \chi),$$

where

$$\Xi(s, \chi) = (\pi/q)^{-\frac{1}{2}(s+\delta)} \Gamma(\tfrac{1}{2}(s+\delta)) L(s, \chi);$$

here $\tau(\chi)$ is the Gaussian sum

$$\tau(\chi) = \sum_{m=1}^{q} \chi(m) e^{2\pi i m/q},$$

so that $|\tau(\chi)| = \sqrt{q}$ and we have $\delta = 0$ if $\chi(-1) = 1$ and $\delta = 1$ if $\chi(-1) = -1$. There is indeed a rich literature on the subject of L-functions and we refer to the recommended books for details.

15.7 Siegel's theorem

Let $L(s, \chi)$ be the Dirichlet L-function introduced in Section 15.4. We referred in Section 15.6 to the celebrated theorem of Siegel asserting that, if χ is a real primitive character with modulus q, then

$$L(1, \chi) \gg q^{-\varepsilon}$$

for any $\varepsilon > 0$, where the implied constant depends only on ε. We give now a proof of this result following an argument of Estermann; as we shall see, there are close similarities with the proof of Lemma 15.2.

We begin by defining χ_1 and χ_2 as real primitive characters with distinct moduli q_1 and q_2. Then $\chi_1 \chi_2$ is a real character with modulus $q_1 q_2$ and, though not necessarily primitive, it is non-principal; for if $\chi_1(n)\chi_2(n) = 1$, that is, $\chi_1(n) = \chi_2(n)$, for all n with $(n, q_1 q_2) = 1$ then χ_1 and χ_2 would induce the same characters $\bmod\, q_1 q_2$ and this is impossible with χ_1 and χ_2 primitive. The proof now rests on a study of the function

$$F(s) = \zeta(s)L(s, \chi_1)L(s, \chi_2)L(s, \chi_1\chi_2).$$

It is in fact the Dedekind zeta-function of a biquadratic field (cf. Sections 12.5 and 12.6); this was noted and critically utilized by Siegel in his original memoir but we shall not need to appeal to the property here.

Now $F(s)$ is analytic for all s except for a simple pole at $s = 1$ with residue

$$\lambda = L(1, \chi_1)L(1, \chi_2)L(1, \chi_1\chi_2).$$

Further, from the Euler products we obtain

$$\log F(s) = \sum_p \sum_{m=1}^{\infty} m^{-1} p^{-ms}(1 + \chi_1(p^m))(1 + \chi_2(p^m))$$

and so

$$F(s) = \sum_{n=1}^{\infty} a_n n^{-s} \quad (\sigma > 1),$$

where $a_n \geq 0$ for all n and $a_1 = 1$. Hence as in the proof of Lemma 15.2 we see that

$$F(s) = \sum_{m=0}^{\infty} b_m (2 - s)^m \quad (|s - 2| < 1),$$

where $b_m \geq 0$ for all m and $b_0 \geq 1$. This gives

$$F(s) - \lambda/(s - 1) = \sum_{m=0}^{\infty} (b_m - \lambda)(2 - s)^m$$

and the equation is valid for $|s - 2| < 2$ since the expression on the left is analytic there.

It is obvious from the definition of $L(s, \chi)$ that

$$|L(s, \chi)| \leq 2q|s| \quad (\sigma \geq \tfrac{1}{2})$$

and so we have $|F(s)| \ll (q_1 q_2)^2$ on the circle $|s - 2| = \tfrac{3}{2}$, where the implied constant is absolute. The same estimate holds for $\lambda/(s - 1)$. Furthermore, by Cauchy's estimates for the coefficients in a power series, we obtain

$$|b_m - \lambda| \ll (q_1 q_2)^2 (\tfrac{2}{3})^m,$$

whence, if $\tfrac{9}{10} < s < 1$, we have

$$\sum_{m=M}^{\infty} |b_m - \lambda|(2 - s)^m \ll \sum_{m=M}^{\infty} (q_1 q_2)^2 (\tfrac{2}{3})^m (2 - s)^m \ll (q_1 q_2)^2 (\tfrac{11}{15})^M.$$

We choose M so that the bound on the right is at most $\tfrac{1}{2}$ and we take the smallest M with this property. Then $M < c \log(q_1 q_2)$ for some absolute constant c. On noting that

$$\sum_{m=0}^{M-1} (2 - s)^m = \left((2 - s)^M - 1\right)/(1 - s)$$

and recalling that $b_0 \geq 1$ and $b_m \geq 0$ when $m \geq 1$ we obtain

$$F(s) \geq \tfrac{1}{2} - \lambda(2 - s)^M/(1 - s).$$

We now choose ε with $0 < \varepsilon < \tfrac{1}{5}c$ freely and we divide into two cases according as there does or does not exist a real primitive character χ such that $L(s, \chi)$ has a real zero $s = \beta$ with $1 - \varepsilon/(2c) < \beta < 1$. In the first case we put $\chi = \chi_1$ and in the second we take χ_1 to be any real primitive character and β to be any number in the above interval. We then define χ_2 as any real primitive character with sufficiently large modulus q_2, the size depending only on the modulus q_1 of χ_1. Plainly $F(\beta) = 0$ in the first case and in the second we have $F(\beta) < 0$ since $\zeta(s) < 0$ for $0 < s < 1$ (see Section 14.10 Exercise (v)), the L-functions in F do not vanish for $\beta \leq s < 1$ and they are positive when $s = 1$ in view of Lemma 15.2 and the Euler product. Thus we get

$$\lambda \gg (1 - \beta)(2 - \beta)^{-M}.$$

But we have

$$\lambda \ll (\log q_1)(\log(q_1 q_2))L(1, \chi_2),$$

whence

$$L(1, \chi_2) \gg (2 - \beta)^{-M}(\log q_1)^{-1}(\log(q_1 q_2))^{-1},$$

the implied constant depending now on β. Since $2 - \beta < 1 + \varepsilon/(2c) < e^{\varepsilon/(2c)}$ we obtain finally $(2 - \beta)^{-M} > (q_1 q_2)^{-\varepsilon/2}$ and, with $\chi_2 = \chi$ and $q_2 = q$, this proves Siegel's theorem.

As a corollary one deduces that if χ is a real non-principal character then for any $\varepsilon > 0$ there exists $c = c(\varepsilon) > 0$ such that $L(s, \chi) \neq 0$ for real $s > 1 - cq^{-\varepsilon}$. For, as can be verified by partial summation using the property that $\sum_{n \leq x} \chi(n)$ is bounded, the relation $L(s, \chi) = \sum \chi(n)/n^s$, already seen for $\sigma > 1$, holds in fact for $\sigma > 0$. Then

$$L'(s, \chi) = -\sum_{n=1}^{\infty} \chi(n)(\log n)n^{-s}$$

and by splitting the sum into two parts according as $n \leq q$ or $n > q$ and applying partial summation again one obtains $|L'(s, \chi)| \ll (\log q)^2$ for $s > 1 - 1/\log q$. Now, if β is a zero of $L(s, \chi)$ with $1 - cq^{-\varepsilon} < \beta < 1$, the mean-value theorem gives

$$L(1, \chi) = L(1, \chi) - L(\beta, \chi) = (1 - \beta)L'(\xi, \chi) \ll (\log q)^2 q^{-\varepsilon},$$

where ξ lies between β and 1. This contradicts Siegel's theorem if χ is primitive and the general result follows from the relation concerning L-functions of induced characters given in Section 15.6.

It is clear from the proof that the implied constant in Siegel's result $L(1, \chi) \gg q^{-\varepsilon}$ depends on the possible existence of a zero β of an L-function near to 1. This is the notorious 'Siegel zero'; it is generally conjectured that no such zero exists but, at present, it cannot be ruled out. Thus the implied constant cannot be explicitly determined and, as we say, Siegel's theorem is ineffective. This has important implications; in particular, as we remarked in Section 15.6, one consequence of Siegel's theorem is the class number estimate $h(d) \gg |d|^{\frac{1}{2} - \varepsilon}$ for an imaginary quadratic field $\mathbb{Q}(\sqrt{d})$ and the implied constant here is again ineffective. Nevertheless, through some deep work of Goldfeld, Gross and Zagier on elliptic curves, a weaker but effective result in this direction has now been established, namely

$$h(d) \gg (\log |d|)^{1 - \varepsilon}.$$

Thus, at least in principle, all the imaginary quadratic fields with any given class number k can now be found explicitly.

The case $k = 1$ was discussed in Section 7.4; it is the Gauss problem of determining all the imaginary quadratic fields with unique factorization and it is known through works of Baker and Stark of the 1960s that there are just nine such fields.[†] The case $k = 2$ was solved in the 1970s, again through works of Baker and Stark; there are in fact just 18 imaginary quadratic fields with class number 2. The case $k = 3$ was the first to be solved by the elliptic curve method, namely by Oesterlé in 1985, and after further results in this context for higher k by Arno and others, Watkins published a paper in 2004 claiming to have covered all $k \leq 100$; the work apparently took some seven months of computation.

15.8 Further reading

The book by Prachar referred to in Section 15.2 is *Primzahlverteilung* (Springer, 1957). The book by Trost referred to in the same section is *Primzahlen* (Birkhäuser, 1953). For further details relating to the first part of Section 15.3, involving the elements of group theory, see the book of Apostol referred to in Section 2.9. The remainder of the section and the rest of the chapter are inspired by the classic work of Davenport referred to in Section 1.7.

Siegel's theorem, the subject of Section 15.7, appeared in *Acta Arith.* **1** (1935), 83–86. Estermann's proof was published in *J. London Math. Soc.* **23** (1948), 275–279. A good account of the topic is given in Chandrasekharan's *Arithmetical Functions*, referred to in Section 2.9. For the work of Watkins see *Math. Comput.* **73** (2004), 907–938; this contains references to the most significant of the earlier papers.

15.9 Exercises

(i) Assuming the prime-number theorem, show that every interval $[a, b]$ with $0 < a < b$ contains a rational p/q with p, q primes.

(ii) Let $M(x) = \sum_{n \leq x} \mu(n)$. Verify that, for $\sigma > 1$,

$$\frac{1}{\zeta(s)} = s \int_1^\infty \frac{M(x)}{x^{s+1}} dx.$$

[†] In 1952, Heegner, a high-school teacher in Berlin, published a paper which claimed to give a solution to the problem based on the theory of elliptic modular functions; the work was not well understood and the experts of the time came to look upon the argument as wrong-headed or at best incomplete. After Stark's work, which was based on the same sphere of ideas, Heegner's paper was re-examined and, with hindsight, it was agreed that it gave a viable approach.

Show that if $M(x) = O(x^{\frac{1}{2}+\varepsilon})$ for every $\varepsilon > 0$ then the equation continues to hold for $\sigma > \frac{1}{2}$ and the Riemann hypothesis is true.

(iii) Let $f(x) \geq 0$ be a real function monotonic increasing for $x \geq 0$. Prove that if

$$\int_0^x f(u)du \sim g(x) \quad \text{as } x \to \infty$$

for a differentiable function $g(x)$ then $f(x) \sim g'(x)$ as $x \to \infty$.

(iv) Let $Q(x) = \sum_{n \leq x} \mu^2(n)$ so that $Q(x)$ is the number of square-free numbers not exceeding x. Show that $\zeta(s)/\zeta(2s) = \sum \mu^2(n)/n^s$ and verify from Section 14.10 Exercise (iv) that

$$\int_0^x Q(u)du \sim \frac{x^2}{2\zeta(2)} \quad \text{as } x \to \infty.$$

Deduce that $Q(x) \sim (6/\pi^2)x$ as $x \to \infty$.

(v) Let k be a positive integer and let $S(x) = \sum_{n \leq x} \sigma_k(n)$ where $\sigma_k(n) = \sum_{d|n} d^k$ as in Section 13.9 Exercise (i). Using the contour of Section 15.1 displaced a distance k to the right, prove that $S(x) \sim \zeta(k+1)x^{k+1}/(k+1)$ as $x \to \infty$.

(vi) Using the fact that $\pi(x) = x/\log x + O(x/(\log x)^2)$ deduce that the nth prime p_n satisfies

$$p_n = n \log n + n \log \log n + O(n).$$

(vii) Verify by partial summation that

$$\sum_{p \leq x} \log p = \pi(x) \log x - \int_2^x \frac{\pi(u)}{u} du.$$

Hence prove that

$$\sum_{p \leq p_n} \log p = n \log n + n \log \log n - n + o(n).$$

(viii) Prove from the Siegel–Walfisz theorem that the least prime in the arithmetical progression $a, a + q, a + 2q, \ldots$, with $(a, q) = 1$, is $\ll e^{q^\varepsilon}$ for any $\varepsilon > 0$, where the implied constant depends only on ε.

(ix) Show that, for $\sigma > 1$,

$$\sum_{p \equiv a \,(\mathrm{mod}\, q)} \frac{1}{p^s} = s \int_1^\infty \frac{\pi(x; q, a)}{x^{s+1}} dx.$$

(x) Prove that

$$\sum_{\substack{p \le x \\ p \equiv a \,(\mathrm{mod}\, q)}} \frac{1}{p} = \frac{\pi(x; q, a)}{x} + \int_2^x \frac{\pi(u; q, a)}{u^2}\, du.$$

(xi) Assuming the Siegel–Walfisz theorem, verify the expression for the latter sum as stated in Section 13.2.

16

The sieve and circle methods

16.1 The Eratosthenes sieve

Eratosthenes observed in ancient Greek times that if, for a given $x \geq 1$, one deletes from the natural numbers $\leq x$ all multiples of 2, of 3, of 5 and so on up to the largest prime $\leq \sqrt{x}$ then, apart from 1, only the primes between $\sqrt{x}$ and x remain. This can be expressed by the following result usually attributed to Legendre.

Theorem 16.1 (Legendre's formula) *Let P denote the product of all primes* $\leq \sqrt{x}$. *Then*

$$\pi(x) - \pi(\sqrt{x}) + 1 = \sum_{d|P} \mu(d)[x/d].$$

Proof The theorem follows from the basic property of the Möbius function, that is, $\sum_{d|n} \mu(d)$ is 0 if $n > 1$ and 1 if $n = 1$. Clearly $\pi(x) - \pi(\sqrt{x}) + 1$ is just the number of $n \leq x$ with $(n, P) = 1$ and so can be expressed as

$$\sum_{n \leq x} \sum_{d|(n,P)} \mu(d) = \sum_{d|P} \mu(d) \sum_{n \leq x, d|n} 1 = \sum_{d|P} \mu(d)[x/d].$$

Alternatively we can apply the inclusion–exclusion principle to the numbers $n \leq x$. Defining $p_1, \ldots, p_k$ as the primes not exceeding $\sqrt{x}$, there are k possible properties corresponding to n being divisible by one of $p_1, \ldots, p_k$. The principle implies that the number of n that are not divisible by any of them is

$$[x] - \sum_r [x/p_r] + \sum_{r>s} [x/p_r p_s] - \sum_{r>s>t} [x/p_r p_s p_t] + \cdots$$

and Legendre's formula follows. □

197

As an application, let us take $x = 50$; then the primes $\leq \sqrt{x}$ are 2,3,5,7 and the right-hand side becomes $50 - 58 + 22 - 2 = 12$; we conclude that there are 11 primes between 8 and 50 and so $\pi(50) = 15$. The process of striking out multiples of primes described above is called the sieve of Eratosthenes and it is the most primitive example of a sieve method. Another, more powerful sieving technique was discovered by Viggo Brun in 1920 and much has followed; indeed techniques of this kind are now used routinely throughout analytic number theory and they have thrown light on many problems, most notably the famous twin-prime conjecture that there are infinitely many primes p such that $p + 2$ is also a prime, and Goldbach's conjecture that every even integer greater than 2 is the sum of two primes. We shall say more about these in Section 16.3.

Legendre's formula cannot be used immediately to give a non-trivial estimate for $\pi(x)$. Indeed if we replace $[x/d]$ by x/d then the right-side becomes

$$x \sum_{d \mid P} \mu(d)/d + O\left(\sum_{d \mid P} |\mu(d)|\right) = x \prod_{p \leq \sqrt{x}} (1 - 1/p) + O(\tau(P)),$$

where τ is the divisor function. By Theorem 13.7, the 'main term' here is $\ll x/\log x$; in fact we see directly that

$$\prod_{p \leq \sqrt{x}} (1 - 1/p)^{-1} \geq \sum_{n \leq \sqrt{x}} (1/n) \geq \log(\sqrt{x})$$

and so the main term is $\leq 2x/\log x$. However, the 'error term' is $O(2^{\pi(\sqrt{x})})$ which is clearly in excess. Nevertheless, if we sieve only up to the bound $y = \log x/\log 4$, taking $P = \prod_{p \leq y} p$, then we obtain the estimate $x/\log y$ for the main term and, since $\pi(y) \leq y$, the error term is given by $O(\sqrt{x})$. We conclude that $\pi(x) - \pi(y) + 1$ is at most $x/\log y + O(\sqrt{x})$ whence, for any $\varepsilon > 0$ and sufficiently large x,

$$\pi(x) \leq (1 + \varepsilon)x/\log\log x.$$

Though significantly weaker than the prime-number theorem or indeed the basic theory of Tchebychev, the estimate is non-trivial and it is indicative of what can be expected by the application of sieve methods.

16.2 The Selberg upper-bound sieve

In the period 1946 to 1951, Selberg developed a new sieve method of great importance for analytic number theory; it arose from earlier ideas of Brun as indicated above. We shall focus entirely on the so-called 'upper-bound' sieve, since it is here that the techniques take their simplest and most elegant form. However, the associated 'lower-bound' sieve lies deeper and gives us some of the most striking applications; we refer to the literature for details.

Let $f(n)$ be a polynomial with integer coefficients. The case $f(n) = n$ will correspond to the sieve of Eratosthenes and $f(n) = n(n+2)$ will relate to the twin-prime conjecture. Let M and N be integers such that $1 \leq M \leq N$. Our object is to estimate the number S of the integers $f(1), f(2), \dots, f(N)$ that are not divisible by any of the primes $p \leq M$.

Plainly S is the number of integers n with $1 \leq n \leq N$ such that $(f(n), P) = 1$, where P is the product of all the primes $p \leq M$. Hence we have

$$S = \sum_{n=1}^{N} \sum_{d \mid (f(n), P)} \mu(d) = \sum_{d \mid P} \mu(d) S(d),$$

where $S(d)$ is the number of $f(n)$ with $1 \leq n \leq N$ that are divisible by d. Now Selberg observed that if $\lambda(n)$ is any real function with $\lambda(1) = 1$ then, for each positive integer k, we have

$$\left(\sum_{d \mid k} \lambda(d) \right)^2 \geq \sum_{d \mid k} \mu(d).$$

Further, if k divides P, then

$$\left(\sum_{d \mid k} \lambda(d) \right)^2 = \sum_{d \mid k} \rho(d),$$

where

$$\rho(d) = \sum_{\substack{d' \mid P \\ \{d', d''\} = d}} \sum_{d'' \mid P} \lambda(d') \lambda(d'')$$

with $\{d', d''\}$ denoting the lowest common multiple of d' and d''. Thus we get

$$S \leq \sum_{n=1}^{N} \sum_{d \mid (f(n), P)} \rho(d) = \sum_{d \mid P} \rho(d) S(d).$$

Now, for every divisor d of P, we define $s(d)$ to be the number of elements $f(n)$ with $1 \leq n \leq d$ that are divisible by d and we put $q(d) = s(d)/d$. Since $f(x)$ is a polynomial we have $f(m) \equiv f(n) \pmod{d}$ whenever $m \equiv n \pmod{d}$ and it follows that $S(d) = [N/d]s(d) + r(d)$, where $0 \leq r(d) \leq s(d)$. This gives $S(d) = Nq(d) + R(d)$, with $|R(d)| \leq dq(d)$, and we have proved the following.

Proposition 16.1 *We have*

$$S \leq N \sum_{d|P} \rho(d)q(d) + \sum_{d|P} |\rho(d)R(d)|.$$

We recall that the discussion so far is valid for any real function $\lambda(n)$ subject only to $\lambda(1) = 1$. The crux of the method lies in the fact that we can select this function such that the first sum on the right in Proposition 16.1 is minimal.

For each divisor d of P we define

$$w(d) = \frac{1}{q(d)} \prod_{p|d} (1 - q(p)).$$

Here we assume that $0 < s(p) < p$ and so $0 < q(p) < 1$ for all $p \leq M$; if $s(p) = 0$ or p for some p we would have either none or all of $f(1), f(2), \ldots, f(N)$ divisible by p and we would simply omit the prime from the definition of P. The functions $w(d)$ and $1/q(d)$ are Möbius inverses in the sense that

$$\frac{1}{q(d)} = \sum_{\delta|d} w(\delta), \qquad w(d) = \sum_{\delta|d} \frac{\mu(d/\delta)}{q(\delta)}.$$

Now, by the definition of $\rho(d)$, the first sum on the right in Proposition 16.1 is

$$\sum_{d|P} \rho(d)q(d) = \sum_{d'|P} \sum_{d''|P} \lambda(d')\lambda(d'')q(\{d', d''\}).$$

By the Chinese remainder theorem the function $s(d)$ is multiplicative. This implies that also $q(d)$ is multiplicative, whence we have

$$q(\{d', d''\}) = q(d')q(d'')/q((d', d''));$$

see Section 16.6 Exercise (i). Hence, by the first of the Möbius relations above, it follows that the double sum can be written as a quadratic form,

$$\sum_{\delta|P} w(\delta)(v(\delta))^2, \quad \text{with} \quad v(\delta) = \sum_{d|P, \delta|d} \lambda(d)q(d).$$

Here the coefficients $w(\delta)$ are independent of λ and the variables $v(\delta)$ depend linearly on λ; we proceed to determine the values of the variables that minimize the form.

By Möbius inversion, we have

$$\lambda(d)q(d) = \sum_{\delta|P, d|\delta} \mu(\delta/d)v(\delta);$$

this defines λ in terms of the $v(\delta)$. Since $q(1) = s(1) = 1$, the condition $\lambda(1) = 1$ becomes

$$\sum_{\delta \mid P} \mu(\delta) v(\delta) = 1.$$

We take $v(\delta) = 0$ for $\delta > M$; then by the method of Lagrange multipliers we find that the desired values of $v(\delta)$ for $\delta \leq M$ are given by

$$v(\delta) = \frac{\mu(\delta)}{W w(\delta)} \quad \text{where} \quad W = \sum_{\delta \mid P, \delta \leq M} \frac{1}{w(\delta)}.$$

This can be seen directly; since δ is square-free, we have $\sum \mu^2(\delta)/w(\delta) = W$, whence the linear condition is satisfied and gives

$$\sum_{\delta \mid P} w(\delta)(v(\delta))^2 = \sum_{\delta \mid P, \delta \leq M} \frac{1}{w(\delta)} \left(w(\delta) v(\delta) - \frac{\mu(\delta)}{W} \right)^2 + \frac{1}{W}.$$

Plainly our values of $v(\delta)$ furnish the minimum $1/W$ of the quadratic form. Hence the following holds.

Proposition 16.2 *We have*

$$S \leq N/W + \sum_{d \mid P} |\rho(d) R(d)|.$$

It remains now only to estimate the second term on the right in Proposition 16.2. By the definition of $\rho(d)$, the expression is at most

$$\sum_{d' \mid P} \sum_{d'' \mid P} |\lambda(d') \lambda(d'') R(\{d', d''\})|.$$

From the estimate $|R(d)| \leq d q(d)$ we see that $|R(\{d', d''\})| \leq d' d'' q(d') q(d'')$. Thus

$$\sum_{d \mid P} |\rho(d) R(d)| \leq \left(\sum_{d \mid P} d |\lambda(d) q(d)| \right)^2.$$

But we have

$$\lambda(d) q(d) = \sum_{\delta \mid P, d \mid \delta} \mu(\delta/d) v(\delta),$$

and, on substituting for the $v(\delta)$ and using the fact that w is multiplicative, this gives

$$|\lambda(d) q(d)| \leq \frac{1}{W} \sum_{\delta \mid P, \delta \leq M, d \mid \delta} \frac{1}{w(\delta)} \leq \frac{1}{W w(d)} \sum_{\delta' \mid P, \delta' \leq M} \frac{1}{w(\delta')} = \frac{1}{w(d)}.$$

for $d \le M$, and it gives $\lambda(d)q(d) = 0$ otherwise. Hence

$$\sum_{d|P} d|\lambda(d)q(d)| \le \sum_{d|P, d\le M} (d/w(d)) \le MW.$$

The second term on the right in Proposition 16.2 is therefore at most $(MW)^2$ and we have shown that the following holds.

Theorem 16.2 *We have* $S \le N/W + (MW)^2$.

16.3 Applications of the Selberg sieve

We now study the estimate for S in Theorem 16.2 when the polynomial $f(n)$ takes some simple forms. Here we recall that

$$W = \sum_{d|P, d\le M} \frac{1}{w(d)} \quad \text{where} \quad w(d) = \frac{1}{q(d)} \prod_{p|d} (1 - q(p))$$

and $q(d) = s(d)/d$ with $s(d)$ the number of elements $f(n)$ with $1 \le n \le d$ that are divisible by d.

Corollary 16.1 (Primes in an interval) *For any integer $y > 0$ and any $\varepsilon > 0$ we have*

$$\pi(x + y) - \pi(y) < (2 + \varepsilon)x/\log x \quad as \quad x \to \infty.$$

Proof We take $f(n) = n + y$. Then $s(d) = 1$ and $q(d) = 1/d$ so that $w(d)$ is Euler's function $\phi(d)$. Thus, noting that $\mu^2(n) = 1$ if n is square-free and 0 otherwise, we obtain

$$W = \sum_{d\le M} \frac{\mu^2(d)}{\phi(d)} = \sum_{d\le M} \mu^2(d) \prod_{p|d} \left(\frac{1}{p} + \frac{1}{p^2} + \dots\right) \ge \sum_{m\le M} \frac{1}{m} > \log M.$$

Further, from the first expression for W above and Theorem 13.7, we see that

$$W \le \prod_{p\le M} \left(1 + \frac{1}{\phi(p)}\right) = \prod_{p\le M} \left(1 - \frac{1}{p}\right)^{-1} \ll \log M.$$

But we have $S \ge \pi(N + y) - \pi(y) - \pi(M)$ whence, taking $N = [x]$, $M = [x^{\frac{1}{2}-\varepsilon}]$ and observing that $\pi(M) < M$, the result follows. $\square$

An improvement in the estimate replacing $2 + \varepsilon$ by 2 was given in 1973 by Montgomery and Vaughan and this is the best result to date.

Corollary 16.2 (Brun–Titchmarsh inequality) *Suppose that $(a, q) = 1$ and $0 < q < x$. Then for any $\varepsilon > 0$ there exists $x_0 = x_0(\varepsilon)$ such that for $x > x_0$ we have*

$$\pi(x; q, a) < \frac{(2 + \varepsilon)x}{\phi(q) \log(x/q)}.$$

Proof We take $f(n) = nq + a$. Then $s(p) = 1$ if $(p, q) = 1$ and $s(p) = 0$ if $(p, q) > 1$. Thus $P = \prod p$ where the product is over all primes $p \leq M$ such that $p \nmid q$. We have $w(d) = \phi(d)$ for $(d, q) = 1$ and thus, as in the proof of Corollary 16.1, we obtain $W \geq \sum 1/m$ where the sum is over all $m \leq M$ with $(m, q) = 1$. But we have

$$\frac{q}{\phi(q)} W = \prod_{p|q} \left(1 + \frac{1}{p} + \frac{1}{p^2} + \cdots\right) W \geq \sum_{m \leq M} \frac{1}{m} > \log M.$$

Further, as before, we have $W \ll \log M$. Furthermore it is clear that $S \geq \pi(Nq + a; q, a) - \pi(M)$ assuming, as we may, that $0 < a < q$. On taking $N = [x/q] + 1$ and $M = [(x/q)^{\frac{1}{2} - \varepsilon}] + 1$ the result follows. $\qquad \square$

Corollary 16.3 (Twin-prime estimate) *The number of primes $p \leq x$ such that $p + 2$ is a prime is $\ll x/(\log x)^2$.*

Proof We take $f(n) = n(n + 2)$. Then $s(p) = 2$ for odd primes p and $s(2) = 1$. Thus $q(p) = 2/p$, $w(p) = \frac{1}{2}p - 1$ and $w(2) = 1$. Hence

$$W = \sum_{d \leq M} \frac{\mu^2(d)}{w(d)} = \sum_{d \leq M} \mu^2(d) \prod_{p|d, p>2} \left(\frac{2}{p} + \frac{2^2}{p^2} + \cdots\right).$$

Since the divisor function τ is multiplicative and $\tau(p^k) = k + 1 \leq 2^k$, we see that the sum on the right is at least

$$\sum_{m \leq M, 2 \nmid m} \frac{\tau(m)}{m} \geq \left(\sum_{m \leq \sqrt{M}, 2 \nmid m} \frac{1}{m}\right)^2 \gg (\log M)^2.$$

Further, by Theorem 13.7, we obtain

$$W \leq \prod_{2 < p \leq M} \left(1 + \frac{1}{w(p)}\right) = \prod_{2 < p \leq M} \left(1 - \frac{2}{p}\right)^{-1} \ll (\log M)^2.$$

Hence, on taking $N = [x]$ and $M = [x^{\frac{1}{4}}]$, Theorem 16.2 gives $S \ll x/(\log x)^2$. But if p is a prime with $M < p \leq N$ such that $p + 2$ is a prime then $p(p + 2)$ is counted in S and, since $\pi(M) < M$, this gives the result. $\qquad \square$

It follows from Corollary 16.3 by partial summation that if p runs through all the primes such that $p, p+2$ are twin primes then $\sum 1/p$ converges.

Corollary 16.4 (Goldbach estimate) *The number of ways that an even integer N can be expressed as a sum of two primes is $\ll N/(\log N)^2$.*

Proof We take $f(n) = n(N-n)$. Then as in Corollary 16.3 we have $w(p) = \frac{1}{2}p - 1$ for odd primes p and $w(2) = 1$. The result follows similarly. □

The proofs of Corollaries 16.3 and 16.4 can be extended to show that, if k is a positive even integer, then the number of primes $p \leq x$ such that $p + k$ is a prime is $\ll (k/\phi(k))x/(\log x)^2$ where the implied constant is absolute. Moreover the same estimate holds for the number of primes $p \leq x$ such that $kp + 1$ is a prime. In these instances we take $f(n) = n(n+k)$ and $f(n) = n(kn+1)$ respectively and we note that then $w(p) = \frac{1}{2}p - 1$ if $p \nmid k$ and $w(p) = p - 1$ if $p \mid k$.

We remark finally that the 'lower bound' sieve gives results in the opposite direction and it was ideas of this kind that enabled Chen Jing-Run to prove that there exist infinitely many primes p such that $p + 2$ has at most two prime factors and that every sufficiently large even integer is the sum of a prime and a number with at most two prime factors. These are the best results to date in the direction of the famous twin-prime and Goldbach conjectures.

16.4 The large sieve

This refers to a general technique for estimating certain exponential sums. The theory dates back to results of Linnik in 1941 and Rényi in 1948. A crucial advance in this context was made by Roth in 1965 and his work was followed shortly afterwards by further important contributions of Bombieri and, independently, of A. I. Vinogradov. They showed that the Riemann hypothesis is true in a certain 'average' sense relating to primes in arithmetical progressions.

The basic result on the large sieve is concerned with the values of the exponential sum

$$S(x) = \sum_{n=1}^{N} c_n \exp(2\pi i n x),$$

where $c_1, \ldots, c_N$ denote any real or complex numbers. It asserts as follows.

Theorem 16.3 *If $x_1, \ldots, x_R$ are real numbers such that $\|x_r - x_s\| > \delta > 0$ for all r, s with $r \neq s$, where $\|x\|$ denotes the distance of x from the nearest integer, then we have*

$$\sum_{r=1}^{R} |S(x_r)|^2 \leq (\delta^{-1} + 2\pi N) \sum_{n=1}^{N} |c_n|^2.$$

Proof We begin by noting that, for any real y,

$$|S(x)|^2 = |S(y)|^2 + \int_y^x \frac{d}{du} |S(u)|^2 \, du,$$

where, as usual, $\int_y^x$ is interpreted as $-\int_x^y$ if $y > x$. On putting $\varepsilon = \frac{1}{2}\delta$, this gives

$$2\varepsilon |S(x)|^2 = \int_{x-\varepsilon}^{x+\varepsilon} |S(y)|^2 \, dy + \int_{x-\varepsilon}^{x+\varepsilon} \int_y^x \frac{d}{du} |S(u)|^2 \, du \, dy.$$

It is readily verified that $|(d/du)|S(u)|| \leq |S'(u)|$, whence the second term on the right is at most

$$2\varepsilon \int_{x-\varepsilon}^{x+\varepsilon} |S(u)S'(u)| \, du.$$

Now, by hypothesis, the intervals $(x_r - \varepsilon, x_r + \varepsilon)$, when translated by integers to $[0, 1)$, are disjoint and hence

$$\sum_{r=1}^{R} |S(x_r)|^2 \leq (2\varepsilon)^{-1} \sum_{r=1}^{R} \int_{x_r-\varepsilon}^{x_r+\varepsilon} |S(y)|^2 \, dy + \sum_{r=1}^{R} \int_{x_r-\varepsilon}^{x_r+\varepsilon} |S(u)S'(u)| \, du$$

$$\leq (2\varepsilon)^{-1} \int_0^1 |S(y)|^2 dy + \int_0^1 |S(u)S'(u)| \, du.$$

The first of these integrals is just $\sum_{n=1}^{N} |c_n|^2$ and, by the Cauchy–Schwarz inequality, we get

$$\left(\int_0^1 |S(u)S'(u)| \, du \right)^2 \leq \left(\int_0^1 |S(u)|^2 \, du \right) \left(\int_0^1 |S'(u)|^2 \, du \right)$$

$$= \left(\sum_{n=1}^{N} |c_n|^2 \right) \left(\sum_{n=1}^{N} |2\pi n c_n|^2 \right).$$

The theorem follows easily. □

As a corollary to Theorem 16.3 we obtain the original result of Roth to the effect that the elements of any set of Z integers between 1 and N inclusive must, in a certain sense, be well distributed among the congruence classes (mod p) for primes $p \leq \sqrt{N}$. More precisely, let $Z(a, p)$ be the number of integers in the set which are congruent to a (mod p). Then the following holds.

Theorem 16.4 *For any positive integer Q, we have*

$$\sum_{p \leq Q} \sum_{a=1}^{p} p(Z(a, p) - Z/p)^2 \ll (Q^2 + N)Z.$$

Proof We take $a_1, a_2, \ldots, a_Z$ as the given set of integers and we put $c_n = 1$ if $n = a_m$ for some m and 0 otherwise. Then

$$S(x) = \sum_{m=1}^{Z} \exp(2\pi i a_m x).$$

Further, we take $x_1, \ldots, x_R$ as the points a/p with $1 \leq a < p$ and $p \leq Q$ so that $\delta = 1/Q^2$. This gives

$$\sum_{p \leq Q} \sum_{a=1}^{p-1} |S(a/p)|^2 = \sum_{r=1}^{R} |S(x_r)|^2.$$

By Theorem 16.3, the latter sum is $\ll (Q^2 + N)Z$. Further, we have

$$\sum_{a=1}^{p-1} |S(a/p)|^2 + Z^2 = \sum_{a=1}^{p} \sum_{n=1}^{Z} \sum_{n'=1}^{Z} \exp(2\pi i (a_n - a_{n'})a/p).$$

The latter expression is just p times the number of pairs n, n' such that $a_n \equiv a_{n'}$ (mod p). Thus, since $Z = \sum_{a=1}^{p} Z(a, p)$, it can be written in the form

$$p \sum_{a=1}^{p} (Z(a, p))^2 = p \sum_{a=1}^{p} (Z(a, p) - Z/p)^2 + Z^2$$

and this establishes the theorem.   □

The main accomplishment arising from this sphere of ideas has been a result, customarily referred to as the Bombieri–Vinogradov large sieve inequality, concerning the theory of primes in arithmetical progressions; see Chapter 15. Let $\psi(x; a, q)$ be the generalized Tchebychev function in this context, that is, $\psi(x; a, q) = \sum \Lambda(n)$ where the sum is taken over all natural numbers $n \leq x$ with $n \equiv a$ (mod q). Then the result in question states that, for each positive A, there is a B such that

$$\sum_{q \leq \sqrt{x}(\log x)^{-B}} \max_{y \leq x} \max_{(a,q)=1} \left| \psi(y; a, q) - \frac{y}{\phi(q)} \right| \ll x(\log x)^{-A}.$$

Now there is a so-called generalized Riemann hypothesis in this context which asserts that the non-trivial zeros not only of $\zeta(s)$ but in fact of all the functions $L(s, \chi)$ as χ runs through the characters mod q lie on the line $\sigma = \frac{1}{2}$. In particular, for $q \le x$, this gives

$$\psi(x; a, q) = \frac{x}{\phi(q)} + O(\sqrt{x}(\log x)^2).$$

The large sieve inequality shows that a result of the latter kind is valid in a certain average sense and it has found wide application.

16.5 The circle method

The Hardy–Littlewood–Vinogradov or circle method was developed during the early part of the last century for solving a wide range of number-theoretical problems of additive type. The work has its origins in studies of Hardy and Ramanujan on the partition function $p(n)$ defined as the number of representations of a positive integer n as a sum of positive integers. It is generated by the formal identity

$$F(x) = \prod_{m=1}^{\infty} (1 - x^m)^{-1} = 1 + \sum_{n=1}^{\infty} p(n)x^n.$$

This gives

$$p(n) = \frac{1}{2\pi i} \int_C \frac{F(z)}{z^{n+1}} dz,$$

where C is any closed contour about the origin which is contained within the unit disc $|z| < 1$. By taking C as a circle centre the origin with radius close to 1 and dividing the path of integration into arcs near to the singularities of $F(z)$, that is, the roots of unity $e^{2\pi i a/q}$ with rational a/q, Hardy and Ramanujan succeeded in 1918 in proving that[†]

$$p(n) \sim \frac{\exp(\pi \sqrt{(2n/3)})}{4n\sqrt{3}} \quad \text{as} \quad n \to \infty.$$

This was the starting point of the circle method. In a series of papers entitled 'Some problems of "Partitio Numerorum"' beginning in 1920, Hardy and Littlewood turned their attention to the famous Waring problem. Waring had conjectured in 1770 that every natural number can be represented as a sum of 4 squares, 9 cubes, 19 biquadrates 'and so on'. This is now interpreted to mean

---

[†] The same result was proved independently in 1920 by J. V. Uspensky.

that, for every integer $k \geq 2$, there exists $s = s(k)$ such that every positive integer n is representable as $x_1^k + \cdots + x_s^k$ with non-negative integers $x_1, \ldots, x_s$. Plainly if the conjecture holds for a particular k then there is a least s for which it is true and we define this minimum value as $g(k)$. The case $k = 2$ of the conjecture is classical (see Section 5.5) and further special cases with $k \leq 10$, $k \neq 9$, were established subsequently. But it was not until 1909 that Hilbert, in a major work, proved the existence of $g(k)$ for every k. Hilbert's proof involved arguments of a combinatorial kind together with certain algebraic identities and though in principle it gave a bound for $g(k)$ it was obvious that this would be very large.[‡]

Hardy and Littlewood gave an independent proof of Hilbert's theorem. They showed that, for $s \geq s_0(k)$, the number $r(n)$ of representations of n in the above form satisfies

$$r(n) \sim \frac{(\Gamma(1+1/k))^s}{\Gamma(s/k)} n^{(s/k)-1} \mathfrak{S}(n) \quad \text{as} \quad n \to \infty,$$

where the function $\mathfrak{S}(n)$, known as the singular series, exceeds a positive constant. In particular this gives $r(n) \to \infty$ as $n \to \infty$ and so there is certainly at least one representation if n is sufficiently large; since every n is trivially a sum of kth powers of 1, this is enough to establish Hilbert's theorem. We define, as customary, $G(k)$ to be the least s such that every sufficiently large n is the sum of s kth powers; then the result implies that $G(k)$ exists and satisfies $G(k) \leq s_0(k)$. The function $G(k)$ is in some ways more fundamental than $g(k)$ since the latter is affected by the need to represent a few small exceptional n; see Section 5.7 Exercise (ix). The argument of Hardy and Littlewood when refined through some later work of Hua gives $s_0(k) = 2^k + 1$ and, for small k, this is the best value to date.

Beginning in 1928 I. M. Vinogradov made important improvements to the circle method. Hardy and Littlewood, in their studies on Waring's problem, had based their arguments on the relation

$$r(n) = \frac{1}{2\pi i} \int_C \frac{(F(z))^s}{z^{n+1}} dz,$$

where C is the circle centre the origin with radius $1 - 1/n$ and

$$F(z) = \sum_{m=0}^{\infty} z^{m^k}.$$

[‡] G. J. Rieger worked out an estimate for $g(k)$ in 1953, namely $(2k+1)^{260(k+3)^{3k+8}}$.

Vinogradov replaced $F(z)$ by a trigonometrical sum

$$f(\alpha) = \sum_{m \leq n^{1/k}} \exp(2\pi i \alpha m^k)$$

and he noted that then

$$r(n) = \int_0^1 (f(\alpha))^s \exp(-2\pi i n \alpha) d\alpha,$$

as is clear since, for any integer j, we have

$$\int_0^1 \exp(2\pi i j \alpha) d\alpha = \begin{cases} 1 & \text{if } j = 0, \\ 0 & \text{otherwise}. \end{cases}$$

Vinogradov obtained an estimate for $G(k)$ of the form $G(k) = O(k \log k)$. It has gone through numerous refinements and the best result to date in this direction, obtained by Wooley in 1992, asserts that

$$G(k) \leq k(\log k + \log \log k + O(1)).$$

A classical result of 1908 due to Hurwitz and Maillet gives the best current lower bound for $G(k)$, namely $G(k) \geq k + 1$; see Section 16.8 Exercise (vi). There has been much further work in this context, particularly for small values of k, but, apart from the classical $G(2) = 4$, the only exact evaluation is $G(4) = 16$ which was obtained by Davenport in 1939. As regards $g(k)$, it has been conjectured that

$$g(k) = 2^k + [(3/2)^k] - 2,$$

and this is known to hold provided that $\|(3/2)^k\| \geq (3/4)^k$ where $\|x\|$ denotes the distance of x from the nearest integer. Mahler has shown that the latter inequality has only finitely many exceptions. Further, through machine calculations, the conjecture has been verified for $k \leq 2 \times 10^5$.[††] Slightly modified expressions for $g(k)$ have been shown to apply in the event that the conjecture is not correct and thus we now have an almost complete evaluation of $g(k)$.

The circle method has been applied widely and has yielded results well beyond the scope of the original Waring problem. Thus, for instance, in 1957 Lewis proved that a cubic form with integer coefficients and sufficiently many variables has a non-trivial integral zero, and Birch extended this a little later to forms of arbitrary odd degree. Hardy and Littlewood themselves applied the circle method to problems involving primes and, in particular, in the third of

---

[††] The case $k = 4$ resisted a complete solution until 1986 when Balasubramanian, Deshouillers and Dress succeeded in showing that g(4) = 19.

their 'Partitio Numerorum' series, they showed that, if the generalized Riemann hypothesis is assumed true (see Section 16.4), then every sufficiently large odd integer is the sum of three primes. In 1937 Vinogradov succeeded in removing the assumption about the Riemann hypothesis and so established the three-prime theorem unconditionally. He also gave an unconditional proof of another of the famous Hardy–Littlewood results, namely that almost all even numbers are the sum of two primes.

16.6 Additive prime number theory

To illustrate the principal features of the circle method, we shall study the equation

$$a_1 p_1 + \cdots + a_s p_s = b,$$

where $s \geq 3$, $a_1, \ldots, a_s$ are non-zero integers, b is an integer and $p_1, \ldots, p_s$ are odd primes. Let n be a positive integer and let $\mathfrak{N}(n, b)$ be the number of solutions of the associated equation

$$a_1 n_1 + \cdots + a_s n_s = b$$

in positive integers $n_1, \ldots, n_s$ not exceeding n. Further, let $\mathfrak{S}(b)$ be the singular series given by

$$\mathfrak{S}(b) = \sum_{q=1}^{\infty} \left(\prod_{j=1}^{s} c_q(a_j) \right) c_q(b)(\phi(q))^{-s},$$

where $c_q(m)$ is Ramanujan's sum, that is,

$$c_q(m) = \sum_{\substack{a=1 \\ (a,q)=1}}^{q} e^{2\pi i a m / q} = \frac{\mu(q/(m,q))\phi(q)}{\phi(q/(m,q))};$$

see Section 2.10 Exercise (vii) and Section 16.8 Exercise (vii). Then we have the following result.

Theorem 16.5 *Assuming that $(a_i, a_j) = 1$ for $i \neq j$, the number of solutions of the above equation in odd primes $p_1, \ldots, p_s$ not exceeding n is given by*

$$\frac{\mathfrak{N}(n, b)\mathfrak{S}(b)}{(\log n)^s} + O\left(\frac{n^{s-1}}{(\log n)^{s+1}} \right),$$

where the constant implied by the O-term depends only on s and $a_1, \ldots, a_s$.

With the hypothesis of the theorem, the singular series reduces to $\mathfrak{S}(b) = \prod_p \chi(p)$ taken over all primes p, where $\chi(p)$ is defined by

$$\chi(p) = 1 + \begin{cases} (-1)^{s+1}(p-1)^{-s} & \text{if } p \nmid a_1 \cdots a_s b, \\ (-1)^{s+1}(p-1)^{2-s} & \text{if } p \mid (a_1 \cdots a_s, b), \\ (-1)^s(p-1)^{1-s} & \text{otherwise.} \end{cases}$$

In particular, we see that the series converges. Further, we observe that $\mathfrak{S}(b) \neq 0$ if $a_1 + \cdots + a_s \equiv b \pmod 2$ and this condition is obviously necessary if there is to be any solution to the given equation in odd primes.

Specializing to the case when $s = 3$, $a_1 = a_2 = a_3 = 1$ and $b = n$ we get

$$\mathfrak{S}(b) = \prod_{p \mid n}(1 - (p-1)^{-2}) \prod_{p \nmid n}(1 + (p-1)^{-3})$$

and if n is odd then

$$\mathfrak{S}(b) > \prod_{p > 2}(1 - (p-1)^{-2}) > \prod_p (1 - p^{-2}) = \frac{6}{\pi^2}.$$

Since also, in this case, $\mathfrak{N}(n, b) \sim \frac{1}{2}n^2$ as $n \to \infty$, Theorem 16.5 shows that the number of reprentations of n as a sum of three primes is asymptotic to $\frac{1}{2}n^2 \mathfrak{S}(b)/(\log n)^3$ and we have the three-prime theorem. Furthermore, on taking $s = 3$, $a_1 = a_3 = 1$, $a_2 = -2$ and $b = 0$, we see that the same asymptotic expression applies to the number of solutions of the equation $p_1 + p_3 = 2p_2$ in primes not exceeding n; in this instance $\mathfrak{S}(b) = 2 \prod_{p > 2}(1 - (p-1)^{-2})$ and we obtain the well-known result, first proved by Chowla and van der Corput, that there exist infinitely many triples of primes in arithmetic progression.

We briefly outline a proof of Theorem 16.5. To begin with we note that the number of solutions in question can be expressed as

$$\int_0^1 S(a_1 \alpha) \cdots S(a_s \alpha) e(-b\alpha) d\alpha,$$

where

$$S(\alpha) = \sum_{2 < p \le n} e(p\alpha)$$

and, for brevity, we have written $e(x)$ for $\exp(2\pi i x)$. For all integers a, q with $1 \le a \le q$, $(a, q) = 1$ and $q \le (\log n)^\kappa$ for some numerical constant $\kappa > 0$ we define the 'major arc' $\mathfrak{M}_{a,q}$ as the interval of real α with

$$\left| \alpha - \frac{a}{q} \right| < \frac{(\log n)^\kappa}{n}.$$

We make the convention that the right-hand half of $\mathfrak{M}_{1,1}$ is translated to the left by an amount 1 and then, for n sufficiently large, the $\mathfrak{M}_{a,q}$ are non-overlapping and contained in $[0, 1)$. We take $\mathfrak{M}$ as the union of all the $\mathfrak{M}_{a,q}$ and we define the 'minor arcs' $\mathfrak{m}$ as the complement of $\mathfrak{M}$ in $[0, 1)$. By Dirichlet's theorem on Diophantine approximation (see Section 6.1), for any real α there exist integers a, q with $1 \le a \le q$, $(a, q) = 1$ and $q \le n/(\log n)^{\kappa}$ such that

$$\left| \alpha - \frac{a}{q} \right| < \frac{(\log n)^{\kappa}}{nq}$$

and if α is in $\mathfrak{m}$ we have $q > (\log n)^{\kappa}$.

Now for each α on $\mathfrak{M}_{a,q}$ we put $\beta = \alpha - a/q$ and we find that then

$$S(a_j\alpha) = (c_q(a_j)/\phi(q))I(a_j\beta) + O\left(ne^{-\lambda\sqrt{(\log n)}}\right)$$

for some numerical constant $\lambda > 0$, where

$$I(\beta) = \sum_{m=2}^{n} e(m\beta)/\log m.$$

Indeed this results on writing

$$S(\alpha) = \sum_{\substack{l=1 \\ (l,q)=1}}^{q} e(la/q) \sum_{\substack{2<p\le n \\ p\equiv l \,(\mathrm{mod}\, q)}} e(p\beta) + \sum_{\substack{2<p\le n \\ p|q}} e(p\alpha)$$

and applying partial summation and the Siegel–Walfisz theorem; see Section 15.6. Hence, on noting that by Abel's lemma

$$|I(\beta)| \ll \min(n/\log n,\, 1/\|\beta\|)$$

where $\|\beta\|$ denotes the distance of β from the nearest integer, we obtain

$$\int_{\mathfrak{M}} S(a_1\alpha) \cdots S(a_s\alpha)e(-b\alpha)d\alpha$$
$$= \mathfrak{S}(b) \int_{-1/2}^{1/2} I(a_1\beta) \cdots I(a_s\beta)e(-b\beta)d\beta + O(n^{s-1}/(\log n)^{s+1}).$$

The last integral evaluates to $\sum(\log m_1 \cdots \log m_s)^{-1}$ summed over all integers $m_1, \ldots, m_s$ with $a_1m_1 + \cdots + a_sm_s = b$ and $2 \le m_j \le n$ and it is readily verified that, within the margin of error above, this is $\mathfrak{N}(n, b)/(\log n)^s$.

To complete the proof of Theorem 16.5 it remains to show that the integral over the minor arcs lies within the same margin of error. This is the deepest aspect of the circle method. Here Vinogradov used a sieving technique to establish a non-trivial estimate for $S(a/q)$ for integers a, q with $1 \le a \le q$,

$(a, q) = 1$ and $(\log n)^{\kappa} < q \le n/(\log n)^{\kappa}$. In our present context this implies that $|S(a_j\alpha)| \ll n/(\log n)^{\nu}$ for all α on $\mathfrak{m}$ and any $\nu > 0$ assuming that κ is sufficiently large. Thus we have

$$\int_{\mathfrak{m}} |S(a_1\alpha) \cdots S(a_s\alpha)| d\alpha \ll (n/(\log n)^{\nu})^{s-2} \int_{\mathfrak{m}} |S(a_1\alpha)S(a_2\alpha)| d\alpha.$$

Now by the Cauchy–Schwarz inequality the last integral is at most

$$\left(\int_0^1 |S(a_1\alpha)|^2 \, d\alpha \right)^{1/2} \left(\int_0^1 |S(a_2\alpha)|^2 \, d\alpha \right)^{1/2}$$

and this evaluates to $\pi(n) - 1 < n$. The desired estimate for the integral over the minor arcs follows at once.

16.7 Further reading

The classic introduction to sieve methods by Halberstam and Richert referred to in Section 1.7 remains one of the best works in the field. A more recent book particularly recommended as an initiation to the subject is that by Cojocaru and Murty, *An Introduction to Sieve Methods and Their Applications* (Cambridge University Press, 2005).

The main books on the circle method are those of Vaughan as cited in Section 5.6 and of Davenport, *Analytic Methods for Diophantine Equations and Diophantine Inequalities* (Cambridge University Press, 2005).

Theorem 16.5 is due to Richert, *J. Reine Angew. Math.* **191** (1953), 179–198. In the case $s = 3$ the problem of bounding small prime solutions has been much studied; see the article by Ming-Chit Liu and Tianze Wang in *A Panorama of Number Theory or the View from Baker's Garden* (ed. G. Wüstholz, Cambridge University Press, 2002), pp. 311–324. The article includes remarks on the connection with the famous theorem of Linnik that the least prime p in the arithmetical progression $a, a + q, a + 2q, \ldots$ with $(a, q) = 1$ satisfies $p \ll q^L$ for some constant L; cf. Section 15.9 Exercise (viii).

The result on the existence of infinitely many triples of primes in arithmetic progression referred to in Section 16.6 was extended to arbitrarily long arithmetic progressions in the primes by Green and Tao in *Annals of Math.* **167** (2008), 481–547. This solved a long-standing open problem in the field.

16.8 Exercises

(i) By considering prime decompositions, verify that if q is a multiplicative function defined on the positive integers then, for all m, n, we have $q(\{m, n\})q((m, n)) = q(m)q(n)$.

(ii) Show that the Brun–Titchmarsh inequality can be extended to give, for any $y > 0$,

$$\pi(x + y; q, a) - \pi(y; q, a) < \frac{(2 + \varepsilon)x}{\phi(q) \log(x/q)} \quad \text{as} \quad x \to \infty.$$

(iii) (Rankin's trick) A natural number n is said to be y-smooth if all of its prime factors are $\leq y$. Let $\Psi(x, y)$ denote the number of y-smooth numbers $\leq x$. Show that, for any $\delta > 0$,

$$\Psi(x, y) \leq \sum_{n \leq x, \, p|n \Rightarrow p \leq y} \left(\frac{x}{n}\right)^{\delta} \leq x^{\delta} \prod_{p \leq y} \frac{1}{1 - p^{-\delta}}.$$

(iv) Verify that, if $\delta > \frac{1}{2} + \varepsilon$ for some given $\varepsilon > 0$, then

$$\Psi(x, y) \ll x^{\delta} \prod_{p \leq y} (1 + 1/p^{\delta}) \leq x^{\delta} \exp(\sum_{p \leq y} 1/p^{\delta})$$

where the implied constant is absolute. Defining $\delta = 1 - (\log 2)/\log y$ and using Theorem 13.6, deduce that $\Psi(x, y) \ll x^{\delta} (\log y)^2$.

(v) Show from Theorem 16.4 that if $Z(a, p) = 0$ for some a and for all $p \leq \sqrt{N}$ then $Z \ll N/\log \log N$. Deduce that this gives $\pi(x) \ll x/\log \log x$.

(vi) Let k be an integer ≥ 2 and let $\mathcal{R}$ be the region in Euclidean k-space defined by

$$0 < x_1 \leq x_2 \leq \cdots \leq x_k \leq X.$$

Show that the volume of $\mathcal{R}$ is $X^k/k!$ and verify that the number of integer points in $\mathcal{R}$ does not exceed the volume. Deduce that the number of integers n with $1 \leq n \leq N$ that are representable as a sum of k kth powers is at most $N/k!$ and show that this gives $G(k) \geq k + 1$.

(vii) Show that, for any integer m, $\sum e^{2\pi i a m/q}$ summed over all integers a with $1 \leq a \leq q$ and $(a, q) = 1$ is another expression for $\sum_{d|(m,q)} d\mu(q/d)$.

(viii) Prove that every sufficiently large even integer n can be represented in the form $n = p_1 + 2p_2 + 3p_3$ with primes p_1, p_2, p_3.

17

Elliptic curves

17.1 Introduction

The subject of elliptic curves has played an important role in the development of the theory of numbers. It has been especially significant in connection with studies on the rational solutions of Diophantine equations and, as indicated in Section 8.4, it has led most famously to a proof of Fermat's last theorem. A brief discussion of elliptic curves was given in Section 8.3 and we broaden this now into a deeper and more advanced exposition.

In a refined geometrical sense, an elliptic curve E over the rationals $\mathbb{Q}$ is a smooth curve of genus 1 with a specified rational point. By the Riemann–Roch theorem, E is then birationally equivalent to an algebraic curve in the projective plane given by a Weierstrass equation

$$y^2 = x^3 + ax + b$$

with a, b in $\mathbb{Q}$ and $4a^3 + 27b^2 \neq 0$ so that, by Section 10.7, the cubic on the right has distinct zeros in $\mathbb{C}$. This can be taken as the definition of E. The Mordell equation discussed in Section 8.3 is the special case with $a = 0$ and, as there, a rational point on E is understood to be either a pair (x, y) of rational numbers satisfying the equation or the point at infinity on E.

More generally an elliptic curve E can be defined with respect to an arbitrary ground field K. One obtains a Weierstrass equation for E of the same shape as above provided that K does not have characteristic 2 or 3; otherwise, with the customary notation, the Weierstrass equation takes the form

$$y^2 + a_1 xy + a_3 y = x^3 + a_2 x^2 + a_4 x + a_6.$$

Elliptic curves over finite fields are of particular interest in connection with cryptography; see the references in Section 9.8.

215

Two elliptic curves E and E' defined over $\mathbb{C}$ are said to be isomorphic if the mapping $x = v^2 x'$, $y = v^3 y'$, where v is a non-zero complex number, takes the Weierstrass equation for E into the Weierstrass equation

$$y'^2 = x'^3 + a'x' + b'$$

for E'. Then $a' = a/v^4$, $b' = b/v^6$, whence elliptic curves over $\mathbb{C}$ are classified up to isomorphism by the ratio $a^3 : b^2$ or, as is more usually stated, they are classified by the j-invariant defined by

$$j = j(E) = 1728(4a^3)/(4a^3 + 27b^2).$$

The classification holds more generally for elliptic curves over any algebraically closed field K, though for fields with characteristic 2 or 3 the formula for j must be modified.

It was noted in Section 8.3 in connection with the Mordell equation that the chord joining any two rational points on the curve intersects the curve again at a rational point, and similarly that the tangent at a rational point intersects again at a rational point. Moreover it was remarked that, as an immediate consequence of the addition formulae for the Weierstrass functions $x = \wp(z)$, $y = \frac{1}{2}\wp'(z)$ that parameterize the curve, the set of all rational points on the curve form a group under the chord and tangent process. Here the identity of the group is the point at infinity O and the inverse of $P = (x, y)$ is $-P = (x, -y)$; the sum of points P and Q is the point $-R$ where R is the third point of intersection of the chord joining P and Q or the tangent at P if $P = Q$ (see Fig. 8.1). We proceed to establish the existence of the group for elliptic curves over $\mathbb{C}$; the main issue here is the verification of the associative law. The same deduction, as will be clear, applies for any subfield of $\mathbb{C}$ and so in particular for $\mathbb{Q}$; in this instance the group is referred to as the Mordell–Weil group of E.

17.2 The Weierstrass $\wp$-function

Let Λ denote the lattice in the complex plane with generators ω_1, ω_2 where ω_1, ω_2 are non-zero complex numbers such that the imaginary part of ω_2/ω_1 is positive. Thus Λ is the set of all complex numbers $\omega = n_1\omega_1 + n_2\omega_2$ with n_1 and n_2 rational integers. By an elliptic function with respect to Λ we mean a meromorphic function f on $\mathbb{C}$ which satisfies $f(z + \omega) = f(z)$ for all ω in Λ. A number ω as here is said to be a period of f and if every period is in Λ then ω_1, ω_2 are said to be a fundamental pair of periods for f.

We note at once that if f is not constant then it must have a pole in $\mathbb{C}$; for otherwise f would be continuous and so bounded in the fundamental parallelogram for Λ, that is, the compact set $u_1\omega_1 + u_2\omega_2$ with $0 \le u_1, u_2 \le 1$. Thus f would be bounded in $\mathbb{C}$ and Liouville's theorem tells us that a bounded entire function is a constant. We shall need the following result.

Theorem 17.1 *Let C denote the closed, positively orientated contour given by the boundary of the fundamental parallelogram. Let f be an elliptic function with respect to Λ and let $z_j (j = 1, \ldots, k)$ be the singular points, that is, zeros or poles, of f within C. Let m_j be the order of z_j and suppose that no singular point of f lies on C. Then $m_1 z_1 + \cdots + m_k z_k$ is in Λ and $m_1 + \cdots + m_k = 0$.*

Proof Let $g(z) = zf'(z)/f(z)$. On noting that, for each j, the point $z = z_j$ is a pole of g with residue $m_j z_j$, we obtain from Cauchy's theorem

$$\int_C g(z)dz = 2\pi i \sum_{j=1}^k m_j z_j.$$

We now separate C into pairs of opposite sides and use periodicity. We have

$$\int_0^{\omega_1} g(z)dz - \int_{\omega_2}^{\omega_1+\omega_2} g(z)dz = \int_0^{\omega_1} (g(u) - g(u + \omega_2))du$$

$$= -\omega_2 \int_0^{\omega_1} (f'(u)/f(u))du.$$

The latter integral is the variation of the amplitude of f as u describes the line from 0 to ω_1 together with a factor i; thus it is an integer multiple of $2\pi i$. The same holds with ω_1 and ω_2 interchanged and the first assertion in the theorem follows. The second assertion is obtained by a similar argument applied to $g(z) = f'(z)/f(z)$. □

On recalling that a meromorphic function has only a finite number of zeros or poles in any bounded region of the complex plane, the fundamental parallelogram for Λ can be translated to a congruent parallelogram such that f has no singular point on the boundary and the theorem continues to hold. It follows in particular that, in any such parallelogram, the number of zeros of f is equal to the number of poles each counted with multiplicity.

The simplest example of a non-constant elliptic function is the Weierstrass ℘-function. It is defined by

$$\wp(z) = \frac{1}{z^2} + \sum_{\omega \ne 0} \left\{ \frac{1}{(z-\omega)^2} - \frac{1}{\omega^2} \right\}$$

with ω in Λ as above. The series converges uniformly on compact sets not containing lattice points since in these sets the summand on the right is $\ll 1/|\omega|^3$ and we have $\sum 1/|\omega|^s$ convergent for any $s > 2$. The latter is a consequence of the fact that, for each $n = 1, 2, \ldots$, the number of lattice points in the annulus R_n given by $n - 1 < |z| \leq n$ is $\ll$ the area $(2n - 1)\pi$ of R_n and thus

$$\sum_{n=1}^{N} \sum_{\omega \in R_n} \frac{1}{|\omega|^s} \ll \sum_{n=1}^{N} \frac{1}{n^{s-1}}.$$

It follows that $\wp(z)$ is a meromorphic function with double poles at each point of Λ and analytic elsewhere. Moreover, since it remains unchanged on replacing ω in the defining sum by $-\omega$, it is an even function, that is, $\wp(z) = \wp(-z)$.

The derivative $\wp'(z)$, expressed by differentiating $\wp(z)$ term by term, is

$$\wp'(z) = -2 \sum_{\omega} \frac{1}{(z - \omega)^3}.$$

This is plainly an odd function, that is, $\wp'(z) = -\wp'(-z)$. Further, both $\wp(z)$ and $\wp'(z)$ are elliptic functions with respect to Λ. This is obvious for $\wp'(z)$ and it follows for $\wp(z)$ on observing that since $\wp'(z + \omega) = \wp'(z)$ we get $\wp(z + \omega) = \wp(z) + c$ for some constant c; substituting $z = -\frac{1}{2}\omega$ we obtain $c = 0$ provided that z is not a period of Λ. The latter is the case when ω is ω_1 or ω_2 and the assertion follows; in fact ω_1, ω_2 is a fundamental pair of periods of $\wp(z)$.

We now establish the basic differential equation satisfied by $\wp(z)$.

Theorem 17.2 *We have*

$$\wp'(z)^2 = 4\wp(z)^3 - g_2 \wp(z) - g_3,$$

where

$$g_2 = 60 \sum_{\omega \neq 0} 1/\omega^4, \quad g_3 = 140 \sum_{\omega \neq 0} 1/\omega^6.$$

Proof The result is obtained by expanding $\wp(z)$ and $\wp'(z)$ as Laurent series about the origin. We have $1/(\omega - z) = (1/\omega) \sum_{j=0}^{\infty} (z/\omega)^j$ and thus

$$1/(z - \omega)^2 = (1/\omega^2) \sum_{j=0}^{\infty} (j + 1)(z/\omega)^j.$$

We shall use classical notation and define G_j $(j = 3, 4, \ldots)$ as the Eisenstein series of order j, that is, $G_j = \sum_{\omega \neq 0} 1/\omega^j$. Then $G_j = 0$ if j is odd and we get

$$\wp(z) - (1/z^2) = \sum_{j=1}^{\infty} (2j+1)G_{2j+2}z^{2j} = 3G_4z^2 + 5G_6z^4 + \cdots.$$

Differentiating term by term gives

$$\wp'(z) + (2/z^3) = \sum_{j=1}^{\infty} 2j(2j+1)G_{2j+2}z^{2j-1} = 6G_4z + 20G_6z^3 + \cdots.$$

It is now easily seen that $4\wp(z)^3$ and $\wp'(z)^2$ are given respectively by

$$(4/z^6) + (36G_4/z^2) + 60G_6 + \cdots \quad \text{and} \quad (4/z^6) - (24G_4/z^2) - 80G_6 + \cdots.$$

It follows that

$$\wp'(z)^2 - 4\wp(z)^3 + 60G_4\wp(z) + 140G_6$$

is an entire elliptic function with a zero at the origin and hence is identically 0. We have $g_2 = 60G_4$ and $g_3 = 140G_6$ and this proves the theorem. $\qquad\square$

It is customary to denote by e_1, e_2, e_3 the values of $\wp(z)$ at the half-periods, that is, $e_j = \wp(\frac{1}{2}\omega_j)$ for $j = 1, 2$ and 3 where $\omega_3 = \omega_1 + \omega_2$. Then the following holds.

Theorem 17.3 *The numbers e_1, e_2, e_3 are distinct and we have*

$$\wp'(z)^2 = 4(\wp(z) - e_1)(\wp(z) - e_2)(\wp(z) - e_3).$$

Proof It is clear that none of $\frac{1}{2}\omega_j$ ($j = 1, 2, 3$) is a pole of $\wp(z)$ and that

$$\wp'(-\tfrac{1}{2}\omega_j) = \wp'(\omega_j - \tfrac{1}{2}\omega_j) = \wp'(\tfrac{1}{2}\omega_j).$$

Thus since $\wp'(z)$ is odd we obtain $\wp'(\frac{1}{2}\omega_j) = 0$. This implies that $\wp(z) - e_j$ has at least a double zero at $z = \frac{1}{2}\omega_j$. But the only poles of the function are double poles at the lattice points and hence from the remark following the proof of Theorem 17.1 there can be no other zeros mod Λ. We conclude that e_1, e_2, e_3 are distinct and the desired factorization follows. $\qquad\square$

The Weierstrass $\wp$-function plays a crucial role in the general theory of elliptic functions. Indeed if Λ denotes the associated lattice then we have the following fundamental result.

Theorem 17.4 *Every even elliptic function with respect to Λ is a rational function of $\wp$. Further, the field of elliptic functions with respect to Λ is generated by $\wp$ and $\wp'$.*

The second property follows immediately from the first on noting that an odd elliptic function with respect to Λ can be expressed as $\wp'(z)f(z)$ with f even and furthermore that every function $g(z)$ is the sum of an even and odd function, namely $\frac{1}{2}(g(z) + g(-z))$ and $\frac{1}{2}(g(z) - g(-z))$. The verification of the first property involves the construction of a rational function of $\wp$ with the same zeros and poles as the given even elliptic function taken with multiplicities and we refer to the literature cited in Section 17.9 for details.

17.3 The Mordell–Weil group

By Theorem 17.2 the functions $x = \wp(z)$, $y = \frac{1}{2}\wp'(z)$ parameterize the elliptic curve E over $\mathbb{C}$ with Weierstrass equation given as in Section 17.1 with $a = -\frac{1}{4}g_2$ and $b = -\frac{1}{4}g_3$. Indeed the polynomial $4x^3 - g_2x - g_3$ has discriminant 16Δ, where $\Delta = g_2^3 - 27g_3^2$, and we proved in Theorem 17.3 that it has distinct zeros. Hence we have $\Delta \neq 0$. We call Δ the discriminant of E and we note that the j-invariant of E is given by $j = 1728g_2^3/\Delta$.

Now suppose that E is an elliptic curve over $\mathbb{C}$ with a Weierstrass parameterization $x = \wp(z)$, $y = \frac{1}{2}\wp'(z)$ as above. We shall say that the point (x, y) has parameter z and we define $-(x, y)$ as the point $(x, -y)$ with parameter $-z$. Plainly the parameters are determined only up to elements of Λ; to specify them uniquely they must be taken as elements of the factor group $\mathbb{C}/\Lambda$ mentioned at the end of the current section and this is implicit in the discussion below.

Let P and Q be points on E with parameters u and v respectively. The chord joining P and Q if $P \neq \pm Q$ and the tangent at P if $P = Q$ are given by equations of the form $y = \lambda x + \nu$ and we see that $f(z) = \wp'(z) - \lambda\wp(z) - \nu$ is an elliptic function with zeros u and v. The poles of f are the lattice points of Λ and these have order -3. Hence, from Theorem 17.1 and the remark following the proof, the sum of the orders of the zeros of f in a parallelogram as specified there is 3. It follows that f has a further zero w such that $u + v + w$ is in Λ; it can be taken as the parameter of the third point of intersection, say R, of the chord or tangent with E and $w = u$ or v if one of u, v is a zero of order 2. We now define $P + Q$ as the point with parameter $u + v$. Then if $P \neq -Q$ we have $P + Q = -R$ and if $P = -Q$ we have $P + Q = O$ where O is the point at infinity on E; it can be taken to have parameter 0. Further, we see that addition is associative since it is certainly so for addition of parameters mod Λ. Thus, on specifying O as the identity element, we obtain an abelian group on E, and, in the case when E is defined over $\mathbb{Q}$, the rational points on E form the Mordell–Weil group referred to earlier.

The structure of the group is intimately related to the well-known addition formulae for the $\wp$-function.

Theorem 17.5 *We have*

$$\wp(u+v) = \left(\frac{\wp'(v) - \wp'(u)}{2(\wp(v) - \wp(u))}\right)^2 - \wp(u) - \wp(v),$$

$$\wp(2u) = \left(\frac{\wp''(u)}{2\wp'(u)}\right)^2 - 2\wp(u),$$

where in the first equation we are assuming that $\wp(u) \neq \wp(v)$ and in the second that $\wp'(u) \neq 0$.

Proof When $P \neq \pm Q$ the coefficient λ in the equation $y = \lambda x + v$ of the chord joining P and Q is given by

$$\lambda = (\wp'(v) - \wp'(u))/(2(\wp(v) - \wp(u))).$$

On taking the limit as $v \to u$ we see that, when $P = Q$, the value of λ in the corresponding equation of the tangent at P is

$$\lambda = \wp''(u)/(2\wp'(u))$$

assuming $\wp'(u) \neq 0$ or equivalently $2P \neq O$. Now the points of intersection of the chord or tangent with the curve are given by the roots of the equation

$$x^3 + ax + b - (\lambda x + v)^2 = 0.$$

The coefficient of x^2 in the polynomial on the left is $-\lambda^2$. Further, as shown above, when $P \neq \pm Q$ the roots are $\wp(u)$, $\wp(v)$, $\wp(-(u+v))$ and when $P = Q$ they are $\wp(u)$ counted twice and $\wp(-2u)$. The theorem follows. $\qquad\square$

Theorem 17.5 gives at once explicit formulae for the group law often used in computations. Purely algebraic proofs can be given but the deduction via the $\wp$-function is immediate and transparent.

Corollary 17.1 *Let E be an elliptic curve over $\mathbb{Q}$ with Weierstrass equation $y^2 = x^3 + ax + b$. Let P and Q be rational points on E with coordinates $(x(P), y(P))$ and $(x(Q), y(Q))$ respectively and let $x(P+Q)$ be the x-coordinate of $P+Q$. If $P \neq \pm Q$ then*

$$x(P+Q) = \left(\frac{y(P) - y(Q)}{x(P) - x(Q)}\right)^2 - x(P) - x(Q).$$

Further, if $2P \neq O$ *then*

$$x(2P) = \left(\frac{3x(P)^2 + a}{2y(P)} \right)^2 - 2x(P).$$

Proof Let $\wp$ be the Weierstrass function that parameterizes E. Then we have $x(P) = \wp(u)$, $y(P) = \frac{1}{2}\wp'(u)$ and $x(Q) = \wp(v)$, $y(Q) = \frac{1}{2}\wp'(v)$ for some u, v. The first assertion is now clear and for the second we note that $2P \neq O$ gives $\wp'(u) \neq 0$ and then from Theorem 17.2 we have $\wp''(u) = 2(3\wp(u)^2 + a)$. $\square$

We remark that the mapping $z \mapsto \wp(z)$ is a homomorphism from the additive group of complex numbers onto E and that the kernel of the mapping is Λ. Thus we have $E \cong \mathbb{C}/\Lambda$. The latter can be viewed as a Riemann surface and topologically it is a torus. More generally, as indicated in Section 17.1, elliptic curves E have been studied with respect to an arbitrary ground field K; the points with coordinates in K together with the point at infinity are called K-rational points on E and the set $E(K)$ of all K-rational points forms a group analogous to the Mordell–Weil group. For the reals we find that $E(\mathbb{R}) \cong \mathbb{Z}/2\mathbb{Z} \times \mathbb{R}/\mathbb{Z}$ or $\mathbb{R}/\mathbb{Z}$ according as $\Delta > 0$ or $\Delta < 0$. For the finite field K with q elements we have $E(K) \cong \mathbb{Z}/k\mathbb{Z} \times \mathbb{Z}/l\mathbb{Z}$ where k and l are positive integers such that $k|(q-1)$ and $k|l$.

We remark also that there is a close connection between elliptic curves and the theory of modular forms. Indeed g_2 and g_3 can be interpreted as modular forms of weight 4 and 6 on $SL_2(\mathbb{Z})$, whence Δ, referred to at the beginning of this section, is a modular form of weight 12 and j is a modular function. It can be expressed in terms of $\tau = \omega_2/\omega_1$ and it then defines, on putting $z = \tau$, the classical elliptic modular function with Fourier expansion

$$j(z) = 1/q + 744 + c_1 q + c_2 q^2 + \cdots,$$

where $q = e^{2\pi i z}$ and $c_1, c_2, \ldots$ are integers. Moreover the theory of modular forms can be applied to establish that j assumes all values in the complex plane as Λ ranges over the set of lattices. Since, from Section 17.1, j classifies the elliptic curves over $\mathbb{C}$ it follows that for any elliptic curve E over $\mathbb{C}$ there is a corresponding lattice Λ such that E has a Weierstrass equation with $a = -\frac{1}{4}g_2$ and $b = -\frac{1}{4}g_3$. This is called the uniformization theorem and it is the converse to the situation with which we started, namely on specifying a lattice Λ we showed there is a corresponding curve E.

17.4 Heights on elliptic curves

The most fundamental result on the arithmetic of elliptic curves is the Mordell–Weil finite basis theorem, which we have already referred to briefly in Section

8.3 and which we shall discuss more fully in the next section. As a preliminary we shall need the notion of the height of points on elliptic curves.

Let E be an elliptic curve defined over $\mathbb{Q}$ with Weierstrass equation $y^2 = f(x)$ where $f(x) = x^3 + ax + b$ as in Section 17.1. Further, let $P = (x, y)$ be a rational point on E other than the point at infinity O. Then on expressing x in lowest terms, that is, $x = p/q$ where p, q are relatively prime integers (possibly $p = 0$, $q = 1$) we define the classical height $h(P)$ of P by

$$h(P) = \log \max(|p|, |q|).$$

Also, by convention, we define $h(O) = 0$. We proceed to verify the following result.

Theorem 17.6 *For any rational points P and Q on E we have*

$$h(P + Q) + h(P - Q) = 2h(P) + 2h(Q) + O(1)$$

where the implied constant depends on a and b but not on P or Q.

Proof First we establish the theorem in the case $P = Q$, that is,

$$h(2P) = 4h(P) + O(1).$$

Let $x(P)$ be the x-coordinate of P. We can assume that $2P \neq O$, for otherwise $x(P)$ is a zero of $f(x)$ and the assertion holds trivially. Then Corollary 17.1 shows that the x-coordinate of $2P$ is $g(x(P))/(4f(x(P)))$ with $g(x) = (f'(x))^2 - 8xf(x)$. Now $g(x)$ has degree 4 and it follows at once on putting $x(P) = p/q$ that $h(2P) \leq 4h(P) + O(1)$.

To prove the converse we note that $f(x)$ has only simple zeros, whence the polynomials $f(x)$ and $g(x)$ have no common factor; the same then applies to the polynomials $\overline{f}(x) = x^4 f(1/x)$ and $\overline{g}(x) = x^4 g(1/x)$. Thus we have $f(x)r(x) + g(x)s(x) = 1$ and $\overline{f}(x)\overline{r}(x) + \overline{g}(x)\overline{s}(x) = 1$ for some polynomials $r(x)$, $s(x)$ and $\overline{r}(x)$, $\overline{s}(x)$ defined over $\mathbb{Q}$ with degrees at most 3. We put $x = x(P) = p/q$ in the first identity and $x = q/p$ in the second, assuming, as we may, that $p \neq 0$. Then writing $F = q^4 f(p/q)$, $G = q^4 g(p/q)$ we obtain $FR + GS = q^7$ and $F\overline{R} + G\overline{S} = p^7$ where

$$R = q^3 r(p/q), \quad S = q^3 s(p/q), \quad \overline{R} = p^3 \overline{r}(q/p), \quad \overline{S} = p^3 \overline{s}(q/p).$$

Now F and G can be expressed as rationals with denominators depending only on the coefficients of f and the same applies to R, S and to $\overline{R}, \overline{S}$. Thus since, by supposition, p and q are relatively prime we see that, apart from a bounded factor, the numerators of F and G are relatively prime and, on noting that the x-coordinate of $2P$ is $G/(4F)$, we get $\log \max(|F|, |G|) = h(2P) + O(1)$. This gives $7h(P) \leq h(2P) + 3h(P) + O(1)$, that is, $4h(P) \leq h(2P) + O(1)$, and the desired expression for $h(2P)$ follows.

Now let P and Q be rational points on E with x-coordinates $x(P)$ and $x(Q)$ respectively. We can suppose that none of P, Q, $P + Q$ and $P - Q$ are O, for otherwise the theorem is trivial or follows from the preceding result. Then by Corollary 17.1 (see Section 17.10 Exercise (ii)) the x-coordinates $x(P + Q)$ and $x(P - Q)$ of $P + Q$ and $P - Q$ satisfy

$$x(P+Q)+x(P-Q) = \frac{2(x(P)x(Q)+a)(x(P)+x(Q))+4b}{(x(P)-x(Q))^2},$$

$$x(P+Q)x(P-Q) = \frac{(x(P)x(Q)-a)^2 - 4b(x(P)+x(Q))}{(x(P)-x(Q))^2}.$$

Thus defining $h(x)$ for rational x like $h(P)$ with $P = (x, y)$, we obtain

$$h(x(P+Q)+x(P-Q)) \le 2h(P) + 2h(Q) + O(1)$$

and $h(x(P+Q)x(P-Q))$ is bounded similarly. Now on writing $x(P+Q) = p/q$ and $x(P-Q) = r/s$ for relatively prime integers p, q and r, s we have $pr, ps + qr, qs$ relatively prime and, as is readily verified,

$$\max(|p|, |q|)\max(|r|, |s|) \le 2\max(|pr|, |ps+rq|, |qs|).$$

On taking logarithms this gives

$$h(P+Q) + h(P-Q) \le 2h(P) + 2h(Q) + O(1).$$

The reverse inequality is obtained by replacing P and Q by $P + Q$ and $P - Q$ respectively and using the result for $h(2P)$ established at the beginning. This proves the theorem. $\qquad\square$

Theorem 17.6 is more than enough for the application needed to establish the Mordell–Weil theorem. However, one can remove the O-term by averaging over the group law and this leads to a particularly elegant result. Let P be any rational point on the elliptic curve E. We define the canonical, or Néron–Tate, height of P by

$$\hat{h}(P) = \lim_{n\to\infty} 4^{-n}h(2^n P).$$

The limit exists since the sequence here is Cauchy. Indeed, for $m \ge n$, we have

$$|4^{-m}h(2^m P) - 4^{-n}h(2^n P)| \le \sum_{j=n+1}^{m} 4^{-j}|(h(2^j P) - 4h(2^{j-1}P))|$$

and, by Theorem 17.6, $h(2P) = 4h(P) + O(1)$; thus the sum on the right is $O(4^{-n})$. The basic property of the canonical height is the following, sometimes referred to as the parallelogram law.

Theorem 17.7 *For any rational points P and Q on E we have*

$$\hat{h}(P + Q) + \hat{h}(P - Q) = 2\hat{h}(P) + 2\hat{h}(Q).$$

Proof Replace P and Q in Theorem 17.6 by $2^n P$ and $2^n Q$ respectively, divide by 4^n, and take the limit as $n \to \infty$. $\qquad\square$

It follows at once by induction that $\hat{h}(nP) = n^2 \hat{h}(P)$ for all integers n. Hence we have $\hat{h}(P) = 0$ if and only if P is a torsion point, that is, a point of finite order in the Mordell–Weil group of E. Further, it is clear, by taking $n = 0$ and letting $m \to \infty$ in the argument above, that the difference between the classical height and the canonical height is bounded. Thus the canonical height shares with the classical height the property that there are only finitely many points with height not exceeding a given bound. In particular any set of rational points on E has an element with least canonical height.

17.5 The Mordell–Weil theorem

Let E be an elliptic curve defined over $\mathbb{Q}$ as in Section 17.1 and let G be the Mordell–Weil group of rational points on E described in Section 17.3. Then the Mordell–Weil theorem asserts the following.

Theorem 17.8 *The group G has a finite basis.*

Proof The proof divides naturally into two parts. First it is shown that the factor group $G/2G$ is finite, customarily termed the weak Mordell–Weil theorem, and Theorem 17.8 is then deduced from it by infinite descent.

We shall assume that a and b in the Weierstrass equation for E are integers; this involves no loss of generality for it suffices to take in place of E a curve E' isomorphic to it as in Section 17.1 with v such that $1/v$ is a denominator for a and b. Then it is readily seen that every point in G other than O has the form $(r/t^2, s/t^3)$ for some integers r, s, t with $(r, t) = (s, t) = 1$. Further, by definition, the Weierstrass equation for E can be written in the form $y^2 = f(x)$ where

$$f(x) = (x - e_1)(x - e_2)(x - e_3)$$

and e_1, e_2, e_3 are distinct algebraic integers with degree at most 3.

Let $e = e_1$ and let $K = \mathbb{Q}(e)$.[†] For rational x we have $x - e$ and $(x - e_2)(x - e_3)$ elements of K, whence, on taking ideals in the field, we obtain

$$[s]^2 = [r - et^2][(r - e_2t^2)(r - e_3t^2)].$$

This gives $[r - et^2] = \mathfrak{a}\mathfrak{b}^2$ for some ideals $\mathfrak{a}, \mathfrak{b}$ in K where $\mathfrak{a}$ is a factor of $[(e - e_2)(e - e_3)]$; for, since $(r, t) = 1$, any common prime ideal factor of the ideals on the right will divide the latter. Standard techniques now show that $r - et^2 = \alpha\beta^2$ for some algebraic integer β in K and some element α in K belonging to a finite set. Indeed from the proof of Theorem 12.4 there exist ideals $\mathfrak{a}'$ and $\mathfrak{b}'$ in the ideal classes inverse to those of $\mathfrak{a}$ and $\mathfrak{b}$ with norms at most $\sqrt{|d|}$ where d denotes the discriminant of K. Then $\mathfrak{a}\mathfrak{a}'$, $\mathfrak{a}'\mathfrak{b}'^2$ and $\mathfrak{b}\mathfrak{b}'$ are principal, say $[\xi]$, $[\eta]$ and $[\zeta]$. We have $[\eta(r - et^2)] = [\xi\zeta^2]$ and thus, by Section 12.1, $r - et^2 = \varepsilon(\xi/\eta)\zeta^2$ for a unit ε in K. But, by Theorem 12.3, that is, Dirichlet's unit theorem, $\varepsilon = \lambda\mu^2$ for some units λ, μ in K with λ in a finite set. Further, by Corollary 11.3, ξ and η have bounded field norms and so, by Corollary 12.1, they can be chosen to belong to finite sets. The desired result follows.

We suppose now that $f(x)$ is irreducible over the rationals so that e has degree 3 and e_2, e_3 are the conjugates of e; we shall indicate the modifications needed to treat the reducible case later. Let K^* denote the multiplicative group of non-zero elements of K and let $\phi : G \to K^*/K^{*2}$ be the map given by $(x, y) \mapsto (x - e)K^{*2}$ and $O \mapsto 1 K^{*2}$. We shall prove that ϕ is a group homomorphism with kernel $2G$. This will suffice to establish the weak Mordell–Weil theorem for, as we showed above, the image of ϕ is finite.

Let $P = (x_1, y_1)$ and $Q = (x_2, y_2)$ be points in G and let $P + Q = (x_3, y_3)$. Then as in the proof of Theorem 17.5 we see that x_1, x_2, x_3 are roots of the equation

$$f(x) - (\lambda x + \nu)^2 = 0,$$

where $y = \lambda x + \nu$ is the chord joining P and Q if $P \neq \pm Q$ and the tangent at P if $P = Q$. Thus the cubic on the left can be written as $(x - x_1)(x - x_2)(x - x_3)$ and we obtain

$$(x_1 - e)(x_2 - e)(x_3 - e) = (\lambda e + \nu)^2.$$

Since e has degree 3 and x_1, x_2, x_3 are rational none of the factors on the left vanishes. Thus, since λ, ν are rational, it follows that

$$(x_3 - e)K^{*2} = (x_1 - e)(x_2 - e)K^{*2}.$$

[†] There can be no confusion here with e, the base for natural logarithms.

This gives $\phi(P+Q)=\phi(P)\phi(Q)$. Plainly the latter holds also when $P=-Q$ and when one or both of P and Q are O and so ϕ is a group homomorphism as asserted.

It remains to consider the kernel. This certainly contains $2G$ since $\phi(2P)=\phi(P)^2$. To prove the converse let $P=(x_1,y_1)$ be a point of the kernel so that

$$x_1 - e = (u + ve + we^2)^2$$

for some rationals u,v,w. Since e has degree 3 it is clear that $w \neq 0$. We put $x_2 = v/w$ and note that

$$(x_2 - e)(u + ve + we^2) = g(e)$$

where $g(x) = \lambda x + v$ with $\lambda = aw - u + vx_2$ and $v = bw + ux_2$. This gives

$$f(x) = g(x)^2 + (x - x_1)(x - x_2)^2;$$

for the expression on the right is a monic polynomial in x of degree 3 with a zero $x = e$ and so must be the minimum polynomial for e. Now the line $y = g(x)$ meets E in points with x-coordinates satisfying $f(x) = g(x)^2$ and thus satisfying $(x - x_1)(x - x_2)^2 = 0$. It is therefore the tangent to the curve at $Q = (x_2, g(x_2))$ and it meets the curve again at $\pm P$. We conclude that $P = 2(\pm Q)$ whence P is contained in $2G$ as required.

The proof of Theorem 17.8 can now be completed readily by infinite descent. Let $\hat{h}$ be the canonical height introduced in Section 17.4. By the weak Mordell–Weil theorem, as R runs through a set of representatives of the elements of $G/2G$, the heights $\hat{h}(R)$ have a maximum value, say M. We show that G is generated by the set S of points P of G with $\hat{h}(P) \leq M$. Indeed suppose the contrary. Then among the points of G not in the subgroup generated by S there is one, say P, of least canonical height. Now $P = R + 2Q$ for some representative R as above and some Q in G and by Theorem 17.7 we have

$$4\hat{h}(Q) = \hat{h}(2Q) = \hat{h}(P - R) \leq 2\hat{h}(P) + 2\hat{h}(R).$$

But since P is not in S we obtain $\hat{h}(R) < \hat{h}(P)$, whence Q is an element of G not in the subgroup generated by S such that $\hat{h}(Q) < \hat{h}(P)$. This contradicts the minimal choice of P and the theorem follows. $\square$

The argument in the case when $f(x)$ is reducible over $\mathbb{Q}$ is similar. One takes $K_i = \mathbb{Q}(e_i)$ ($i = 1, 2, 3$) and one defines maps $\phi_i : G \to K_i^* / K_i^{*2}$ as for ϕ above with $e = e_i$ but specifying

$$(e_i, 0) \mapsto (e_i - e_j)(e_i - e_k)K_i^{*2}$$

with distinct suffixes i, j, k for the $(e_i, 0)$ in G. Then the map ϕ given by

$$\phi(P) = (\phi_1(P), \phi_2(P), \phi_3(P))$$

is a group homomorphism with kernel $2G$ and the result follows as before. Note here that if $P = (x_1, y_1)$ is in the kernel of ϕ and $y_1 \neq 0$ then $x_1 - e_i = u_i^2$ with u_i in K_i and, with appropriate choices of sign, the monic polynomial with zeros u_i $(i = 1, 2, 3)$ has rational coefficients. Now there is a point $Q = (x_2, g(x_2))$ in G, where $g(x) = \lambda x + \nu$ with rational λ and ν, such that $f(x) - g(x)^2$ factorizes as $(x - x_1)(x - x_2)^2$ and thus $P = 2(\pm Q)$. Indeed the factorization is valid if and only if, for all i,

$$g(e_i)^2 = (x_1 - e_i)(x_2 - e_i)^2;$$

this certainly holds if $\lambda e_i + \nu = u_i(x_2 - e_i)$, that is, if

$$g(x_1) - \lambda u_i^2 = u_i(x_2 - x_1 + u_i^2),$$

and thus it suffices to define λ, ν and x_2 by an identity in t, namely

$$t^3 + \lambda t^2 + (x_2 - x_1)t - g(x_1) = (t - u_1)(t - u_2)(t - u_3).$$

Another way of dealing with the reducible case, well known in the subject, is through the theory of isogenies; see Section 17.8 and the books cited in Section 17.9.

17.6 Computing the torsion subgroup

Theorem 17.8 shows that $G \cong T \times \mathbb{Z}^r$ where T is the torsion subgroup consisting of the elements of G of finite order and r is a non-negative integer termed the Mordell–Weil rank of E. Now any finite subgroup of $\mathbb{C}/\Lambda$, where Λ is the lattice for E, has at most two generators and we have $E \cong \mathbb{C}/\Lambda$. Thus it follows that there is a finite set of parameters $u_1, \ldots, u_r$, a basis ω, ω' for Λ and a pair q, q' of positive integers whose product qq' is the order of T such that the rational points on E are given by $x = \wp(u)$, $y = \frac{1}{2}\wp'(u)$ with

$$u = m_1 u_1 + \cdots + m_r u_r + (m/q)\omega + (m'/q')\omega'$$

where $m_1, \ldots, m_r, m, m'$ are integers with $0 \leq m < q$ and $0 \leq m' < q'$.

In the case when the coefficients a and b in the Weierstrass equation for E are integers there is an effective criterion for computing T due to Lutz and Nagell as follows.

Theorem 17.9 *Let (x, y) be a point of T other than O. Then x, y are integers and either $y = 0$ or y^2 divides $4a^3 + 27b^2$.*

For the proof of Theorem 17.9 we refer to the literature in Section 17.9. It depends on the theory of reduction of E mod p for a prime p. Let $\mathbb{F}_p$ be the finite field with p elements and for any integer r let $\tilde{r}$ be the representative of r in $\mathbb{F}_p$. Further, for any rational $u = r/s$ where r, s are relatively prime integers with p not dividing s let $\tilde{u} = \tilde{r}/\tilde{s}$; since $\tilde{s}$ is invertible $\tilde{u}$ exists and is an element of $\mathbb{F}_p$. Now replacing a and b in the equation for E by $\tilde{a}$ and $\tilde{b}$ yields a curve $\tilde{E}$ defined over $\mathbb{F}_p$ and termed the reduced curve of E mod p. We define a map $E \to \tilde{E}$ by $(x, y) \mapsto (\tilde{x}, \tilde{y})$ if the denominators of x and y in lowest terms are not divisible by p and by $(x, y) \mapsto \tilde{O}$, the point at infinity in $\tilde{E}$, otherwise. The reduced curve $\tilde{E}$ is an elliptic curve and it is said to be non-singular if p is odd and does not divide $4a^3 + 27b^2$. In this instance the points on $\tilde{E}$ form a group $\tilde{G}$ analogous to the Mordell–Weil group G of E and the map defines a group homomorphism from G to $\tilde{G}$. Further, the torsion subgroup T of G maps injectively into $\tilde{G}$, whence the order of T divides the order of $\tilde{G}$, that is, the number of points on $\tilde{E}$. This gives another useful criterion for the computation of T.

Example 17.1 Consider the elliptic curve $y^2 = x^3 + x + 1$. We have $4a^3 + 27b^2 = 31$ and, since the latter is prime, the only possible non-trivial torsion point is $P = (0, 1)$. But a simple calculation shows that $2P = (\frac{1}{4}, -\frac{9}{8})$ and the coordinates here are not integers. Hence P is a point of infinite order and the torsion subgroup is trivial, that is, it consists of O only.

Example 17.2 The elliptic curve $y^2 = x^3 + 4x$ has torsion subgroup isomorphic to $\mathbb{Z}/4\mathbb{Z}$; for here $4a^3 + 27b^2 = 4^4$ and so the only possible torsion points (x, y) occur with x, y integers and $y = 0$ or y dividing 16. Now $y = \pm1$ and $y = \pm2$ are plainly impossible and since any common factor of x and $x^2 + 4$ divides 4 we see that $y = \pm8$ and $y = \pm16$ can also be excluded. Thus, apart from O, the only possible torsion points are $(0, 0)$ and $(2, \pm4)$. It is simple to verify that $2(2, 4) = (0, 0)$ and the assertion follows.

Example 17.3 Consider the elliptic curve $y^2 = x^3 + 10$. The reduced curve mod p is non-singular for all primes $p > 5$. It is easy to calculate that $x^3 + 10$ is a square mod 7 for all x mod 7 except $x = 0$ (in fact Fermat's theorem of Section 3.3 gives $x^3 \equiv \pm1$ for $x \not\equiv 0$ mod 7) and that it is a square mod 11 for all x mod 11 except $x = 0$, $x = \pm2$, $x = 4$ and $x = -5$. Thus, including O, the reduced curve mod 7 has 13 points and mod 11 it has 12 points. It follows that the order of the torsion subgroup divides 13 and 12 and so the group is trivial. Note that the curve contains the point $(-1, 3)$ and so there is at least one generator of the Mordell–Weil group of infinite order.

Example 17.4 Let l be an integer not divisible by 3 and consider the curve $y^2 = x^3 + lx$. By taking separately the cases $l \equiv \pm 1 \pmod 3$ we see that the reduced curve mod 3 has order 4. Now if $l = -k^2$ for an integer k then the curve includes, apart from O, the points $(0, 0)$ and $(\pm k, 0)$ and these are torsion points. We conclude that the torsion subgroup is isomorphic to $(\mathbb{Z}/2\mathbb{Z})^2$.

A celebrated theorem of Mazur based on deep studies on the arithmetic of modular curves establishes that, for an elliptic curve E defined over $\mathbb{Q}$, the only possibilities for the torsion subgroup are

$$T \cong \mathbb{Z}/n\mathbb{Z} \text{ with } 1 \le n \le 10 \text{ or } n = 12,$$

$$T \cong \mathbb{Z}/2\mathbb{Z} \times \mathbb{Z}/2n\mathbb{Z} \text{ with } 1 \le n \le 4.$$

Each of the possibilities does in fact occur for some E.

17.7 Conjectures on the rank

Computing the subgroup of G generated by the basis elements of infinite order is a different matter. Indeed the proof of the weak Mordell–Weil theorem is not effective and to the present day no general algorithm is known for finding $G/2G$. Nevertheless, in many instances, direct search for rational points on E yields a sharp bound for r from below and infinite descent yields an equally sharp bound for r from above; thus the Mordell–Weil group G can be explicitly determined. There has been extensive work on the problem of calculating r and it has led to important conjectures in the subject, which we now describe.

Let E be an elliptic curve defined over $\mathbb{Q}$ with Weierstrass equation as in Section 17.1. We shall assume that a and b are integers and that there are no primes p such that p^4 divides a and p^6 divides b; these properties can be realized by taking in place of E an isomorphic elliptic curve with Weierstrass coefficients $v^4 a$ and $v^6 b$ for a suitable rational v. If p is a prime other than 2 or 3 then the curve E is said to have good reduction at p if p does not divide $4a^3 + 27b^2$ and bad reduction at p otherwise. In other words E has good or bad reduction at p according as the reduced curve $\widetilde{E}$ is or is not non-singular. It is customary to write the number of points on $\widetilde{E}$ as $p + 1 - a_p$. Then, for complex s, we define

$$L_p(E, s) = (1 - a_p p^{-s} + p^{1-2s})^{-1} \quad \text{or} \quad L_p(E, s) = (1 - a_p p^{-s})^{-1}$$

according as E has good or bad reduction at p. Similar definitions apply in the cases $p = 2$ and $p = 3$ but it is then necessary to work with the more

general form of the Weierstrass equation as indicated in Section 17.1. The
local functions above are combined to give a global function

$$L(E, s) = \prod_p L_p(E, s)$$

known as the Hasse–Weil L-function for E. Now there is a famous theorem of
Hasse asserting that $|a_p| \leq 2\sqrt{p}$ and it follows readily from this that $L(E, s)$
is analytic in the half-plane $\mathrm{Re}(s) > \frac{3}{2}$. Hasse conjectured that the L-function
satisfies a functional equation relating $L(E, s)$ to $L(E, 2 - s)$ and that, as a
consequence, the function can be analytically continued throughout the complex plane. This was proved by Breuil, Conrad, Diamond and Taylor as an out-
come of the work of Taylor and Wiles on Fermat's last theorem as described
in Section 8.4.

In the early 1960s, as a result of extensive machine computations, Birch and
Swinnerton-Dyer formulated the following conjecture; the analytic continuity
of the L-function was known at the time only when E has complex multiplication (see Section 17.8) and the property in general was simply assumed.

Conjecture 17.1 (Birch–Swinnerton-Dyer) *The order of the zero of* $L(E, s)$
at $s = 1$ *is equal to the Mordell–Weil rank* r *of* E.

The conjecture has been verified in a large number of numerical cases. Moreover, through works of Coates–Wiles, Gross–Zagier, Kolyvagin and others,
some initial steps have been taken in the direction of the general assertion;
most notably it has been shown to be valid when the order of the zero in question is 0 or 1. There is a more explicit version of the Birch–Swinnerton-Dyer
conjecture giving an exact expression for $L(E, s)/(s - 1)^r$ as $s \to 1$; we refer
to the literature cited in Section 17.9 for details.

As an application, we discuss briefly the congruent number problem; it dates
back to the ancient Greeks. By a congruent number we mean a positive rational number k such that there exists a right-angled triangle with rational side-
lengths and area k. The basic problem is to determine whether a given rational
number is a congruent number. It will be seen at once that there is no loss in
assuming that k is a square-free integer; for when the sides of a right-angled
triangle are scaled by a factor s the area becomes scaled by a factor s^2. An
equivalent formulation of the problem is to determine for which k there exists an x with each of x, $x + k$ and $x - k$ the square of a rational; see Section
17.10 Exercise (vii). For instance, the triangle with side-lengths $\frac{3}{2}$, $\frac{20}{3}$ and $\frac{41}{6}$
has area 5; thus $k = 5$ is a congruent number and $x = (\frac{41}{12})^2$ has the asserted
property.

The relation to elliptic curves is given by the fact that a square-free integer $k > 0$ is a congruent number if and only if the curve E with Weierstrass equation $y^2 = x^3 - k^2 x$ has a point (x, y) with x, y rational and with $y \neq 0$. Indeed the rational points on E with $y \neq 0$ are in bijective correspondence with the non-zero rational triples X, Y, Z satisfying $X^2 + Y^2 = Z^2$ and $\frac{1}{2} XY = k$; the correspondence is given by

$$X = (x^2 - k^2)/y, \quad Y = 2kx/y, \quad Z = (x^2 + k^2)/y$$

which inverts to $x = k(X + Z)/Y$ and $y = 2kx/Y$. Now by Section 17.10 Exercise (iv) the non-trivial torsion points on E are $(0, 0)$ and $(\pm k, 0)$ and since these have $y = 0$ we see that k is a congruent number if and only if the rank r of E is at least 1. Further, the curve E has complex multiplication (see again Section 17.8) and for curves with this property Coates and Wiles showed in 1977 that if $r \geq 1$ then, in accordance with Conjecture 17.1, the Hasse–Weil L-function for E vanishes at $s = 1$. Combining the result with theorems of Shimura and Waldspurger on modular forms, Tunnell succeeded in 1983 in deriving an easily verifiable criterion for k to be a congruent number; namely, he proved that if an odd square-free integer $k > 0$ is congruent then the number of solutions to $2x^2 + y^2 + 8z^2 = k$ with x, y, z integers and with z odd is the same as the number with z even. A similar result was given for k even in terms of solutions to $4x^2 + y^2 + 8z^2 = \frac{1}{2} k$. Furthermore Tunnell showed that if Conjecture 17.1 holds for E then, in both cases, the converse is also valid.

We remark that the work of Goldfeld, Gross and Zagier on the Gauss class number problem which we mentioned briefly in Section 15.7 has a close connection with Conjecture 17.1. Indeed, in 1976 Goldfeld demonstrated that the problem could be reduced to showing that there exists an elliptic curve for which the associated Hasse–Weil L-function has a triple zero at $s = 1$. Since it was known that there certainly exist elliptic curves with rank at least 3 it followed that a verification of a special case of the Birch–Swinnerton-Dyer conjecture would suffice in principle for a solution. For a time it seemed that Goldfeld had merely related one intractable problem with another but, remarkably, in 1986 Gross and Zagier succeeded in giving an explicit elliptic curve with the desired property. This led to the work of Oesterlé, Arno and others on the complete determination of all the imaginary quadratic fields with given class numbers as referred to in Section 15.7.

17.8 Isogenies and endomorphisms

Let E and E' be elliptic curves defined over $\mathbb{C}$ and let Λ and Λ' be their associated lattices with bases ω_1, ω_2 and ω_1', ω_2' respectively. Then E and E' are said to be isogenous if there exists an $\alpha \neq 0$ in $\mathbb{C}$ such that $\alpha \Lambda \subseteq \Lambda'$; here $\alpha \Lambda$ signifies the lattice with basis $\alpha \omega_1, \alpha \omega_2$. An equivalent formulation is to say that E and E' are isogenous if and only if there exists an $\alpha \neq 0$ in $\mathbb{C}$ such that

$$\alpha \omega_1 = p\omega_1' + q\omega_2', \qquad \alpha \omega_2 = r\omega_1' + s\omega_2'$$

for some integers p, q, r, s. When the curves have this property the map ϕ from $\mathbb{C}/\Lambda$ to $\mathbb{C}/\Lambda'$ given by $z \mapsto \alpha z$ is called an isogeny from E to E'. It is clear that isogenies are group homomorphisms. Moreover the kernel of an isogeny ϕ is finite; it has precisely $|ps - qr|$ elements and this is called the degree of ϕ. Further, on noting that if $\overline{\wp}(z)$ is the Weierstrass function corresponding to Λ' then $\overline{\wp}(\alpha z)$ and $\overline{\wp}'(\alpha z)$ are elliptic functions with respect to Λ, one deduces from Theorem 17.4 that ϕ furnishes a mapping from points (x, y) on E to points $(\xi(x), \eta(x, y))$ on E' for some rational functions ξ and η, and it turns out that the degree of ϕ is the maximum of the degrees of the numerator and denominator of ξ when expressed as a quotient of polynomials in lowest terms.

Example 17.5 Consider the elliptic curve $y^2 = x^3 + ax$ where a is a non-zero complex number. Let Λ be the associated lattice with basis ω_1, ω_2 and let $\wp(z)$ be the Weierstrass function that parameterizes the curve. We shall assume that the basis is chosen so that $z = \frac{1}{2}\omega_1$ yields the point $(0, 0)$ and we shall determine the isogenous curve with lattice Λ' generated by $\frac{1}{2}\omega_1, \omega_2$.

Clearly $\wp(z)$ has a double pole at each point of Λ and, by Theorem 17.3, a double zero at each point of Λ' other than the points of Λ. Thus $\wp'(z)/\wp(z)$ has a simple pole with residue ± 2 at each point of Λ'. Now let $\overline{\wp}(z)$ be the Weierstrass function corresponding to Λ'. Then $\overline{\wp}(z) - \frac{1}{4}(\wp'(z)/\wp(z))^2$ is a bounded entire function and therefore a constant c, say. We have

$$\overline{\wp}(z) - c = (\wp(z)^2 + a)/\wp(z) \quad \text{and so} \quad \overline{\wp}'(z) = \wp'(z)(1 - a/\wp(z)^2).$$

Hence, on putting $x = \wp(z)$, $y = \frac{1}{2}\wp'(z)$ and similarly $x' = \overline{\wp}(z)$, $y' = \frac{1}{2}\overline{\wp}'(z)$ we get $x' - c = (y/x)^2$ and $y' = y(1 - a/x^2)$. But we have

$$x^2(1 - a/x^2)^2 = (y/x)^4 - 4a,$$

whence $y'^2 = (x' - c)^3 - 4a(x' - c)$. By Theorem 17.2 we see that $c = 0$ and thus the desired curve is $y'^2 = x'^3 - 4ax'$.

Note that the rational functions here are $\xi(x) = (x^2 + a)/x$ and $\eta(x, y) = (x^2 - a)y/x^2$ and that the degree of the isogeny is 2 in agreement with earlier remarks. The argument applies more generally to the elliptic curve $y^2 = x^3 +$

$bx^2 + ax$ with complex a, b satisfying $a \neq 0$ and $a' = b^2 - 4a \neq 0$. Taking now $x = \wp(z) - \wp(\frac{1}{2}\omega_1)$ we find that the curve is isogenous to

$$y'^2 = x'^3 - 2bx'^2 + a'x';$$

we obtain $\xi(x) = (x^2 + bx + a)/x$ and the same $\eta(x, y)$ as above.

Example 17.5 together with Exercises (ix), (x) and (xi) in Section 17.10 can be used to calculate the rank of some elliptic curves; we recall that if G is the Mordell–Weil group of E then $G \cong T \times \mathbb{Z}^r$ where T is the torsion subgroup and r is termed the rank of E. Let E be the elliptic curve $y^2 = x^3 + ax$ where a is an integer, let E' be the curve $y^2 = x^3 - 4ax$ and let G and G' be the Mordell–Weil groups of E and E' respectively. Further, let ϕ and ϕ' be the maps $G \to G'$ and $G' \to G$ induced by the isogenies defined in Exercise (ix). Then $\phi'\phi(G) = 2G$.

Now let ψ be the homomorphism defined in Exercise (x) and let ψ' be the corresponding homomorphism for E'. Then the kernels of ψ and ψ' are $\phi'(G')$ and $\phi(G)$, whence their images are isomorphic to $G/\phi'(G')$ and $G'/\phi(G)$ respectively. Further, Exercise (xi) shows that these images are finite; we shall suppose that they have n and n' elements respectively. We have

$$G/\phi'\phi(G) = G/2G \cong (T/2T) \times (\mathbb{Z}/2\mathbb{Z})^r$$

and this gives $2^{r+2} = nn'$. Indeed the sequence

$$0 \to \ker(\phi) \to \ker(\phi'\phi) \to \ker(\phi')$$
$$\to \operatorname{coker}(\phi) \to \operatorname{coker}(\phi'\phi) \to \operatorname{coker}(\phi') \to 0$$

is exact, where, by definition,

$$\operatorname{coker}(\phi) = G'/\phi(G), \quad \operatorname{coker}(\phi'\phi) = G/\phi'\phi(G), \quad \operatorname{coker}(\phi') = G/\phi'(G').$$

By the first isomorphism theorem for groups applied to each of the above maps one deduces that the group orders satisfy

$$|\ker(\phi)||\ker(\phi')||\operatorname{coker}(\phi'\phi)| = |\ker(\phi'\phi)||\operatorname{coker}(\phi)||\operatorname{coker}(\phi')|.$$

But, since T is a finite group, the kernel and cokernel of multiplication by 2 on T have the same order whence $|\operatorname{coker}(\phi'\phi)| = 2^r |\ker(\phi'\phi)|$. On noting that $|\ker(\phi)| = |\ker(\phi')| = 2$ and that $n = |\operatorname{coker}(\phi')|$, $n' = |\operatorname{coker}(\phi)|$ the assertion follows.

To compute n and n' we use the fact that if (x, y) is a rational point on E with $x \neq 0$ then $x = kt^2$ with t in $\mathbb{Q}$ and with k a square-free divisor of a, positive or negative; see Exercise (xi). Thus $(y/(kt))^2 = kt^4 + l$ with $kl = a$, that is,

$$w^2 = ku^4 + lv^4$$

where $(u, v, w) = (t, 1, y/(kt))$. Conversely if there exist rationals u, v, w satisfying the equation and $uv \neq 0$ then $x = k(u/v)^2$ and $y = kuw/v^3$ give a rational point on E with $x \neq 0$. This holds similarly for E' with a replaced by $-4a$. Further, we note that a solution in rationals u, v, w yields also a solution in integers u, v, w with $(u, v) = 1$ on multiplying through by a suitable denominator and cancelling common factors.

Example 17.6 Consider the elliptic curve $y^2 = x^3 + 5x$. In this case $k = \pm 1$ or $k = \pm 5$. But if $k < 0$ then $l < 0$ and there is no non-trivial real solution u, v, w. Thus since ψ maps $(0, 0)$ to $5\,\mathbb{Q}^{*2}$ it follows that this generates the image and $n = 2$. For the isogenous curve $y^2 = x^3 - 20x$ we have the possible generators given by -1, 2 and 5. Hence we must check which of -1, ± 2, ± 5 and ± 10 correspond to rational points. Now since ± 2 are quadratic nonresidues mod 5 we see that there is no solution to $w^2 = \pm(2u^4 - 10v^4)$ in integers u, v, w with $(u, v) = 1$. Further, we note that ψ' maps $(0, 0)$ to $-5\,\mathbb{Q}^{*2}$. Thus it remains only to consider $k = 5$, that is, the equation $w^2 = 5u^4 - 4v^4$, and here we have the solution $(1, 1, 1)$. We conclude that the image of ψ' is generated by $-1\,\mathbb{Q}^{*2}$ and $5\,\mathbb{Q}^{*2}$ and we have $n' = 4$. This gives $r = 1$.

The argument here applies more generally with $a = p$ where p is any prime with $p \equiv 5 \pmod 8$ and it gives again $r = 1$ provided that $w^2 = pu^4 - 4v^4$ is soluble in integers; some examples are $p = 13$ and $p = 37$ when there are solutions $(1, 1, 3)$ and $(5, 3, 151)$ respectively.

By an endomorphism of E we mean either an isogeny ϕ from E to itself or the zero map $z \mapsto 0$. If ϕ is an isogeny as here then the equations at the beginning of the section give

$$(p - \alpha)(s - \alpha) - qr = 0 \quad \text{and} \quad \alpha = p + q\omega_2/\omega_1,$$

whence α is either a rational integer or an integer in an imaginary quadratic field. When the latter occurs the quotient ω_2/ω_1 is an element of the field and the curve E is said to have complex multiplication. Further, by the definitions of g_2 and g_3 in Theorem 17.2, it is clear that the periods of lattices associated with isomorphic elliptic curves differ only by a scaling factor v, with v as in Section 17.1, and so complex multiplication is a property of the isomorphism class. Furthermore the endomorphisms of E form a ring $\mathrm{End}(E)$. It is given by the set of α in $\mathbb{C}$ such that $\alpha \Lambda \subseteq \Lambda$, where $\alpha \Lambda$ signifies the origin when $\alpha = 0$, and it has the property that $\text{End}(E)$ is isomorphic either to $\mathbb{Z}$ or to an order in the imaginary quadratic field $K = \mathbb{Q}(\omega_2/\omega_1)$, that is, a subring of the algebraic integers $\mathcal{O}_K$ of K of the form $\mathbb{Z} + f\mathcal{O}_K$ for an integer $f > 0$.

To give some examples we shall need the following result on periods of elliptic curves expressed in Legendre normal form, that is, $y^2 = f(x)$ where $f(x) = x(x-1)(x-\lambda)$ with λ in $\mathbb{C}$ and not 0 or 1.

Theorem 17.10 *The periods of the lattice associated with an elliptic curve in Legendre normal form can be taken as*

$$\omega_1 = \int_{-\infty}^{0} \frac{dt}{\sqrt{f(t)}}, \quad \omega_2 = \int_{1}^{\infty} \frac{dt}{\sqrt{f(t)}}.$$

The proof depends on the mapping $z \mapsto \wp(z)$ of the fundamental parallelogram defined by the lattice in the complex plane onto the Riemann sphere and we refer to the literature mentioned in the following section for details. At first sight it may appear that the hypothesis in Theorem 17.10 that the curve has Legendre normal form is restrictive but in fact this is not so. For if E is any elliptic curve over $\mathbb{C}$ with Weierstrass equation $y^2 = (x - e_1)(x - e_2)(x - e_3)$, say, then on putting $x - e_1 = v^2 x'$ and $y = v^3 y'$ with $v = \sqrt{(e_2 - e_1)}$ we obtain $y'^2 = f(x')$ where $\lambda = (e_3 - e_1)/(e_2 - e_1)$. Thus every elliptic curve over $\mathbb{C}$ is isomorphic to an elliptic curve which, after translating the x-coordinate by $-\frac{1}{3}(1 + \lambda) = e_1/(e_2 - e_1)$, assumes Legendre normal form.

We remark that changing the basis ω_1, ω_2 of the lattice associated with E to any two distinct pairs from $\omega_1, \omega_2, \omega_1 + \omega_2$ permutes e_1, e_2, e_3 and changes λ to one of $\lambda, 1 - \lambda, 1 - 1/\lambda$ or one of their reciprocals. Further, the j-invariant of a curve in Legendre normal form is $2^8(\lambda^2 - \lambda + 1)^3/(\lambda(\lambda - 1))^2$ (see Section 17.10 Exercise (xii)) and this remains unchanged by any of the six substitutions. Furthermore the latter are distinct except when λ maps to -1 or $-\varrho$, where $\varrho = \frac{1}{2}(-1 + \sqrt{(-3)})$ is a primitive cube-root of unity, and then $j = 1728$ and $j = 0$ respectively.

Example 17.7 Consider the elliptic curve E given by $y^2 = x^3 - x$. We take $e_1 = -1$, $e_2 = 1$ and $e_3 = 0$ so that $\lambda = \frac{1}{2}$ and $v = \sqrt{2}$. Then putting $u = 2t - 1$ and noting that ω_1, ω_2 become scaled by a factor $\sqrt{2}$ we obtain as a basis

$$\omega_1 = \int_{-\infty}^{-1} \frac{du}{\sqrt{(u^3 - u)}}, \quad \omega_2 = \int_{1}^{\infty} \frac{du}{\sqrt{(u^3 - u)}}.$$

On substituting $-u$ for u in the second integral we see that both integrals are the same except for a factor i. Hence $\omega_2 = i\omega_1$ and E has complex multiplication. Since elliptic curves of the form $y^2 = x^3 + ax$, where a is a non-zero complex number, are all isomorphic to E over $\mathbb{C}$, they too have complex multiplication. Note that the j-invariant in this case is $j = 1728$.

Example 17.8 Consider the elliptic curve E given by $y^2 = x^3 - 1$. We take $e_1 = \varrho^2$, $e_2 = \varrho$ and $e_3 = 1$ so that $\lambda = 1 + \varrho^2$ and $v^2 = \varrho(1 - \varrho)$. Then putting $u = v^2 t + \varrho^2$ we get as a basis for the lattice

$$\omega_1 = \int_{-\infty}^{\varrho^2} \frac{du}{\sqrt{(u^3 - 1)}}, \quad \omega_2 = \int_{\varrho}^{\infty} \frac{du}{\sqrt{(u^3 - 1)}},$$

where the paths of integration are on the line joining ϱ^2 and ϱ, that is, on the imaginary axis translated by $-\frac{1}{2}$. But since we are dealing with simply connected regions we can vary the paths so that they lie on the lines through the origin and ϱ^2 and ϱ respectively. Then on substituting ϱu for u in the second integral we obtain the first multiplied by $-\varrho$. Thus $\omega_2 = -\varrho\omega_1$ and since ϱ is quadratic it follows that E has complex multiplication. Further, elliptic curves of the form $y^2 = x^3 + b$ for a non-zero complex number b are all isomorphic to E over $\mathbb{C}$ and so they too have complex mutiplication. Note that the j-invariant in this case is $j = 0$.

17.9 Further reading

The books *The Arithmetic of Elliptic Curves* (Springer, 2009) and *Advanced Topics in the Arithmetic of Elliptic Curves* (Springer, 1994) by J. H. Silverman give comprehensive treatments of the subject and are currently regarded as standard references. A more elementary text, however, and closer in spirit to our work here, is that of Silverman and Tate, *Rational Points on Elliptic Curves* (Springer, 1992). Cassels' short *Lectures on Elliptic Curves* (Cambridge University Press, 1991) covers a deceptively large range and is especially instructive.

The congruent number problem referred to in Section 17.7 is the central theme of Koblitz's *Introduction to Elliptic Curves and Modular Forms* (Springer, 1993). The topic of complex multiplication referred to in Section 17.8 is a particular feature of Cox's *Primes of the Form $x^2 + ny^2$* (Wiley, 1989). For proofs of Theorems 17.4 and 17.10 and for a good treatment of the subject in general, see Husemöller's *Elliptic Curves* (Springer, 2004). Computational aspects of the field are covered well by Schmitt and Zimmer's *Elliptic Curves: A Computational Approach* (de Gruyter, 2003) and by Cremona's *Algorithms for Modular Elliptic Curves* (Cambridge University Press, 1997). For the method discussed in Section 17.8 of determining the rank of certain curves, see the first of the books of Silverman, above; the method is called descent via two-isogeny. For applications to cryptography see the books by Washington and by Blake, Seroussi and Smart referred to in Section 9.8.

We remark finally that there is now much important interplay between elliptic curves and transcendence theory. In particular, there is an elliptic isogeny theorem due to Masser and Wüstholz which has found significant application in Diophantine geometry, and there is a so-called elliptic logarithm method for solving Diophantine equations which has found widespread usage. We refer to the book of Baker and Wüstholz cited in Section 6.8, to that of Schmitt and Zimmer, mentioned above, and to Smart's *The Algorithmic Resolution of Diophantine Equations* (Cambridge University Press, 1998).

17.10 Exercises

(i) Let E be the elliptic curve defined by $y^2 + y = x^3 + x^2 - 2x$. Show that the transformation $x' = x + \frac{1}{3}$, $y' = y + \frac{1}{2}$ gives the Weierstrass equation for E. For the rational points $P = (0, 0)$ and $Q = (1, 0)$ on E compute $2P, 2Q, P + Q, P - Q$ and $2P - Q$.

(ii) Let E be an elliptic curve with Weierstrass equation $y^2 = x^3 + ax + b$ and let $P = (x_1, y_1)$ and $Q = (x_2, y_2)$ be points on E with $x_1 \neq x_2$. Show that $((y_1 - y_2)/(x_1 - x_2))^2$ and $((y_1 + y_2)/(x_1 - x_2))^2$ are zeros of

$$(x_1 - x_2)^2 t^2 - 2(x_1^3 + x_2^3 + a(x_1 + x_2) + 2b)t + (x_1^2 + x_1 x_2 + x_2^2 + a)^2$$

as a polynomial in t. Show further that the x-coordinates of $P + Q$ and $P - Q$ are zeros of the polynomial in u obtained on putting $t = u + x_1 + x_2$. Thus verify the formulae in Section 17.4.

(iii) Let $f(x)$ be a polynomial with integer coefficients and let E be the curve $y^2 = f(x)$. Further, let p be an odd prime and let $\widetilde{E}$ be the reduction of E mod p as in Section 17.6. Show that the number of points on $\widetilde{E}$ is

$$1 + \sum_{x=0}^{p-1} \left(\left(\frac{f(x)}{p} \right) + 1 \right).$$

Deduce that if $f(-x) = -f(x)$ identically and if $p \equiv 3 \pmod 4$ then $\widetilde{E}$ has $p + 1$ points.

(iv) Let E be the elliptic curve $y^2 = x^3 - k^2 x$ where k is a positive integer. Show that the order q of the torsion subgroup of E divides $p + 1$ for all primes $p \equiv 3 \pmod 4$ with p not dividing $2k$. Show further, as in Section 17.6 Example 17.4, that q is divisible by 4.

Let q' be the greatest factor of q not divisible by 3. By considering primes $p \equiv q' + 3 \pmod q$, deduce from Dirichlet's theorem on arithmetical progressions that $q = 4$ and that the torsion subgroup of E is isomorphic to $(\mathbb{Z}/2\mathbb{Z})^2$.

(v) Show that the curve $y^2 = x^3 + 5$ has infinitely many solutions over the rationals.

(vi) Let E be the elliptic curve $y^2 = x^3 - 43x + 166$. Verify by reduction mod 3 that the order of the torsion subgroup T of E divides 7. Hence show that $T \cong \mathbb{Z}/7\mathbb{Z}$ and that T is generated by $(3, 8)$.

(vii) By taking $Z = 2\sqrt{x}$, show that if k is a positive rational then the positive rationals X, Y, Z such that $X^2 + Y^2 = Z^2$ and $\frac{1}{2}XY = k$ are in bijective correspondence with the rationals x such that x, $x + k$ and $x - k$ are rational squares.

(viii) Use Tunnell's criterion to show that the least congruent number k with $k \equiv 1 \pmod 8$ is 41.

(ix) Let E be the elliptic curve $y^2 = x^3 + ax$ and let ϕ be the isogeny that takes E to the curve E' given by $y'^2 = x'^3 - 4ax'$ as in Section 17.8 Example 17.5. By considering the lattice with basis $\frac{1}{2}\omega_1, \frac{1}{2}\omega_2$, show that there is an isogeny ϕ' from E' to E that takes (x', y') to (x, y) where $x = \frac{1}{4}(y'/x')^2$ and $y = \frac{1}{8}(x'^2 + 4a)y'/x'^2$. Verify that the composition $\phi'\phi$ maps each point P on E to $2P$.

(x) Let E be an elliptic curve as in Exercise (ix) but now defined over $\mathbb{Q}$ and let G be the Mordell–Weil group of E. Show that the map ψ taking G to $\mathbb{Q}^*/\mathbb{Q}^{*2}$ given by $(x, y) \mapsto x\,\mathbb{Q}^{*2}$ if $x \neq 0$, $(x, y) \mapsto a\,\mathbb{Q}^{*2}$ if $x = 0$ and $O \mapsto 1\,\mathbb{Q}^{*2}$ is a group homomorphism.

By considering the substitution $x' = 2x + 2y/t$ and $y' = 2tx'$ with $x = t^2$ for some rational $t \neq 0$, prove that the kernel of ψ is $\phi'(G')$ where G' is the Mordell–Weil group of E' and ϕ' is the induced map $G' \to G$ arising from the isogeny.

(xi) Let E be the elliptic curve $y^2 = x^3 + ax + b$ with a, b integers. Verify the assertion in Section 17.5 that if (x, y) is a rational point on E then $x = r/t^2$ and $y = s/t^3$ for some integers r, s, t with $(r, t) = (s, t) = 1$.

Now let $b = 0$. Show that if p is a prime dividing r which occurs to an odd power in the canonical factorization of r then p divides a. Hence verify that the image of the map ψ in Exercise (x) is contained in the subgroup of $\mathbb{Q}^*/\mathbb{Q}^{*2}$ generated by $-1\,\mathbb{Q}^{*2}$ and $p\,\mathbb{Q}^{*2}$ where p runs through the prime factors of a.

(xii) Let E be the elliptic curve given in Legendre normal form by $y^2 = f(x)$ with $f(x) = x(x - 1)(x - \lambda)$. Show that the discriminant of f is $(\lambda(\lambda - 1))^2$. Show further that the Weierstrass equation for E is $y^2 = x^3 + ax + b$ with $3a = -(\lambda^2 - \lambda + 1)$. Hence verify the formula for the j-invariant of E quoted in Section 17.8.

Bibliography

S. Alaca and K. S. Williams, *Introductory Algebraic Number Theory* (Cambridge University Press, 2004)

T. M. Apostol, *Introduction to Analytic Number Theory* (Springer-Verlag, 1976)

E. Artin and J. Tate, *Class Field Theory* (W. A. Benjamin, 1967; AMS Chelsea Publishing, 2nd revised edn, 2008)

P. Bachmann, *Niedere Zahlentheorie* (Teubner, 1902; reprint, AMS Chelsea Publishing, 1968)

A. Baker, *Transcendental Number Theory* (Cambridge Mathematical Library, Cambridge University Press, 3rd edn, 1990)

A. Baker and G. Wüstholz, *Logarithmic Forms and Diophantine Geometry* (New Mathematical Monographs **9**, Cambridge University Press, 2007)

W. E. H. Berwick, *Integral Bases* (Cambridge University Press, 1927)

I. F. Blake, G. Seroussi and N. P. Smart, *Elliptic Curves in Cryptography* (LMS Lecture Note Series **265**, Cambridge University Press, 2000)

Z. I. Borevich and I. R. Shafarevich, *Number Theory* (Academic Press, 1966)

J. W. S. Cassels, *An Introduction to Diophantine Approximation* (Cambridge University Press, 1957)

J. W. S. Cassels, *An Introduction to the Geometry of Numbers* (Springer-Verlag, 2nd edn, 1971)

240

J. W. S. Cassels, *Rational Quadratic Forms* (Academic Press, 1978)

J. W. S. Cassels, *Local Fields* (LMS Student Text Series **3**, Cambridge University Press, 1986)

J. W. S. Cassels, *Lectures on Elliptic Curves* (LMS Student Text Series **24**, Cambridge University Press, 1991)

J. W. S. Cassels and A. Fröhlich, eds, *Algebraic Number Theory* (Academic Press, 1967)

K. Chandrasekharan, *Introduction to Analytic Number Theory* (Grundlehren Math. Wiss. **148**, Springer-Verlag, 1968)

K. Chandrasekharan, *Arithmetical Functions* (Grundlehren Math. Wiss. **167**, Springer-Verlag, 1970)

H. Cohen, *Number Theory*, Vols I, II (Graduate Texts in Mathematics **239**, **240**, Springer-Verlag, 2007)

A. C. Cojocaru and M. R. Murty, *An Introduction to Sieve Methods and Their Applications* (LMS Student Text Series **66**, Cambridge University Press, 2005)

D. A. Cox, *Primes of the Form $x^2 + ny^2$: Fermat, Class Field Theory and Complex Multiplication* (Wiley, 1989)

J. E. Cremona, *Algorithms for Modular Elliptic Curves* (Cambridge University Press, 2nd edn, 1997)

H. Davenport, *Multiplicative Number Theory* (Graduate Texts in Mathematics **74**, revised by H. L. Montgomery, Springer-Verlag, 3rd edn, 2000)

H. Davenport, *Analytic Methods for Diophantine Equations and Diophantine Inequalities* (Campus Publications, 1963; Cambridge Mathematical Library, 2nd revised edn prepared by T. D. Browning, Cambridge University Press, 2005)

H. Davenport, *The Higher Arithmetic* (Cambridge University Press, 8th edn, 2008)

L. E. Dickson, *History of the Theory of Numbers* (Washington, 1920; reprint, Dover, 2005)

H. M. Edwards, *Riemann's Zeta Function* (Academic Press, 1974; reprint, Dover, 2001)

W. and F. Ellison, *Prime Numbers* (Wiley, Hermann, 1985)

J. Esmonde and M. R. Murty, *Problems in Algebraic Number Theory* (Graduate Texts in Mathematics **190**, Springer-Verlag, 2nd edn, 2004)

A. Fröhlich and M. J. Taylor, *Algebraic Number Theory* (Cambridge Studies in Advanced Mathematics **27**, Cambridge University Press, 1991)

C. F. Gauss, *Disquisitiones Arithmeticae* (Springer-Verlag, translated into English from original 1801 Latin edn, 1986)

H. Halberstam and H. E. Richert, *Sieve Methods* (Academic Press, 1974)

G. H. Hardy and E. M. Wright, *An Introduction to the Theory of Numbers* (Oxford University Press, 6th edn, 2008)

E. Hecke, *Lectures on the Theory of Algebraic Numbers* (Graduate Texts in Mathematics **77**, Springer-Verlag, translated from original 1923 German edn, 1981)

D. Husemöller, *Elliptic Curves* (Graduate Texts in Mathematics **111**, Springer-Verlag, 2nd edn, 2004)

M. N. Huxley, *The Distribution of Prime Numbers: Large Sieves and Zero-Density Theorems* (Oxford Mathematical Monographs, Clarendon Press, 1972)

A. E. Ingham, *The Distribution of Prime Numbers* (Cambridge Mathematical Library, Cambridge University Press, 2nd edn, 1990)

H. Iwaniec and E. Kowalski, *Analytic Number Theory* (AMS Colloquium Publications **53**, American Mathematical Society, 2004)

A. Ivić, *The Riemann Zeta-Function* (Wiley, 1985)

A. A. Karatsuba and S. M. Voronin, *The Riemann Zeta-Function* (de Gruyter, 1992)

A. Khintchine, *Kettenbrüche* (Teubner, 1956)

N. Koblitz, *Introduction to Elliptic Curves and Modular Forms* (Graduate Texts in Mathematics **97**, Springer-Verlag, 1993)

N. Koblitz, *A Course in Number Theory and Cryptography* (Graduate Texts in Mathematics **114**, Springer-Verlag, 2nd edn, 1994)

E. Landau, *Einführung in die Elementare und Analytische Theorie der Algebraischen Zahlen und der Ideale* (Teubner, 1918)

E. Landau, *Foundations of Analysis* (Chelsea Publishing, 1951)

E. Landau, *Handbuch der Lehre von der Verteilung der Primzahlen* (Chelsea Publishing, 2nd edn, 1953)

E. Landau, *Elementary Number Theory* (Chelsea Publishing, 1958; reprint, American Mathematical Society, 1999)

S. Lang, *Algebraic Number Theory* (Graduate Texts in Mathematics **110**, Springer-Verlag, 2nd edn, 1994)

D. A. Marcus, *Number Fields* (Springer-Verlag, 1995)

H. L. Montgomery and R. C. Vaughan, *Multiplicative Number Theory I: Classical Theory* (Cambridge Studies in Advanced Mathematics **97**, Cambridge University Press, 2006)

L. J. Mordell, *Diophantine Equations* (Academic Press, 1969)

M. R. Murty, *Problems in Analytic Number Theory* (Springer-Verlag, 2nd edn, 2008)

T. Nagell, *Introduction to Number Theory* (Wiley, 1951; reprint, AMS Chelsea Publishing, 2001)

W. Narkiewicz, *Elementary and Analytic Theory of Algebraic Numbers* (Springer-Verlag, 3rd edn, 2004)

J. Neukirch, *Algebraic Number Theory* (Grundlehren Math. Wiss. **322**, Springer-Verlag, 1999)

I. Niven, H. S. Zuckerman and H. L. Montgomery, *An Introduction to the Theory of Numbers* (Wiley, 5th edn, 1991)

S. J. Patterson, *An Introduction to the Riemann Zeta-Function* (Cambridge Studies in Advanced Mathematics **14**, Cambridge University Press, 1988)

O. Perron, *Die Lehre von den Kettenbrüchen* (Teubner, 1913)

K. Prachar, *Primzahlverteilung* (Springer-Verlag, 1957)

P. Ribenboim, *13 Lectures on Fermat's Last Theorem* (Springer-Verlag, 1979)

P. Ribenboim, *The New Book of Prime Number Records* (Springer-Verlag, 1996)

H. Riesel, *Prime Numbers and Computer Methods for Factorization* (Progress in Mathematics **126**, Birkhäuser, 2nd edn, 1994)

W. M. Schmidt, *Diophantine Approximation* (Lecture Notes in Mathematics **785**, Springer-Verlag, 1980)

S. Schmitt and H. G. Zimmer, *Elliptic Curves: A Computational Approach* (Studies in Mathematics **31**, with an appendix by A. Pethő, de Gruyter, 2003)

R. Schoof, *Catalan's Conjecture* (Springer-Verlag, 2008)

J.-P. Serre, *Local Fields* (Graduate Texts in Mathematics **67**, Springer-Verlag, 1979)

J. H. Silverman, *Advanced Topics in the Arithmetic of Elliptic Curves* (Graduate Texts in Mathematics **151**, Springer-Verlag, 1994)

J. H. Silverman, *The Arithmetic of Elliptic Curves* (Graduate Texts in Mathematics **106**, Springer-Verlag, 2nd edn, 2009)

J. H. Silverman and J. Tate, *Rational Points on Elliptic Curves* (Undergraduate Texts in Mathematics, Springer-Verlag, 1992)

T. Skolem, *Diophantische Gleichungen* (Springer-Verlag, 1938; reprint, Chelsea Publishing, 1950)

N. P. Smart, *The Algorithmic Resolution of Diophantine Equations* (LMS Student Text Series **41**, Cambridge University Press, 1998)

H. M. Stark, *An Introduction to Number Theory* (MIT Press, 1978)

I. Stewart and D. Tall, *Algebraic Number Theory and Fermat's Last Theorem* (A. K. Peters, 3rd edn, 2002)

E. C. Titchmarsh, *The Theory of the Riemann Zeta-Function* (Oxford University Press, 2nd edn revised by D. R. Heath-Brown, 1986)

E. Trost, *Primzahlen* (Birkhäuser, 1953)

R. C. Vaughan, *The Hardy–Littlewood Method* (Cambridge University Press, 2nd edn, 1997)

I. M. Vinogradov, *An Introduction to the Theory of Numbers* (Pergamon Press, 1961)

L. C. Washington, *Elliptic Curves: Number Theory and Cryptography* (Chapman & Hall/CRC, 2nd edn, 2008)

A. Weil, *Basic Number Theory* (Grundlehren Math. Wiss. **144**, Springer-Verlag, 3rd edn, 1974; reprinted in the Classics in Mathematics series, 1995)

Index

Printed in the United States
By Bookmasters